Lecture Notes in Mathematics

Edited by A. Dold and B. Eckmann

Subseries: Nankai Institute of Mathematics, Tianjin, P.R. China
vol. 7
Adviser: S.S. Chern, B.-j. Jiang

1353

Richard S. Palais

Chuu-lian Terng

Critical Point Theory and Submanifold Geometry

Springer-Verlag

Berlin Heidelberg New York London Paris Tokyo

Authors

Richard S. Palais
Department of Mathematics, Brandeis University
Waltham, MA 02254, USA

Chuu-liang Terng
Department of Mathematics, Northeastern University
Boston, MA 02115, USA

Mathematics Subject Classification (1980): 49-02, 49 F 15, 53-02, 53 C 40, 53 C 42, 58-02, 58 E 05

ISBN 3-540-50399-4 Springer-Verlag Berlin Heidelberg New York
ISBN 0-387-50399-4 Springer-Verlag New York Berlin Heidelberg

Printing and binding: Druckhaus Beltz, Hemsbach/Bergstr.
2146/3140-543210

To Shiing-shen Chern

Scholar, Teacher, Friend

Preface

This book is divided into two parts. Part I is a modern introduction to the very classical theory of submanifold geometry. We go beyond the classical theory in at least one important respect; we study submanifolds of Hilbert space as well as of Euclidean spaces. Part II is devoted to critical point theory, and here again the theory is developed in the setting of Hilbert manifolds. The two parts are inter-related through the Morse Index Theorem, that is, the fact that the structure of the set of critical points of the distance function from a point to a submanifold can be described completely in terms of the local geometric invariants of the submanifold.

Now it is perfectly standard and natural to study critical point theory in infinite dimensions; one of the major applications of critical point theory is to the Calculus of Variations, where an infinite dimensional setting is essential. But what is the rationale for extending the classical theory of submanifolds to Hilbert space? The elementary theory of Riemannian Hilbert manifolds was developed in the 1960's, including for example the existence of Levi-Civita connections, geodesic coordinates, and some local theory of submanifolds. But Kuiper's proof of the contractibility of the group of orthogonal transformations of an infinite dimensional Hilbert space was discouraging. It meant that one could not expect to obtain interesting geometry and topology from the study of Riemannian Hilbert manifolds with the seemingly natural choice of structure group, and it was soon realized that a natural Fredholm structure was probably necessary for an interesting theory of infinite dimensional Riemannian manifolds. However, for many years there were few interesting examples to inspire further work in this area. The recent development of Kac-Moody groups and their representation theory has changed this picture. The coadjoint orbits of these infinite dimensional groups are nice submanifolds of Hilbert space with natural Fredholm structures. Moreover they arise in the study of gauge group actions and have a rich and interesting geometry and topology. Best of all from our point of view, they are isoparametric (see below) and provide easily studied explicit models that suggest good assumptions to make in order to extend classical Euclidean submanifold theory to a theory of submanifolds of Hilbert space.

One of the main goals of part I is to help graduate students get started doing research in Riemannian geometry. As a result we have tried to make it a reasonably self-contained source for learning the techniques of the subject. We do assume that the reader is familiar with the elementary theory of differentiable manifolds, as presented for example in Lang's book [La], and the basic theory of Riemannian geometry as in Hicks' book [Hk], or selected parts of Spivak's [Sp]. But in Chapter 1 we give a review of finite dimensional Riemannian geometry, with emphasis on the techniques of computation. We use Cartan's moving frame method, always trying to emphasize the intrinsic meaning behind seemingly non-invariant computations. We also give many exercises that are meant as an introduction to a variety of interesting research topics. The local geometry of submanifolds of R^n is treated in Chapter 2. In Chapter 3 we apply the local theory to study Weingarten surfaces in R^3 and S^3. The focal structure of submanifolds and its relation to the critical point structure of distance and height functions are explained in Chapter 4. The remaining chapters in part I are devoted to two problems, the understanding of which is a natural step towards developing a more general theory of submanifolds:

(1) Classify the submanifolds of Hilbert space that have the "simplest local invariants", namely the so-called isoparametric submanifolds. (A submanifold is called isoparametric if its normal curvature is zero and the principal curvatures along any parallel normal field are constant).

(2) Develop the relationship between the geometry and the topology of isoparametric submanifolds.

Many of these "simple" submanifolds arise from representation theory. In particular the generalized flag manifolds (principal orbits of adjoint representations) are isoparametric and so are the principal orbits of other isotropy representations of symmetric spaces. In fact it is now known that all homogeneous isoparametric submanifolds arise in this way, so that they are effectively classified. But there are also many non-homogeneous examples. In fact, problem (1) is far from solved, and the ongoing effort to better understand and classify isoparametric manifolds has given rise to a beautiful interplay between Riemannian geometry, algebra, transformation group theory, differential equations, and Morse theory.

In Chapter 5 we develop the basic theory of proper Fredholm Riemannian group actions (for both finite and infinite dimensions). In Chapter 6 we study the geometry of finite dimensional isoparametric submanifolds. In Chapter 7 we develop the basic theory of proper Fredholm submanifolds of Hilbert space (the condition "proper Fredholm" is needed in order to use the techniques of differential topology and Morse theory on Hilbert manifolds). Finally, in chapter 8, we use the Morse theory developed in part II to study the homology of isoparametric submanifolds of Hilbert space.

Part II of the book is a self-contained account of critical point theory on Hilbert manifolds. In Chapters 9 we develop the standard critical point theory for non-degenerate functions that satisfy Condition C: the deformation theorems, minimax principal, and Morse inequalities. We then develop the theory of linking cycles in Chapters 10; this is used in Chapter 8 of Part I to compute the homology of isoparametric submanifolds of Hilbert space. In Chapter 11, we apply our abstract critical point theory to the Calculus of Variations. We treat first the easy case of geodesics, where the abstract theory fits like a glove. We then consider a model example of the more complex "multiple integral" problems in the Calculus of Variations; the so-called Yamabe Problem, that arises in the conformal deformation of a metric to constant scalar curvature. Here we illustrate some of the major techniques that are required to make the abstract theory work in higher dimensions.

This book grew out of lectures we gave in China in May of 1987. Over a year before, Professor S.S. Chern had invited the authors to visit the recently established Nankai Mathematics institute in Tianjin, China, and lecture for a month on a subject of our choice. Word had already spread that the new Institute was an exceptionally pleasant place in which to work, so we were happy to accept. And since we were just then working together on some problems concerning isoparametric submanifolds, we soon decided to give two inter-related series of lectures. One series would be on isoparametric submanifolds; the other would be on aspects of Morse Theory, with emphasis on our generalization to the isoparametric case of the Bott-Samelson technique for calculating the homology and cohomology of certain orbits of group actions. At Professor Chern's request we started to write up our lecture notes in advance, for eventual publication as a volume in a new Nankai Institute sub-series of the Springer Verlag Mathematical Lecture Notes. Despite all good intentions, when we arrived in Tianjin in May of 1987 we each had only about a week's worth of lectures written up, and just rough notes for the rest. Perhaps it was for the best! We were completely surprised by the nature of the audience that greeted us. Eighty graduate students and young faculty, interested in geometry, had come to Tianjin from all over China to participate in our mini courses. From the beginning this was as bright and enthusiastic a group of students as we have lectured to anywhere. Moreover, before we arrived, they had received considerable background preparation for our lectures and were soon clamoring for us to pick up the pace. Perhaps we did not see as much of the wonderful city of Tianjin as we had hoped, but nevertheless we spent a very happy month talking to these students and scrambling to prepare appropriate lectures. One result was that the scope of these notes has been considerably expanded from what was originally planned. For example, the Hilbert space setting for the part on Morse Theory reflects the students desire to hear about the infinite dimensional aspects of the theory. And the part on isoparametric submanifolds was expanded to a general exposition of the modern theory of submanifolds of space forms, with material on orbital ge-

ometry and tight and taut immersions. We would like to take this opportunity
to thank those many students at Nankai for the stimulation they provided.

We will never forget our month at Nankai or the many good friends we
made there. We would like to thank Professor and Mrs. Chern and all of the
faculty and staff of the Mathematics Institute for the boundless effort they put
into making our stay in Tianjin so memorable.

After the first draft of these notes was written, we used them in a differential geometry seminar at Brandeis University. We would like to thank the
many students who lectured in this seminar for the errors they uncovered and
the many improvements that they suggested.

Both authors would like to thank The National Science Foundation for its
support during the period on which we wrote and did research on this book.
We would also like to express our appreciation to our respective Universities,
Brandeis and Northeastern, for providing us with an hospitable envoironment
for the teaching and research that led up to its publication.

And finally we would both like to express to Professor Chern our grat-
itude for his having been our teacher and guide in differential geometry. Of
course there is not a geometer alive who has not benefited directly or indi-
rectly from Chern, but we feel particularly fortunate for our many personal
contacts with him over the years.

Contents

Part II.   Critical Point Theory.

Part I.   Submanifold Theory.

Chapter 1.
Preliminaries.

In this chapter we review some basic facts concerning connections and the existence theory for systems of first order partial differential equations. These are basic tools for the study of submanifold geometry. A connection is defined both globally as a differential operator (Koszul's definition) and locally as connection 1-forms (Cartan's formulation). While the global definition is better for interpreting the geometry, the local definition is easier to compute with. A first order system of partial differential equations can be viewed as a system of equations for differential 1-forms, and the associated existence theory is referred to as the Frobenius theorem.

1.1 Connections on a vector bundle.

Let M be a smooth manifold, ξ a smooth vector bundle of rank k on M, and $C^\infty(\xi)$ the space of smooth sections of ξ.

1.1.1. Definition. A *connection* for ξ is a linear operator

$$\nabla : C^\infty(\xi) \to C^\infty(T^*M \otimes \xi)$$

such that

$$\nabla(fs) = df \otimes s + f\nabla(s)$$

for every $s \in C^\infty(\xi)$ and $f \in C^\infty(M)$. We call $\nabla(s)$ the *covariant derivative* of s.

If ξ is trivial, i.e., $\xi = M \times R^k$, then $C^\infty(\xi)$ can be identified with $C^\infty(M, R^k)$ by $s(x) = (x, f(x))$. The differential of maps gives a trivial connection on ξ, i.e., $\nabla s(x) = (x, df_x)$. The collection of all connections on ξ can be described as follows. We call k smooth sections $s_1, \ldots, s_k$ of ξ a *frame field* of ξ if $s_1(x), \ldots, s_k(x)$ is a basis for the fiber ξ_x at every $x \in M$. Then every section of ξ can be uniquely written as a sum $f_1 s_1 + \ldots + f_k s_k$, where f_i are uniquely determined smooth functions on M. A connection ∇ on ξ is uniquely determined by $\nabla(s_1), \ldots, \nabla(s_k)$, and these can be completely arbitrary smooth sections of the bundle $T^*M \otimes \xi$. Each of the sections $\nabla(s_i)$ can be written uniquely as a sum $\sum \omega_{ij} \otimes s_j$, where (ω_{ij}) is an arbitrary $n \times n$ matrix of smooth real-valued one forms on M. In fact, given $\nabla(s_1), \ldots, \nabla(s_k)$ we can define ∇ for an arbitrary section by the formula

$$\nabla(f_1 s_1 + \cdots + f_k s_k) = \sum(df_i \otimes s_i + f_i \nabla(s_i)).$$

(Here and in the sequel we use the convention that $\sum$ always stands for the summation over all indices that appear twice).

Suppose U is a small open subset of M such that $\xi|U$ is trivial. A frame field $s_1, \ldots, s_k$ of $\xi|U$ is called a *local frame field* of ξ on U.

It follows from the definition that a connection ∇ is a local operator, that is, if s vanishes on an open set U then ∇s also vanishes on U. In fact, since $s(p) = 0$ and $ds_p = 0$ imply $\nabla s(p) = 0$, ∇ is a first order differential operator ([Pa3]).

Since a connection is a local operator, it makes sense to talk about its restriction to an open subset of M. If a collection of open sets U_α covers M such that $\xi|U_\alpha$ is trivial, then a connection ∇ on ξ is uniquely determined by its restrictions to the various U_α. Let $s_1, \ldots, s_k$ be a local frame field on U_α, then there exists unique $n \times n$ matrix of smooth real-valued one forms (ω_{ij}) on U_α such that $\nabla(s_i) = \sum \omega_{ij} \otimes s_j$.

Let $GL(k)$ denote the Lie group of the non-singular $k \times k$ real matrices, and $gl(k)$ its Lie algebra. If s_i and s_i^* are two local frame fields of ξ on U, then there is a uniquely determined smooth map $g = (g_{ij}) : U \to GL(k)$ such that $s_i^* = \sum g_{ij} s_j$. Let $g^{-1} = (g^{ij})$ denote the inverse of g, so that $s_i = \sum g^{ij} s_j^*$. Suppose

$$\nabla s_i = \sum \omega_{ij} \otimes s_j, \qquad \nabla s_i^* = \sum \omega_{ij}^* \otimes s_j^*.$$

Let $\omega = (\omega_{ij})$ and $\omega^* = (\omega_{ij}^*)$. Since

$$\nabla s_i^* = \nabla(\sum g_{im} s_m) = \sum dg_{im} s_m + g_{im} \nabla s_m$$

$$= \sum_m (dg_{im} + \sum_k g_{ik} \omega_{km}) s_m$$

$$= \sum_j (\sum_m dg_{im} g^{mj} + \sum_{m,k} g_{ik} \omega_{km} g^{mj}) s_j^*$$

$$= \sum_j \omega_{ij}^* s_j^*,$$

we have

$$\omega^* = (dg) g^{-1} + g \omega g^{-1}.$$

Given an open cover U_α of M and local frame fields $\{s_i^\alpha\}$ on U_α, suppose $s_i^\alpha = \sum(g_{ij}^{\alpha\beta}) s_j^\beta$ on $U_\alpha \cap U_\beta$. Let $g^{\alpha\beta} = (g_{ij}^{\alpha\beta})$. Then a connection on ξ is defined by a collection of $gl(k)$–valued 1-forms ω^α on U_α, such that on $U_\alpha \cap U_\beta$ we have $\omega^\beta = (dg^{\alpha\beta})(g^{\alpha\beta})^{-1} + g^{\alpha\beta} \omega^\alpha (g^{\alpha\beta})^{-1}$.

Identify $T^*M \otimes \xi$ with $L(TM, \xi)$, and let $\nabla_X s$ denote $(\nabla s)(X)$. For $X, Y \in C^\infty(TM)$ and $s \in C^\infty(\xi)$ we define

$$K(X, Y)(s) = -(\nabla_X \nabla_Y - \nabla_Y \nabla_X - \nabla_{[X,Y]})(s). \qquad (1.1.1)$$

It follows from a direct computation that

$$K(Y, X) = -K(X, Y),$$

$$K(fX, Y) = K(X, fY) = fK(X, Y),$$

$$K(X, Y)(fs) = fK(X, Y)(s).$$

Hence K is a smooth section of $L(\xi \otimes \bigwedge^2 TM, \xi) \simeq L(\xi, \bigwedge^2 T^*M \otimes \xi)$.

1.1.2. Definition. This section K of the vector bundle $L(\xi, \bigwedge^2 T^*M \otimes \xi)$ is called the *curvature* of the connection ∇.

Recall that the bracket operation on vector fields and the exterior differentiation on p forms are related by

$$dw(X_0, \ldots, X_p) = \sum_i (-1)^i X_i \omega(X_0, \ldots, \hat{X}_i, \ldots, X_p)$$
$$+ \sum_{i<j} (-1)^{i+j} \omega([X_i, X_j], X_0, \ldots, \hat{X}_i, \ldots, \hat{X}_j, \ldots, X_p). \qquad (1.1.2)$$

Suppose $s_1, \ldots, s_k$ is a local frame field on U, and $\nabla s_i = \sum \omega_{ij} \otimes s_j$. Then there exist 2-forms Ω_{ij} such that

$$K(s_i) = \sum \Omega_{ij} \otimes s_j.$$

Since

$$-K(X, Y)(s_i) = \nabla_X \nabla_Y s_i - \nabla_Y \nabla_X s_i - \nabla_{[X,Y]} s_i$$
$$= \nabla_X (\sum \omega_{ij}(Y) s_j) - \nabla_Y (\sum \omega_{ij}(X) s_j)$$
$$- \sum \omega_{ij}([X, Y]) s_j$$
$$= \sum (X(\omega_{ij}(Y)) - Y(\omega_{ij}(X)) - \omega_{ij}([X, Y])) s_j$$
$$+ \sum (\omega_{ij}(Y) \omega_{jk}(X) - \omega_{ij}(X) \omega_{jk}(Y)) s_k$$
$$= \sum (d\omega_{ij} - \sum \omega_{ik} \wedge \omega_{kj})(X, Y) s_j,$$

we have

$$-\Omega_{ij} = d\omega_{ij} - \sum \omega_{ik} \wedge \omega_{kj}.$$

Thus K can is locally described by the $k \times k$ matrix $\Omega = (\Omega_{ij})$ of 2-forms just as ∇ is defined locally by the matrix $\omega = (\omega_{ij})$ of 1-forms. In matrix notation, we have

$$-\Omega = d\omega - \omega \wedge \omega. \tag{1.1.3}$$

Let $g = (g_{ij}) : U \to GL(k)$ be a smooth map and let $\omega = (dg)g^{-1}$. Then ω is a $gl(k)-$ valued 1-form on U, satisfying the so-called *Maurer-Cartan equation*

$$d\omega = \omega \wedge \omega.$$

Conversely, given a $gl(k)-$ valued 1-form on U with $d\omega = \omega \wedge \omega.$, it follows from Frobenius theorem (cf. 1.4) that given any $x_0 \in U$ and $g_0 \in GL(k)$ there is a neighborhood U_0 of x_0 in U and a smooth map $g = (g_{ij}) : U_0 \to GL(k)$ such that $g(x_0) = g_0$ and $(dg)g^{-1} = \omega$. Thus $d\omega = \omega \wedge \omega$ is a necessary and sufficient condition for being able to solve locally the system of first order partial differential equations:

$$dg = \omega g. \tag{1.1.4}$$

Let e_i denote the ith row of the matrix g and $\omega = (\omega_{ij})$. Then (1.1.4) can be rewritten as

$$de_i = \sum_j \omega_{ij} \otimes e_j.$$

1.1.3. Definition. A smooth section s of $\xi|U$ is *parallel* with respect to ∇ if $\nabla s = 0$ on U.

1.1.4. Definition. A connection is *flat* if its curvature is zero.

1.1.5. Proposition.   *The connection ∇ on ξ is flat if and only if there exist local parallel frame fields.*

PROOF.   Let s_i and $\omega = (\omega_{ij})$ be as before. Suppose $\Omega = 0$, then ω satisfies the Maurer- Cartan equation $d\omega = \omega \wedge \omega$. So locally there exists a $GL(k)-$ valued map $g = (g_{ij})$ such that $(dg)g^{-1} = \omega$. Let $g^{-1} = (g^{ij})$, and $s_i^* = \sum g^{ij}s_j$. Then $\nabla s_i^* = \sum \omega_{ij}^* \otimes s_j^*$, and

$$\omega^* = d(g^{-1})g + g^{-1}\omega g$$
$$= -g^{-1}(dg)g^{-1}g + g^{-1}(dg)g^{-1}g = 0 \,\dot{}$$

So s_i^* is a parallel frame.   ∎

1.1.6. Definition. A connection ∇ on ξ is called globally flat if there exists a parallel frame field defined on the whole manifold M.

1.1.7. Example. Let ξ be the trivial vector bundle $M \times R^k$, and ∇ the trivial connection on ξ given by the differential of maps. Then a section $s(x) = (x, f(x))$ is parallel if and only if f is a constant map, so ∇ is globally flat.

1.1.8. Remarks.

(1) If ξ is not a trivial bundle then no connection on ξ can be globally flat.

(ii) A flat connection need not be globally flat. For example, let M be the Möbius band $[0, 1] \times R/ \sim$ (where $(0, t) \sim (1, -t)$)). Then the trivial connection on $[0, 1] \times R$ induces a flat connection on TM. But since TM is not a product bundle this connection is not globally flat.

Given $x_0 \in M$, a smooth curve $\alpha : [0, 1] \to M$ such that $\alpha(0) = x_0$ and $v_0 \in \xi_{x_0}$ (the fiber of ξ over x_0), then the following first order ODE

$$\nabla_{\alpha'(t)} v = 0, \quad v(0) = v_0, \tag{1.1.5}$$

has a unique solution. A solution of (1.1.5) is called a parallel field along α, and $v(1)$ is called the parallel translation of v_0 along α to $\alpha(1)$. Let $P(\alpha) : \xi_{x_0} \to \xi_{x_0}$ be the map defined by $P(\alpha)(v_0) = v(1)$ for closed curve α such that $\alpha(0) = \alpha(1) = x_0$. The set of all these $P(\alpha)$ is a subgroup of $GL(\xi_{x_0})$, that is called the *holonomy group* of ∇ with respect to x_0. It is easily seen that ∇ is globally flat if and only if the holonomy group of ∇ is trivial.

1.1.9. Definition. A local frame s_i of vector bundle ξ is called parallel at a point x_0 with respect to the connection ∇, if $\nabla s_i(x_0) = 0$ for all i.

1.1.10. Proposition. *Let ∇ be a connection on the vector bundle ξ on M. Given $x_0 \in M$, then there exist an open neighborhood U of x_0 and a frame field defined on U, that is parallel at x_0.*

PROOF. Let s_i be a local frame field, $\nabla s_i = \sum_j \omega_{ij} \otimes s_j$, and $\omega = (\omega_{ij})$. Let $x_1, \ldots, x_n$ be a local coordinate system near x_0, and $\omega = \sum_i f_i(x) \, dx_i$, for some smooth $gl(k)$ valued maps f_i. Let $a_i = f_i(x_0)$. Then $a_i \in gl(k)$, and $g^{-1}dg + \omega = 0$ at x_0, where $g(x) = \exp(\sum_i x_i a_i)$. So we have $dg \, g^{-1} + g \omega g^{-1} = 0$ at x_0, i.e., $s_i^* = \sum g_{ij} s_j$ is parallel at x_0, where $g = (g_{ij})$. ∎

Let $O(m, k)$ denote the Lie group of linear isomorphism that leave the following bilinear form on R^{m+k} invariant:

$$(x, y) = \sum_{i=1}^{m} x_i y_i - \sum_{j=1}^{k} x_{m+j} y_{m+j}.$$

So an $(m+k) \times (m+k)$ matrix A is in $O(m,k)$ if and only if

$$A^t E A = E, \quad \text{where} \quad E = \text{diag}(1,\ldots,1,-1,\ldots,-1),$$

and its Lie algebra is:

$$o(m,k) = \{A \in gl(m+k) \mid A^t E + EA = 0\}.$$

1.1.11. Definition. A rank $(m+k)$ vector bundle ξ is called an $O(m,k)-$ bundle (an orthogonal bundle if $k = 0$) if there is a smooth section g of $S^2(\xi^*)$ such that $g(x)$ is a non-degenerate bilinear form on ξ_x of index k for all $x \in M$. A connection ∇ on ξ is said to be *compatible* with g if

$$X(g(s,t)) = g(\nabla_X s, t) + g(s, \nabla_X t),$$

for all $X \in C^\infty(TM), s, t \in C^\infty(\xi)$.

Suppose $s_1,\ldots,s_{m+k}$ is a local frame field, $g(s_i,s_j) = g_{ij}$, and

$$\nabla s_i = \sum_j \omega_{ij} \otimes s_j.$$

Then ∇ is compatible with g if and only if

$$\omega G + G\omega^t = dG,$$

where $\omega = (\omega_{ij})$ and $G = (g_{ij})$. In particular, if $G = E$ as above, then

$$\omega E + E\omega^t = 0, \tag{1.1.6}$$

i.e., ω is an $o(m,k)-$ valued 1-form on M.

The collection of all connections on ξ does not have natural vector space structure. However it *does* have a natural affine structure. In fact if ∇_1 and ∇_2 are two connections on ξ and f is a smooth function on M then the linear combination $f\nabla_1 + (1-f)\nabla_2$ is again a well-defined connection on ξ, and $\nabla_1 - \nabla_2$ is a smooth section of $L(\xi, T^*M \otimes \xi)$.

Next we consider connections on induced vector bundles. Given a smooth map $\varphi : N \to M$ we can form the induced vector bundle $\varphi^*\xi$. Note that there are canonical maps

$$\varphi^* : C^\infty(\xi) \to C^\infty(\varphi^*\xi),$$

$$\varphi^* : C^\infty(T^*M) \to C^\infty(T^*N).$$

So there is also a canonical map

$$\varphi^* : C^\infty(T^*M \otimes \xi) \to C^\infty(T^*N \otimes \varphi^*\xi).$$

1.1.12. Lemma. *To each connection ∇ on ξ there corresponds a unique connection $\varphi^*\nabla$ on the induced bundle $\varphi^*\xi$ so that*

$$(\varphi^*\nabla)(\varphi^*s) = \varphi^*(\nabla s).$$

For example, given a local frame field $s_1, \ldots, s_k$ over an open subset U of M with $\nabla(s_i) = \sum \omega_{ij} \otimes s_j$, then

$$(\varphi^*\nabla)(\varphi^*s_i) = \sum \varphi^*\omega_{ij} \otimes \varphi^*s_j,$$

i.e., the connection 1-form for $\varphi^*\nabla$ is $\varphi^*\omega_{ij}$.

Suppose ∇_1 and ∇_2 are connections on the vector bundles ξ_1 and ξ_2 over M. Then there is a natural connection ∇ on $\xi_1 \otimes \xi_2$ that satisfies the usual "product rule", i.e.,

$$\nabla(s_1 \otimes s_2) = \nabla_1(s_1) \otimes s_2 + s_1 \otimes \nabla_2(s_2)$$

1.2 Levi-Civita Connections.

Let M be an n-dimensional smooth manifold, and g a smooth metric on M, i.e., $g \in C^\infty(S^2T^*M)$, such that $g(x)$ is positive definite for all $x \in M$ (or equivalently, TM is an orthogonal bundle). Suppose ∇ is a connection on TM, and given vector fields X and Y on M let

$$T(X,Y) = \nabla_X Y - \nabla_Y X - [X,Y].$$

It follows from a direct computation that we have

$$T(fX,Y) = T(X,fY) = fT(X,Y), \qquad T(X,Y) = -T(Y,X).$$

So T is a section of $\bigwedge^2 T^*M \otimes TM$, called the *torsion tensor* of ∇.

1.2.1. Definition. A connection ∇ on TM is said to be *torsion free* if its torsion tensor T is zero.

Let $e_1, \ldots, e_n$ be a local orthonormal tangent frame field on an open subset U of M, i.e., $e_1(x), \ldots, e_n(x)$ forms an orthonormal basis for TM_x for all $x \in M$. We denote by $\omega_1, \ldots, \omega_n$ the 1-forms in U dual to $e_1, \ldots, e_n$, i.e., satisfying $\omega_i(e_j) = \delta_{ij}$. Suppose

$$\nabla e_i = \sum \omega_{ij} \otimes e_j.$$

It follows from (1.1.6) that ∇ is compatible with g if and only if $\omega_{ij} + \omega_{ji} = 0$. The torsion is zero if and only if

$$[e_i, e_j] = \sum (\omega_{jk}(e_i) - \omega_{ik}(e_j)) e_k. \tag{1.2.1}$$

Then (1.1.2) and (1.2.1) imply that

$$d\omega_k = \sum \omega_l \wedge \omega_{lk}.$$

Let c_{ijk}, and γ_{ijk} be the coefficients of $[e_i, e_j]$ and ω_{ij} respectively, i.e.,

$$[e_i, e_j] = \sum c_{ijk} e_k,$$

and

$$\omega_{ij} = \sum \gamma_{ijk} \omega_k.$$

Then we have:

$$\gamma_{ijk} = -\gamma_{jik}, \qquad \gamma_{jki} - \gamma_{ikj} = c_{ijk}.$$

This system of linear equations for the γ_{ijk} has a unique solution that is easily found explicitly; namely

$$\gamma_{ijk} = \frac{1}{2}(-c_{ijk} + c_{jki} + c_{kij}).$$

Equivalently, $\nabla_Z X$ is determined by the following equation:

$$\begin{aligned} g(\nabla_Z X, Y) = \frac{1}{2}\{&g([Y, Z], X) + g([Z, X], Y) - g([X, Y], Z) \\ &+ X(g(Y, Z)) + Y(g(Z, X)) - Z(g(X, Y))\} \end{aligned} \tag{1.2.2}$$

for all smooth vector field Y on M. So we have:

1.2.2. Theorem. *There is a unique connection ∇ on a Riemannian manifold (M, g) that is torsion free and compatible with g. This connection* is called the *Levi-Civita connection* of g. If $e_1, \ldots, e_n$ is a local orthonormal frame field of TM and $\omega_1, \ldots, \omega_n$ is its dual coframe, then the Levi-Civita connection 1-form ω_{ij} of g are characterized by the following "structure equations":

$$d\omega_i = \sum \omega_j \wedge \omega_{ji}, \qquad \omega_{ij} + \omega_{ji} = 0,$$

or equivalently

$$d\omega_i = \sum \omega_{ij} \wedge \omega_j, \qquad \omega_{ij} + \omega_{ji} = 0. \tag{1.2.3}$$

1.2.3. Definition. The curvature of the Levi-Civita connection of (M, g) is called the *Riemann tensor* of g.

Let $\omega = (\omega_{ij})$ be the Levi-Civita connection 1-form of g, and $\Omega = (\Omega_{ij})$ the Riemann tensor. It follows from (1.1.3) that we have

$$d\omega - \omega \wedge \omega = -\Omega. \tag{1.2.4}$$

This is called the *curvature equation*. Write

$$\Omega_{ij} = \frac{1}{2} \sum_{k \neq l} R_{ijkl} \omega_k \wedge \omega_l \tag{1.2.5}$$

with $R_{ijkl} = -R_{ijlk}$. It is easily seen that

$$R_{klij} = g(K(e_i, e_j)(e_k), e_l).$$

Next we will derive the first Bianchi identity. Taking the exterior derivative of (1.2.3) and using (1.2.4), we get

$$\sum_j \Omega_{ij} \wedge \omega_j = \frac{1}{2} \sum_{j,k,l} R_{ijkl} \omega_k \omega_l \omega_j = 0,$$

which implies the first Bianchi identity

$$R_{ijkl} + R_{iklj} + R_{iljk} = 0. \tag{1.2.6}$$

If the dimension of M is 2 and $\Omega_{12} = K\omega_1 \wedge \omega_2$, then K is a well-defined smooth function on M, called the *Gaussian curvature* of g. The curvature equation (1.2.4) gives

$$d\omega_{12} = -K\omega_1 \wedge \omega_2.$$

Let M be a Riemannian n-manifold, E a linear 2-plane of TM_p and v_1, v_2 is an orthonormal basis of E. Then $g(K(v_1, v_2)(v_1), v_2)$ is independent of the choice of v_1, v_2 and depends only on E; it is called the *sectional curvature* $K(E)$ of the 2-plane E with respect to g. In fact $K(E)$ is equal to the Gaussian curvature of the surface $\exp_p(B)$ at p with induced metric from M, where B is a small disk centered at the origin in E. The metric g is said to have *constant sectional curvature* c if $K(E) = c$ for all two planes. It is easily seen that g has constant sectional curvature c if and only if

$$\Omega_{ij} = c \omega_i \wedge \omega_j.$$

The metric g has positive sectional curvature if $K(E) > 0$ for all two planes E.

The *Ricci curvature*,

$$\mathrm{Ric} = \sum r_{ij} \omega_i \otimes \omega_j,$$

of g is defined by the following contraction of the Riemann tensor Ω:

$$r_{ij} = \sum_k R_{ikjk}.$$

The *scalar curvature*, μ, of g is the trace of the Ricci curvature, i.e.,

$$\mu = \sum_i r_{ii}.$$

It is easily seen that Ric is a symmetric 2-tensor. We say that Ric is positive, negative, non-positive, or non-negative if it has the corresponding property as a quadratic form, e.g, $\mathrm{Ric} > 0$ if $\mathrm{Ric}(X, X) > 0$ for all non-zero tangent vector X. The metric g is called an *Einstein metric*, if the Ricci curvature $\mathrm{Ric} = cg$ for some constant c.

The study of constant scalar curvature metrics and Einstein metrics plays very important role in geometry, partial differential equations and physics, for example see [Sc1],[KW] and [Be].

1.2.4. Example. Suppose $g = A^2(x, y)dx^2 + B^2(x, y)dy^2$ is a metric on an open subset U of R^2. Set

$$\omega_1 = A dx, \quad \omega_2 = B dy, \quad \omega_{12} = p\omega_1 + q\omega_2.$$

Then using the structure equations:

$$d\omega_1 = \omega_{12} \wedge \omega_2, \quad d\omega_2 = \omega_1 \wedge \omega_{12},$$

we can solve p and q explicitly. Let f_x denote $\frac{\partial f}{\partial x}$. We have

$$\omega_{12} = -\frac{A_y}{B} \, dx + \frac{B_x}{A} \, dy,$$

$$K = \frac{-1}{AB} \left[\left(\frac{A_y}{B} \right)_y + \left(\frac{B_x}{A} \right)_x \right].$$

1.2.5. Example. Let $M = R^n$, and $g = dx_1^2 + \ldots + dx_n^2$ the standard metric. A smooth vector field u of R^n can be identified as a smooth map $u = (u_1, \ldots, u_n) : R^n \to R^n$. Then the constant vector fields $e_i(x) = (0, \ldots, 1, \ldots, 0)$ with 1 at the ith place form an orthonormal frame of TR^n, and $\omega_i = dx_i$ are the dual coframe. It is easily seen that $\omega_{ij} = 0$ is the solution of the structure equations (1.2.3). So $\nabla e_i = 0$, and the curvature forms are

$$\Omega = -d\omega + \omega \wedge \omega = 0.$$

If $u = (u_1, \ldots, u_n)$ is a vector field, then $u = \sum u_i e_i$ and

$$\nabla u = \sum du_i \otimes e_i = (du_1, \ldots, du_n),$$

i.e., the covariant derivative of the tangent vector field u is the same as the differential of u as a map.

Exercises.

1. Using the first Bianchi identity and the fact that R_{ijkl} is antisymmetric with respect to ij and kl, show that $R_{ijkl} = R_{klij}$, i.e., if we identify T^*M with TM via the metric, then the Riemann tensor Ω is a self-adjoint operator on $\bigwedge^2 TM$. (Note that if g has positive sectional curvature then Ω is a positive operator, but the converse is not true.)

2. Show that Ricci curvature tensor is a section of $S^2(T^*M)$, i.e., $r_{ij} = r_{ji}$.

3. Suppose (M, g) is a Riemannian 3-manifold. Show that the Ricci curvature Ric determines the Riemann curvature Ω. In fact since Ric is symmetric, there exists a local orthonormal frame e_1, e_2, e_3 such that Ric $= \sum \lambda_i \omega_i \otimes \omega_i$. Then R_{ijkl} can be solved explicitly in terms of the λ_i from the linear system $\sum_k R_{ikjk} = \lambda_i \delta_{ij}$.

4. Let (M^n, g) be a Riemannian manifold with $n \geq 3$. Suppose that for all 2-plane E_x of TM_x we have $K(E_x) = c(x)$, depending only on x. Show that $c(x)$ is a constant, i.e., independent of x.

5. Let G be a Lie group, V a linear space, and $\rho : G \to GL(V)$ a group homomorphism, i.e., a representation. Then V is called a linear G-space

and we let gv denote $\rho(g)(v)$. A linear subspace V_0 of V is G-invariant if $g(V_0) \subseteq V_0$ for all $g \in G$.

  (i) Let V_1, V_2 be linear G-spaces and $T : V_1 \to V_2$ a linear equivariant map, i.e., $T(gv) = gT(v)$ for all $g \in G$ and $v \in V_1$. Show that both $\mathrm{Ker}(T)$ and $\mathrm{Im}(T)$ are G-invariant linear subspaces.

  (ii) If V is a linear G-space given by ρ then the dual V^* is a linear G-space given by ρ^*, where $\rho^*(g)(\ell)(v) = \ell(\rho(g^{-1})(v))$.

  (iii) Suppose V is an inner product and $\rho(G) \subseteq O(V)$. If V_0 is an invariant linear subspace of V then $V_0^{\perp}$ is also invariant.

  (iv) With the same assumption as in (iii), if we identify V^* with V via the inner product then $\rho^* = \rho$.

6. Let M be a smooth (Riemannian) n-manifold, and $F(M)$ ($F_0(M)$) the bundle of (orthonormal) frames on M, i.e., the fiber $F(M)_x$ ($F_0(M)_x$) over $x \in M$ is the set of all (orthonormal) bases of TM_x.

  (i) Show that $F(M)$ is a principal $GL(n)$-bundle.

  (ii) Show that $F_0(M)$ is a principal $O(n)$-bundle,

  (iii) Show that the vector bundle associated to the representation $\rho = id : GL(n) \to GL(n)$ is TM.

  (iv) Find the $GL(n)$-representations associated to the tensor bundles of M, $S^2 TM$ and $\bigwedge^p TM$.

7. Let $v_1, \ldots, v_n$ be the standard basis of $\boldsymbol{R}^n$, and

$$V = \left\{ \sum x_{ijkl} v_i \otimes v_j \otimes v_k \otimes v_l \mid \right.$$
$$\left. x_{ijkl} + x_{jikl} = x_{ijkl} + x_{ijlk} = x_{ijkl} + x_{iklj} + x_{iljk} = 0 \right\}.$$

Let $r : V \to S^2(\boldsymbol{R}^n)$ be defined by $r(x) = \sum x_{ikjk} v_i \otimes v_j$.

  (i) Show that V is an $O(n)$-invariant linear subspace of $\otimes^4 \boldsymbol{R}^n$, and the Riemann tensor Ω is a section of the vector bundle associated to V.

  (ii) Show that r is an $O(n)$-equivariant map, $V = \mathrm{Ker}(r) \oplus S^2(\boldsymbol{R}^n)$ as $O(n)$-spaces, and the Ricci tensor is a section of the vector bundle associated to $S^2(\boldsymbol{R}^n)$, i.e., $S^2 TM$. The projection of Riemann tensor Ω onto the vector bundle associated to $\mathrm{Ker}(r)$ is called the *Weyl tensor* (For detail see [Be]).

  (iii) Write down the equivariant projection of V onto $\mathrm{Ker}(r)$ explicitly. (This gives a formula for the Weyl tensor).

8. Let $M = \boldsymbol{R}^n$ and $g = a_1^2(x)\, dx_1^2 + \ldots + a_n^2(x)\, dx_n^2$. Find the Levi-Civita connection 1-form of (M, g).

9. Let $\boldsymbol{H}^n = \{(x_1, \ldots, x_n) \mid x_n > 0\}$, and $g = (dx_1^2 + \ldots + dx_n^2)/x_n^2$. Show that the sectional curvature of $(\boldsymbol{H}^n, g)$ is -1.

1.3   Covariant derivative of tensor fields.

Let (M, g) be a Riemannian manifold, and ∇ the Levi-Civita connection on TM. There is a unique induced connection ∇ on T^*M by requiring

$$X(\omega(Y)) = (\nabla_X \omega)(Y) + \omega(\nabla_X Y). \tag{1.3.1}$$

Let $e_1, \ldots, e_n$ be a local orthonormal frame field on M, and $\omega_1, \ldots, \omega_n$ its dual coframe. Suppose

$$\nabla e_i = \sum \omega_{ij} \otimes e_j,$$

$$\nabla \omega_i = \sum \tau_{ij} \otimes \omega_j.$$

Then (1.3.1) implies that $\tau_{ij} = -\omega_{ji} = \omega_{ij}$, i.e.,

$$\nabla \omega_i = \sum \omega_{ij} \otimes \omega_j. \tag{1.3.2}$$

So ∇ can be naturally extended to any tensor bundle $T_s^r = \otimes^r T^*M \otimes^s TM$ of type (r,s) as in section 1.1.

For $r > 0$, $s > 0$, let $C_q^p : T_s^r \to T_{s-1}^{r-1}$ denote the linear map such that

$$C_q^p(\omega_{i_1} \otimes \ldots \otimes \omega_{i_r} \otimes e_{j_1} \otimes \ldots \otimes e_{j_s})$$
$$= \omega_{i_1} \otimes \ldots \omega_{i_{p-1}} \otimes \omega_{i_{p+1}} \otimes \ldots \omega_{i_r} \otimes e_{j_1} \otimes \ldots e_{j_{q-1}} \otimes e_{j_{q+1}} \otimes \ldots e_{j_s}.$$

These linear maps C_q^p are called *contractions*. If we make the standard identification of T_1^1 with $L(TM, TM)$, then for $t = \sum t_{ij}\omega_i \otimes e_j$, we have

$$C_1^1(t) = \sum t_{ii} = \mathrm{tr}(t).$$

Since T^*M can be naturally identified with TM via the metric, the contraction operators are defined for any tensor bundles. For example if $t = \sum t_{ij}\omega_i \otimes \omega_j$, then $C(t) = \sum t_{ii}$ defines a contraction. The induced connections on the tensor bundles commute with tensor product and contractions.

In the following we will demonstrate how to compute the covariant derivatives of tensor fields. Let f be a smooth function on M, and

$$\nabla f = \sum f_i \omega_i = df. \tag{1.3.3}$$

Since $\nabla(df)$ is a section of T_0^2, it can be written as a linear combination of $\{\omega_i \otimes \omega_j\}$:

$$\nabla(df) = \sum f_{ij}\omega_j \otimes \omega_i, \tag{1.3.4}$$

where $\nabla_{e_j}(df) = \sum f_{ij}\omega_i$. Using the product rule, we have

$$\nabla(df) = \sum df_i \otimes \omega_i + f_i \nabla\omega_i$$
$$= \sum_i df_i \otimes \omega_i + \sum_{i,j} f_i\omega_{ij} \otimes \omega_j \qquad (1.3.5)$$
$$= \sum_i df_i \otimes \omega_i + \sum_{i,m} f_m\omega_{mi} \otimes \omega_i.$$

Compare (1.3.4) and (1.3.5), we obtain

$$\sum_j f_{ij}\omega_j = df_i + \sum_m f_m\omega_{mi}. \qquad (1.3.6)$$

Taking the exterior derivative of (1.3.3) and using (1.2.3), (1.3.6), we obtain

$$0 = \sum df_i \wedge \omega_i + \sum f_i\omega_{ij} \wedge \omega_j$$
$$= \sum(\sum f_{ij}\omega_j - f_j\omega_{ji}) \wedge \omega_i + \sum f_i\omega_{ij} \wedge \omega_j$$
$$\doteq \sum_{ij} f_{ij}\omega_j \wedge \omega_i,$$

which implies that $f_{ij} = f_{ji}$. So we have

1.3.1. Proposition. *If $f : M \to R$ is a smooth function then $\nabla^2 f$ is a smooth section of $S^2 T^* M$.*

The *Laplacian* of f is defined to be the trace of $\nabla^2 f$, i.e.,

$$\Delta f = \sum_i f_{ii}.$$

Now suppose that $u = \sum u_{ij}\omega_i \otimes \omega_j$ is a smooth section of $\otimes^2 T^* M$, and

$$\nabla u = \sum u_{ijk}\omega_k \otimes \omega_i \otimes \omega_j,$$

where

$$\nabla_{e_k}(u) = \sum u_{ijk}\omega_i \otimes \omega_j.$$

Since

$$\nabla u = \sum du_{ij} \otimes \omega_i \otimes \omega_j + u_{ij}\nabla\omega_i \otimes \omega_j + u_{ij}\omega_i \otimes \nabla\omega_j,$$

and (1.3.2), we have

$$\sum_k u_{ijk}\omega_k = du_{ij} + \sum_m u_{im}\omega_{mj} + \sum_m u_{mj}\omega_{mi}. \qquad (1.3.7)$$

For example, if u is the metric tensor g, then we have $u_{ij} = \delta_{ij}$ and by (1.3.7) we see that $u_{ijk} = 0$, i.e., $\nabla g = 0$ or g is parallel.

In the following we derive the formula for the covariant derivative of the Riemann tensor and the second Bianchi identity. Let $\Omega = \sum R_{ijkl}\omega_i \otimes e_j \otimes \omega_k \otimes \omega_l$ be the Riemann tensor of g. Set

$$\nabla\Omega = \sum R_{ijklm}\omega_m \otimes \omega_i \otimes e_j \otimes \omega_k \otimes \omega_l,$$

where

$$\nabla_{e_m}\Omega = \sum R_{ijklm}\omega_i \otimes e_j \otimes \omega_k \otimes \omega_l.$$

Using an argument similar to the above we find

$$\sum_m R_{ijklm}\omega_m = dR_{ijkl} + \sum_m R_{mjkl}\omega_{mi} + \sum_m R_{imkl}\omega_{mj}$$
$$+ \sum_m R_{ijml}\omega_{mk} + \sum_m R_{ijkm}\omega_{ml}.$$

(1.3.8)

Taking the exterior derivative of (1.2.4) and using (1.3.8) we have

$$\sum_{k,l,m} R_{ijklm}\omega_k \wedge \omega_l \wedge \omega_m = 0.$$

So we obtain the second Bianchi identity :

$$R_{ijklm} + R_{ijlmk} + R_{ijmkl} = 0. \qquad (1.3.9)$$

Let u be a smooth section of tensor bundle T_r^s. Then $\nabla^2 u$ is a section of $T_r^s \otimes T^*M \otimes T^*M$. The *Laplacian* of u, Δu, is the section of T_r^s defined by contracting on the last two indices of $\nabla^2 u$. For example, if

$$u = \sum u_{ij}\omega_i \otimes \omega_j, \quad \nabla^2 u = \sum_{ijkl} \omega_i \otimes \omega_j \otimes \omega_k \otimes \omega_l,$$

then $(\Delta u)_{ij} = \sum_k u_{ijkk}$.

Exercises.

1. Let μ, Ric be the scalar and Ricci curvature of g respectively, $d\mu = \sum \mu_k \omega_k$ and ∇ Ric $= \sum r_{ijk}\omega_i \otimes \omega_j \otimes \omega_k$. Show that $\mu_k = 2\sum_i r_{iki}$.

2. Suppose (M, g) is an n-dimensional Riemannian manifold, and its Ricci curvature Ric satisfies the condition that Ric $= fg$ for some smooth function f on M. If $n > 2$, then f must be a constant, i.e., g is Einstein.

3. Let f be a smooth function on M, and $\nabla^3 f = \sum f_{ijk}\omega_i \otimes \omega_j \otimes \omega_k$. Show that

$$f_{ijk} = f_{ikj} + \sum_m f_m R_{mijk}.$$

4. Let $\varphi : R \to R$ and $u : M \to R$ be smooth functions. Show that

$$\Delta(\varphi(u)) = \varphi'(u)\Delta u + \varphi''(u)\|\nabla u\|^2.$$

6. Let (M^n, g) be an orientable Riemannian manifold, $f : M \to R$ a smooth function, and $df = \sum_i f_i \omega_i$. Show that

   (i) there is a unique linear operator $* : \bigwedge^p T^*M \to \bigwedge^{n-p} T^*M$ such that

$$\omega \wedge *\tau = \langle \omega, \tau \rangle dv$$

   for all p-forms ω and τ, where dv is the volume form of g.

   (ii)

$$*df = \sum_i (-1)^{i-1} f_i \omega_1 \wedge \ldots \omega_{i-1} \wedge \omega_{i+1} \ldots \wedge \omega_n.$$

   (iii)

$$\int_M \Delta f\, dv = \int_{\partial M} *df,$$

   In the following we assume that $\partial M = \emptyset$, show that

   (iv)

$$\int_M f \Delta f\, dv = - \int_M \|\nabla f\|^2 dv,$$

   (v) if $\Delta f = \lambda f$ for some $\lambda \geq 0$ then f is a constant.

1.4 Vector fields and differential equations.

A time independent system of ordinary differential equations (ODE) for n functions $\alpha = (\alpha_1, \ldots, \alpha_n)$ of one real variable t is given by a smooth map $f : U \to R^n$ on an open subset U of R^n. Corresponding to this ODE we have the following "initial value problem" : Given $x_0 \in U$, find $\alpha : (-t_0, t_0) \to U$ for some $t_0 > 0$ such that

$$\begin{cases} \alpha'(t) = f(\alpha(t)), \\ \alpha(0) = x_0. \end{cases} \tag{1.4.1}$$

The map f is a local vector field on $\boldsymbol{R}^n$ and the solutions of (1.4.1) are called the integral curves of the vector field f. As a consequence of the existence and uniqueness theorem of ODE, we have

1.4.1. Theorem. *Suppose M is a compact, smooth manifold, and X is a smooth vector field on M. Then there exists a unique family of diffeomorphisms $\varphi_t : M \to M$ for all $t \in \boldsymbol{R}$ such that*
(i) $\varphi_0 = id, \quad \varphi_{s+t} = \varphi_s \circ \varphi_t$,
(ii) let $\alpha(t) = \varphi_t(x_0)$, then α is the unique solution for the ODE system
$$\begin{cases} \alpha'(t) = X(\alpha(t)), \\ \alpha(0) = x_0. \end{cases}$$

The map $t \mapsto \varphi_t$ from the additive group $\boldsymbol{R}$ to the group $\mathrm{Diff}(M)$ of the diffeomorphisms of M is a group homomorphism, and is called the one-parameter subgroup of diffeomorphisms generated by the vector field X. Conversely, any group homomorphism $\rho : \boldsymbol{R} \to \mathrm{Diff}(M)$ arises this way, namely it is generated by the vector field X, where

$$X(x_0) = \frac{d}{dt}\bigg|_{t=0} (\rho(t)(x_0)).$$

In fact, $\mathrm{Diff}(M)$ is an infinite dimensional Fréchet Lie group and $C^\infty(TM)$ is its Lie algebra.

It follows from Theorem 1.4.1 that if X is a vector field on M such that $X(p) \neq 0$, then there exists a local coordinate system (U, x), $x = (x_1, \ldots, x_n)$ around p such that $X = \frac{\partial}{\partial x_1}$. It is obvious that $[\frac{\partial}{\partial x_i}, \frac{\partial}{\partial x_j}] = 0$, for all i, j. This is also a sufficient condition for any k vector fields being part of coordinate vector fields, i.e.,

1.4.2. Theorem. *Let $X_1, \ldots, X_k$ be k smooth tangent vector fields on an n-dimensional manifold M such that $X_1(x), \ldots, X_k(x)$ are linearly independent for all x in a neighborhood U of p. Suppose $[X_i, X_j] = 0, \ \forall \ 1 \leq i < j \leq k$ on U. Then there exist $U_0 \subset U$ and a coordinate system (x, U_0) around p such that $X_i = \frac{\partial}{\partial x_i}$ for all $1 \leq i \leq k$.*

The following first order system of partial differential equations (PDE) for u,

$$\frac{\partial u}{\partial x_i} = P_i(x_1, \ldots, x_n), \qquad i = 1, \ldots, n, \tag{1.4.2}$$

is equivalent to $\tau = du$ for some u, i.e., τ is an exact 1-form, where

$$\tau = \sum_i P_i(x_1, \ldots, x_n) dx_i.$$

So by the Poincaré Lemma, (1.4.2) is solvable if and only if τ is closed, i.e., $d\tau = 0$, or equivalently

$$\frac{\partial P_i}{\partial x_j} = \frac{\partial P_j}{\partial x_i}, \quad \text{for all} \quad i \neq j.$$

For the more general first order PDE:

$$\frac{\partial u}{\partial x_i} = P_i(x_1, \ldots, x_n, u(x)), \tag{1.4.3}$$

the solvability condition is that "the mixed second order partial derivatives are independent of the order of derivatives". But to check this condition for a complicated system can be tedious, and the Frobenius theorem gives a systematic way to determine whether a system is solvable, that can be stated either in terms of vector fields or differential forms.

1.4.3. Frobenius Theorem.   *Let $X_1, \ldots, X_k$ be k smooth tangent fields on an n-dimensional manifold M such that $X_1(x), \ldots, X_k(x)$ are linearly independent for all x in a neighborhood U of p. Suppose*

$$[X_i, X_j] = \sum_{l=1}^{k} f_{ijl} X_l, \quad \forall \quad i \neq j, \tag{1.4.4}$$

on U, for some smooth functions f_{ijl}. Then there exist an open neighborhood U_0 of p and a local coordinate system (x, U_0) such that the span of $\frac{\partial}{\partial x_1}, \ldots, \frac{\partial}{\partial x_k}$ is equal to the span of $X_1, \ldots, X_k$.

A rank k *distribution* E on M is a smooth rank k subbundle of TM. It is *integrable* if whenever $X, Y \in C^\infty(E)$, we have $[X, Y] \in C^\infty(E)$. Given a rank k distribution, locally there exist k smooth vector fields $X_1, \ldots, X_k$ such that E_x is the span of $X_1(x), \ldots, X_k(x)$. The vector fields $X_1, \ldots, X_k$ satisfies condition (1.4.4) if and only if E is integrable. A submanifold N of M is called an integral submanifold of E, if $TN_x = E_x$ for all $x \in N$. Then Theorem 1.4.3. can be restated as:

1.4.4. Theorem.   *If E is a smooth, integrable, rank k distribution of M, then there exists a local coordinate system (x, U) such that*

$$\{q \in U \mid x_{k+1}(q) = c_{k+1}, \ldots, x_n(q) = c_n\}$$

are integral submanifolds for E.

The space $\mathcal{A}$ of all differential forms is an anti-commutative ring under the standard addition and the wedge product. An ideal $\wp$ of $\mathcal{A}$ is called *d-closed* if $d\wp \subseteq \wp$. Given a rank k distribution E, locally there also exist

$(n - k)$ linearly independent 1-forms $\omega_{k+1}, \ldots, \omega_n$ such that $E_x = \{u \in TM_x \mid \omega_{k+1}(x) = \cdots = \omega_n(x) = 0\}$. Using (1.1.2), Theorem 1.4.3. can be formulated in terms of differential forms.

1.4.5. Theorem. *Let $\omega_1, \ldots, \omega_m$ be linearly independent 1-forms on M^n, and $\wp$ the ideal in the ring $\mathcal{A}$ of differential forms generated by $\omega_1, \ldots, \omega_m$. Suppose $\wp$ is d-closed. Then given $x_0 \in M$ there exists a local coordinate system (x, U) around x_0 such that $dx_1, \ldots, dx_m$ generates $\wp$.*

1.4.6. Corollary. *With the same assumption as in Theorem 1.4.5, given $x_0 \in M$, there exists an $(n-m)$-dimensional submanifold N of M through x_0 such that $i^*\omega_j = 0$ for all $1 \leq j \leq m$, where $i : N \rightarrow M$ is the inclusion.*

Let $\omega_1, \ldots, \omega_k$ be linearly independent 1-forms on M^n, and $\wp$ the ideal generated by $\omega_1, \ldots, \omega_k$. Then locally we can find smooth 1-forms $\omega_{k+1}, \ldots, \omega_n$ such that $\omega_1, \ldots, \omega_n$ are linear independent. We may assume that

$$d\omega_i = \sum_{j < l} f_{ijl} \omega_j \wedge \omega_l,$$

for some smooth functions f_{ijl}. Then it is easily seen that $\wp$ is d-closed if and only if one of the following conditions holds:

(i) $f_{ijl} = 0$ if $i \leq k$ and $j, l > k$,

(ii) $d\omega_i = 0 \mod (\omega_1, \ldots, \omega_k)$ for all $i \leq k$.

1.4.7. Example. In order to solve (1.4.3), we consider the following 1-form on $R^n \times R$:

$$\omega = dz - \sum_i P(x, z) dx_i.$$

Let $\wp$ be the ideal generated by ω. Then the condition that $\wp$ is d-closed is equivalent to one of the following:

(i) there exists a 1-form τ such that $d\omega = \omega \wedge \tau$,

(ii) $\omega \wedge d\omega = 0$.

If $\wp$ is integrable then there is a smooth function $f(x, z)$ such that $f(x, z) = c$ defines integrable submanifolds of $\wp$. Since df never vanishes and is proportional to ω, $\frac{\partial f}{\partial z} \neq 0$. So it follows from the Implicit Function Theorem that locally there exists a smooth function $u(x)$ such that $f(x, u(x)) = c$. So u is a solution of (1.4.3). In particular, the first order system for $g : U \rightarrow GL(n)$:

$$dg = \omega g,$$

is solvable if and only if $d\omega = \omega \wedge \omega$.

Exercises.

1. Let $\{X_1, X_2\}$ be a local frame field around p on the surface M. Show that there exists a local coordinate system (x_1, x_2) around p such that X_i is parallel to $\frac{\partial}{\partial x_i}$.

1.5   Lie derivative of tensor fields.

Let $\varphi : M \to N$ be a diffeomorphism. Then the pull back φ^* on vector fields and 1-forms are defined as follows:

$$\varphi^* : C^\infty(TN) \to C^\infty(TM), \quad \varphi^*(X)_p = (d\varphi_p)^{-1}(X(\varphi(p)),$$

$$\varphi^* : C^\infty(T^*N) \to C^\infty(T^*M), \quad \varphi^*(\omega)_p = \omega_{\varphi(p)} \circ d\varphi_p.$$

Hence φ^* is defined for any tensor fields by requiring that

$$\varphi^*(t_1 \otimes t_2) = \varphi^*(t_1) \otimes \varphi^*(t_2),$$

for any two tensor fields t_1 and t_2.

Let X be a vector field on M, and φ_t the one-parameter subgroup of M generated by X. Then the *Lie derivative* of a tensor field u with respect to X is defined to be

$$L_X u = \left. \frac{\partial}{\partial t} \right|_{t=0} (\varphi_t^* u). \qquad (1.5.1)$$

Let $\mathcal{T}(M)$ denote the direct sum of all the tensor bundles of M. Then L_X is a linear operator on $\mathcal{T}(M)$, that has the following properties (for proof see [KN] and [Sp]):

(i) If $u \in C^\infty(T_s^r(M))$, then $L_X u \in C^\infty(T_s^r(M))$.

(ii) L_X commute with the tensor product and contractions, i.e.,

$$L_X(u_1 \otimes u_2) = (L_X u_1) \otimes u_2 + u_1 \otimes (L_X u_2),$$

$$L_X(C(u)) = C(L_X u),$$

for any contraction operator C.

(iii) $L_X f = Xf = df(X)$, for any smooth function f.

(iv) $L_X Y = [X, Y]$, for any vector field Y.

The *interior derivative*, i_X, is the linear operator

$$i_X : C^\infty \left(\bigwedge^p T^*M \right) \to C^\infty \left(\bigwedge^{p-1} T^*M \right),$$

defined by

$$i_X(\omega)(X_1,\ldots,X_{p-1}) = \omega(X,X_1,\ldots,X_{p-1}).$$

Then on differential forms we have

$$L_X = i_X d + d i_X.$$

Let (M,g) be a Riemannian manifold, $e_1,\ldots,e_n$ a local orthonormal frame field, and $\omega_1,\ldots,\omega_n$ its dual coframe. Suppose $X = \sum X_i e_i$, and $\nabla X = \sum X_{ij} e_i \otimes \omega_j$. Using the fact that L_X commutes with contractions, we can easily show that

$$(L_X \omega_i) = \sum_j \omega_{ij}(X)\omega_j + X_{ij}\omega_j.$$

So we have

$$\begin{aligned}
L_X g &= L_X \left(\sum \omega_i \otimes \omega_i \right) \\
&= \sum (L_X \omega_i) \otimes \omega_i + \omega_i \otimes (L_X \omega_i). \\
&= \sum_{ij} (X_{ij} + X_{ji})\omega_i \otimes \omega_j
\end{aligned} \qquad (1.5.2)$$

A diffeomorphism $\varphi : M \to M$ is called an *isometry* if $\varphi^* g = g$ for all t, or equivalently $d\varphi_x : TM_x \to TM_{\varphi(x)}$ is a linear isometry for all $x \in M$. If φ_t is a one-parameter subgroup of isometries of M, and X is its vector field, then $\varphi_t^* g = g$ and by definition of $L_X g$ we have $L_X g = 0$. So by (1.5.2), we have

$$X_{ij} + X_{ji} = 0.$$

Any vector field satisfying this condition is called a *Killing vector field* of M. Conversely, if X is a Killing vector field on a complete manifold (M,g), then the 1-parameter subgroup φ_t generated by X consists of isometries.

Exercises.

1. Find all isometries of (R^n, g), where g is the standard metric.
2. If ξ is a Killing vector field and v a smooth tangent vector field on M, then $\langle \nabla_v \xi, v \rangle = 0$.
3. Let X be a smooth Killing vector field on the closed Riemannian manifold M. Show that
   (i)

$$\frac{1}{2}\Delta(\|X\|^2) = -\operatorname{Ric}(X,X) + \|\nabla X\|^2.$$

(ii)

$$\int_M \mathrm{Ric}(\nabla X, \nabla X)\,dv = \int_M \|\nabla X\|^2\,dv.$$

(iii) If $\mathrm{Ric} \leq 0$ (i.e., $\mathrm{Ric}(X, X) \leq 0$ for all vector field X) then the dimension of the group of isometries of M is 0.

Chapter 2.
Local Geometry of Submanifolds.

Given an immersed submanifold M^n of the simply connected space form $N^{n+k}(c)$ there are three basic local invariants associated to M: the first and second fundamental forms and the normal connection. These three invariants are related by the Gauss, Codazzi and Ricci equations, and they determine the isometric immersion of M into $N^{n+k}(c)$ uniquely up to isometries of $N^{n+k}(c)$.

2.1 Local invariants of submanifolds.

Let M be an n-dimensional submanifold of an (n+p)-dimensional Riemannian manifold (N, g), and $\bar{\nabla}$ the Levi-Civita connection of g. Let $TM_x^{\perp}$ denote the orthogonal complement of TM_x in TN_x, and $\nu(M)$ the normal bundle of M in N, i.e., $\nu(M)_x = (TM_x)^{\perp}$. In this section we will derive the three basic local invariants of submanifolds: the first and second fundamental forms, the induced normal connection, and we will derive the equations that relate them.

Let $i : M \to N$ denote the inclusion. The *first fundamental form*, I, of M is the induced metric i^*g, i.e., the inner product I_x on TM_x is the restriction of the inner product g_x to TM_x.

Let $v \in C^{\infty}(\nu(M))$ and let $A_v : TM_{x_0} \to TM_{x_0}$ denote the linear map defined by $A_v(u) = -((\bar{\nabla}_u v)(x_0))^T$, the projection of $(\bar{\nabla}_u v)(x_0)$ onto TM_{x_0}. Since

$$\bar{\nabla}_u(fv) = df(u)v + f\bar{\nabla}_u v, \quad \text{for} \quad f \in C^{\infty}(M, R),$$

and $df(u)v$ is a normal vector, we have

$$A_{fv}(u) = fA_v(u).$$

In particular, if v_1, v_2 are two normal fields on M such that $v_1(x_0) = v_2(x_0)$, then $A_{v_1}(u) = A_{v_2}(u)$ for $u \in TM_{x_0}$. So we have associated to each normal vector $v_0 \in \nu(M)_{x_0}$ a linear operator A_{v_0} on TM_{x_0}, that is called the *shape operator* of M in the normal direction v_0.

2.1.1. Proposition. *The shape operator $A_{v_0} : TM_{x_0} \to TM_{x_0}$ is self-adjoint, i.e., $g(A_{v_0}(u_1), u_2) = g(u_1, A_{v_0}(u_2))$.*

PROOF. Let v be a smooth normal field on M defined on a neighborhood U of x_0 such that $v(x_0) = v_0$, and X_i smooth tangent vector field on U such that $X_i(x_0) = u_i$. Let $\langle\ ,\ \rangle$ denote the inner product g_x on TN_x. Then

$$\langle A_v(X_1), X_2 \rangle = -\langle (\bar{\nabla}_{X_1}(v))^T, X_2 \rangle = -\langle (\bar{\nabla}_{X_1}(v)), X_2 \rangle$$
$$= -X_1(\langle v, X_2 \rangle)\ +\ \langle v, \bar{\nabla}_{X_1} X_2 \rangle$$
$$= \langle v, \bar{\nabla}_{X_1} X_2 \rangle.$$

Similarly, we have

$$\langle A_v(X_2), X_1 \rangle = \langle v, \bar{\nabla}_{X_2} X_1 \rangle,$$

so

$$\langle A_v(X_1), X_2 \rangle - \langle A_v(X_2), X_1 \rangle = \langle v, [X_1, X_2] \rangle.$$

Then the proposition follows from the fact that $[X_1, X_2]$ is a tangent vector field. ∎

By identifying T^*M with TM via the induced metric, the shape operator A_v corresponds to a smooth section of $S^2(T^*M) \otimes \nu(M)$, called the *second fundamental form* of M, and denoted by II. Explicitly,

$$\langle II(u_1, u_2), v \rangle = \langle A_v(u_1), u_2 \rangle.$$

The third invariant of M is the induced *normal connection* ∇^ν on $\nu(M)$, defined by $(\nabla^\nu)_u(v) = (\bar{\nabla}_u v)^\nu$, the orthogonal projection of $\bar{\nabla}_u v$ onto $\nu(M)$.

In the following we will write the above local invariants in terms of moving frames. A local orthonormal frame field $e_1, \ldots, e_{n+p}$ in N is said to be *adapted* to M if, when restricted to M, $e_1, \ldots, e_n$ are tangent to M. From now on, we shall agree on the following index ranges:

$$1 \leq A, B, C \leq (n+p), \quad 1 \leq i, j, k \leq n, \quad (n+1) \leq \alpha, \beta, \gamma \leq (n+p).$$

Let $\omega_1, \ldots, \omega_{n+p}$ be the dual coframe on N. Then the first fundamental form on M is

$$I = \sum_i \omega_i \otimes \omega_i.$$

The structure equations of N are

$$d\omega_A = \sum \omega_{AB} \wedge \omega_B, \quad \omega_{AB} + \omega_{BA} = 0, \tag{2.1.1}$$

and the curvature equation is

$$d\omega_{AB} = \sum_C \omega_{AC} \wedge \omega_{CB} - \Theta_{AB}, \tag{2.1.2}$$

$$\Theta_{AB} = \frac{1}{2} \sum_{C,D} K_{ABCD} \, \omega_C \wedge \omega_D, \quad K_{ABCD} = -K_{ABDC},$$

where ω_{AB} and Θ_{AB} are the Levi-Civita connection and the Riemann curvature tensor of g respectively.

For a differential form τ on N, we still use τ to denote $i^*\tau$, where $i : M \to N$ is the inclusion. Restricting ω_α to M, i.e., applying i^* to ω_α, we have

$$\omega_\alpha = 0. \tag{2.1.3}$$

Using (2.1.3), and applying i^* to (2.1.1), we obtain

$$d\omega_i = \sum \omega_{ij} \wedge \omega_j, \quad \omega_{ij} + \omega_{ji} = 0, \tag{2.1.4}$$

$$d\omega_\alpha = \sum \omega_{\alpha i} \wedge \omega_i = 0. \tag{2.1.5}$$

Note that (2.1.4) implies that the connection 1-form $\{\omega_{ij}\}$ is the Levi-Civita connection ∇ of the induced metric I on M. Set

$$\omega_{i\alpha} = \sum_j h_{i\alpha j} \omega_j. \tag{2.1.6}$$

Then (2.1.5) becomes

$$\sum_{i,j} h_{i\alpha j} \omega_i \wedge \omega_j = 0,$$

which implies that

$$h_{i\alpha j} = h_{j\alpha i}.$$

Note that

$$A_{e_\alpha}(e_i) = -(\bar{\nabla}_{e_i} e_\alpha)^T = -\sum_j \omega_{\alpha j}(e_i) e_j = \sum_j h_{i\alpha j} e_j.$$

So the second fundamental form of M is

$$II = \sum_{i,j,\alpha} h_{i\alpha j} \omega_i \otimes \omega_j \otimes e_\alpha$$

$$= \sum_{i,\alpha} \omega_{i\alpha} \otimes \omega_i \otimes e_\alpha.$$

It follows from the definition of the normal connection that

$$\nabla^\nu(e_\alpha) = \sum_\beta \omega_{\alpha\beta} \otimes e_\beta.$$

Restricting the curvature equations (2.1.2) of N to M, we have

$$d\omega_{ij} = \sum_k \omega_{ik} \wedge \omega_{kj} + \sum_\alpha \omega_{i\alpha} \wedge \omega_{\alpha j} - \Theta_{ij}, \qquad (2.1.7)$$

$$d\omega_{i\alpha} = \sum_k \omega_{ik} \wedge \omega_{k\alpha} + \sum_\beta \omega_{i\beta} \wedge \omega_{\beta\alpha} - \Theta_{i\alpha}, \qquad (2.1.8)$$

$$d\omega_{\alpha\beta} = \sum_\gamma \omega_{\alpha\gamma} \wedge \omega_{\gamma\beta} + \sum_i \omega_{\alpha i} \wedge \omega_{i\beta} - \Theta_{\alpha\beta}. \qquad (2.1.9)$$

Then (2.1.7) and (2.1.9) imply that the Riemann curvature tensor Ω of the induced metric I and the curvature Ω^ν of the normal connection ∇^ν (called the normal curvature of M) are:

$$\Omega_{ij} = \sum_\alpha \omega_{i\alpha} \wedge \omega_{j\alpha} + \Theta_{ij}, \qquad (2.1.10)$$

$$\Omega^\nu_{\alpha\beta} = \sum_i \omega_{i\alpha} \wedge \omega_{i\beta} + \Theta_{\alpha\beta}, \qquad (2.1.11)$$

respectively. Equations (2.1.7)-(2.1.9) are called the *Gauss, Codazzi,* and *Ricci* equations of the submanifold M.

Henceforth we assume that (N, g) has constant sectional curvature c, i.e.,

$$\Theta_{AB} = c \, \omega_A \wedge \omega_B.$$

So the Gauss, Codazzi and Ricci equations (2.1.7)-(2.1.9) for the submanifold M are

$$d\omega_{ij} = \sum_k \omega_{ik} \wedge \omega_{kj} + \sum_\alpha \omega_{i\alpha} \wedge \omega_{\alpha j} - c \omega_i \wedge \omega_j, \qquad (2.1.12)$$

$$d\omega_{i\alpha} = \sum_k \omega_{ik} \wedge \omega_{k\alpha} + \sum_\beta \omega_{i\beta} \wedge \omega_{\beta\alpha}, \qquad (2.1.13)$$

$$d\omega_{\alpha\beta} = \sum_\gamma \omega_{\alpha\gamma} \wedge \omega_{\gamma\beta} + \sum_i \omega_{\alpha i} \wedge \omega_{i\beta}. \qquad (2.1.14)$$

And (2.1.10) and (2.1.11) become

$$\Omega_{ij} = \sum_\alpha \omega_{i\alpha} \wedge \omega_{j\alpha} + c \, \omega_i \wedge \omega_j, \qquad (2.1.15)$$

$$\Omega^\nu_{\alpha\beta} = \sum_i \omega_{i\alpha} \wedge \omega_{i\beta}, \qquad (2.1.16)$$

Let

$$\Omega_{ij} = \frac{1}{2} \sum_{k,l} R_{ijkl} \omega_k \wedge \omega_l, \qquad \text{with} \quad R_{ijkl} + R_{ijlk} = 0,$$

$$\Omega^{\nu}_{\alpha\beta} = \frac{1}{2} \sum_{k,l} R^{\nu}_{\alpha\beta kl} \omega_k \wedge \omega_l, \qquad \text{with} \quad R^{\nu}_{\alpha\beta kl} + R^{\nu}_{\alpha\beta lk} = 0.$$

Using $\omega_{i\alpha} = \sum_j h_{i\alpha j}\omega_j$, we have

$$R_{ijkl} = \sum_{\alpha} (h_{i\alpha k} h_{j\alpha l} - h_{i\alpha l} h_{j\alpha k}) + c(\delta_{ik}\delta_{jl} - \delta_{il}\delta_{jk}), \qquad (2.1.17)$$

$$R^{\nu}_{\alpha\beta kl} = \sum_{i} (h_{i\alpha k} h_{i\beta l} - h_{i\alpha l} h_{i\beta k}.) \qquad (2.1.18)$$

By identifying T^*M with TM via the induced metric, then the Ricci equation (2.1.6) becomes $\Omega^{\nu}_{\alpha\beta} = [A_\alpha, A_\beta]$. So we have

2.1.2. Proposition. *Suppose (N, g) has constant sectional curvature, and M is a submanifold of N. Then the normal curvature Ω^{ν} of M measures the commutativity of the shape operators. In fact, $\Omega^{\nu}(u, v) = [A_u, A_v]$.*

A normal vector field v is *parallel* if $\nabla^{\nu} v = 0$. The normal bundle $\nu(M)$ is *flat* if ∇^{ν} is flat. Then it follows from Proposition 1.1.5 that $\nu(M)$ is flat if one of the following equivalent conditions holds:
(i) The normal curvature Ω^{ν} is zero.
(ii) Given $x_0 \in M$, there exist a neighborhood U of x_0 and a parallel normal frame field on U.

The normal bundle $\nu(M)$ is called *globally flat* if ∇^{ν} is globally flat, or equivalently, there exists a global parallel normal frame on M.

Since there are connections ∇ on TM and ∇^{ν} on $\nu(M)$, there exists a unique connection ∇ on the vector bundle $\otimes^2 T^*M \otimes \nu(M)$ that satisfies the "product rule", i.e.,

$$\nabla_X(\theta \otimes \tau \otimes v) = (\nabla_X \theta) \otimes \tau \otimes v + \theta \otimes (\nabla_X \tau) \otimes v + \theta \otimes \tau \otimes (\nabla_X v).$$

Set

$$\nabla II = \sum_{i,j,k\alpha} h_{i\alpha jk} \omega_i \otimes \omega_j \otimes \omega_k \otimes e_\alpha,$$

where

$$\nabla_{e_k} II = \sum_{i,j,k,\alpha} h_{i\alpha jk} \omega_i \otimes \omega_j \otimes e_\alpha.$$

Using an argument similar to that in section 1.3, we have

$$\sum_k h_{i\alpha jk}\omega_k = dh_{i\alpha j} + \sum_m h_{m\alpha j}\omega_{mi} + \sum_m h_{i\alpha m}\omega_{mj} + \sum_\beta h_{i\beta j}\omega_{\beta\alpha}.$$

$$(2.1.19)$$

Taking the exterior derivative of (2.1.6), we obtain

$$dw_{i\alpha} = d\left(\sum_j h_{i\alpha j}\omega_j\right)$$

$$= \sum_j dh_{i\alpha j}\omega_j + \sum_{j,k} h_{i\alpha j}\omega_{jk} \wedge \omega_k.$$

$$(2.1.20)$$

By the Codazzi equation (2.1.13), we have

$$dw_{i\alpha} = \sum_j \omega_{ij} \wedge \omega_{j\alpha} + \sum_\beta \omega_{i\beta} \wedge \omega_{\beta\alpha}$$

$$= \sum_{j,k} h_{j\alpha k}\omega_{ij} \wedge \omega_k + \sum_{\beta,j} h_{i\beta j}\omega_j \wedge \omega_{\beta\alpha}$$

$$= \sum_j \left(\sum_k h_{k\alpha j}\omega_{ik} - \sum_\beta h_{i\beta j}\omega_{\beta\alpha}\right) \wedge \omega_j.$$

$$(2.1.21)$$

Equating (2.1.19) and (2.1.20), we get

$$\sum_j (dh_{i\alpha j} + \sum_k \{h_{k\alpha j}\omega_{ki} + h_{i\alpha k}\omega_{kj}\} + \sum_\beta h_{i\beta j}\omega_{\beta\alpha}) \wedge \omega_j = 0.$$

So by (2.1.19), we have

$$\sum_{j,k} h_{i\alpha jk}\omega_j \wedge \omega_k = 0,$$

i.e., $h_{i\alpha jk} = h_{i\alpha kj}$. Since $h_{i\alpha j} = h_{j\alpha i}$, $h_{i\alpha jk} = h_{j\alpha ik}$, so we have

2.1.3. Proposition.   *Suppose (N, g) has constant sectional curvature c, and M is an immersed submanifold of N. Then ∇II is a section of $S^3 T^* M \otimes \nu(M)$, i.e., $h_{i\alpha jk}$ is symmetric in i, j, k.*

Although all our discussion above have been for embedded submanifolds, they hold equally well for immersions. For, locally an immersion $f : M \to N$ is an embedding, and we can naturally identify $TM_x \simeq T(f(M))_{f(x)}$.

The *principal curvatures* of an immersed submanifold M along a normal vector v are the eigenvalues of the shape operator A_v. The *mean curvature* vector H of M in N is the trace of II, i.e.,

$$H = \sum_\alpha H_\alpha e_\alpha, \quad \text{where} \quad H_\alpha = \sum_i h_{i\alpha i}.$$

The mean curvature vector of an immersion $f : M \to N$ is the gradient of the area functional at f. To be more precise, for any immersion $f : M \to N$, we let $dv(f^*g)$ be the volume element given by the induced metric f^*g, and define

$$A(f) = \int_M dv(f^*g),$$

to be the volume of the immersion f. A compact deformation of an immersion f_0 is a smooth family of immersions $\{f_t : M \to N\}$ such that there exists a relatively compact open set U of M with $f_t|(M \setminus U) = f_0|(M \setminus U)$. Then the deformation vector field

$$\xi = \left.\frac{\partial f_t}{\partial t}\right|_{t=0}$$

is a section of $f_0^*(TN)$ with compact support. It is well-known (cf. Exercise 4 below) that

$$\left.\frac{d}{dt}\right|_{t=0} A(f_t) = -\int_M \langle H, \xi \rangle \, dv_0, \qquad (2.1.22)$$

where dv_0 is the volume form of f_0^*g and H is the mean curvature vector of the immersion f_0. The immersion f_0 is called a *minimal*, if its mean curvature vector $H = 0$ or equivalently

$$\left.\frac{d}{dt}\right|_{t=0} A(f_t) = 0,$$

for all compact deformations f_t. The study of minimal immersions plays a very important role in differential geometry, for example see [Lw2], [Os], [Ch5] and [Bb].

Exercises.

1. Let M be the graph of a smooth function $u : R^n \to R$, i.e., $M = \{(x, u(x)) \mid x \in R^n\}$. Find I, II and H for M in R^{n+1}.
2. Suppose $\alpha(s) = (f(s), g(s))$ is a smooth curve in the $yz-$plane, parametrized by arc length. Let M be the surface of revolution generated by

the curve α, i.e., M is the surface of $\mathbf{R}^3$ obtained by rotating the curve α around the z-axis.

   (i) Find I, II for M.

   (ii) Find a curve α such that M has constant Gaussian curvature.

   (iii) Find a curve α such that M has constant mean curvature.

3. Let $\gamma : [0, \ell] \to \mathbf{R}^n$ be an immersion parametrized by arc length.

   (i) If $n = 2$, then $I = ds^2$ and $II = k(s)\, ds^2$, where $k(s)$ is the curvature of the plane curve.

   (ii) For generic immersions, show that we can choose an orthonormal frame field e_A on γ such that

$$\omega_{AB} = \begin{cases} 0, & \text{if } |A - B| \neq 1; \\ k_i(s)\, ds, & \text{if } (A, B) = (i, i+1); \\ -k_i(s)\, ds, & \text{if } (AB) = (i+1, i), \end{cases}$$

i.e., (ω_{AB}) is anti-symmetric and tridiagonal. (When $n = 3$, this frame e_A is the Frenet frame for curves in $\mathbf{R}^3$ and k_1, k_2 are the curvature and torsion of γ respectively, for more on the theory of curves see [Ch4], [Do]).

4. Let A denote the area functional for immersions of M into N.

   (i) If $\varphi : M \to M$ is a diffeomorphism, then $A(f \circ \varphi) = A(f)$, i.e., A is invariant under the group of diffeomorphisms of M.

   (ii) Show that $\nabla A(f)$ has to be a normal field along f.

   (iii) It suffices to show (2.1.22) for normal deformations, i.e., we may assume that ξ is a normal field for the immersion f.

   (iv) Prove (2.1.22) for normal deformations.

5. Let M be an immersed submanifold of $\mathbf{R}^m$, $p \in M$, $u \in TM_p$ and $v \in \nu(M)_p$ unit vectors. Let E be the plane spanned by u, v, and σ the curve given by the intersection of M and $p + E$. Show that $\langle II(u, u), v \rangle$ is equal to the curvature the curve σ at p.

6. Let M^n be an immersed submanifold of $N^{n+k}(c)$.

   (i) If we identify T^*M with TM via the metric then

$$\text{Ric} = HA - A^2 + (n-1)c\, I,$$

$$\mu = H^2 - \|II\|^2 + cn(n-1).$$

   (ii) If M is minimal in $\mathbf{R}^{n+k}$ then $\text{Ric}(M) \leq 0$.

2.2   Totally umbilic submanifolds.

A submanifold M of N is called *totally geodesic* (*t.g.*) if its second fundamental form is identically zero. A smooth curve α of N is called a *geodesic* if as a submanifold of N it is totally geodesic. It is easily seen that if e_A is an adapted frame for M then M is t.g. if and only if $\omega_{i\alpha} = 0$ for all $1 \leq i \leq n$ and $n+1 \leq \alpha \leq n+p$.

2.2.1. Proposition.   *Let γ be a smooth curve on N. Then the following statements are equivalent:*
(i) γ is a geodesic,
(ii) the tangent vector field γ' is parallel along γ,
(iii) the mean curvature of γ as a submanifold of N is zero.

PROOF.   We may assume that $\gamma(s)$ is parametrized by its arc length and $e_1(\gamma(s)) = \gamma'(s)$. Then γ is a geodesic if and only if $\omega_{1i}(\gamma') = 0$ for all $1 < i \leq n$, (ii) is equivalent to

$$0 = \nabla_{\gamma'}\gamma' = (\nabla_{e_1}e_1)(\gamma(s)) = \sum_{i=2}^{n} \omega_{1i}(\gamma'(s))e_i,$$

and (iii) gives

$$H = \sum_{i>1} \omega_{1i}(\gamma')e_i = 0.$$

So these three statements are equivalent.   ∎

2.2.2. Proposition.   *A submanifold M of a Riemannian manifold N is totally geodesic if and only if every geodesic of M (with respect to the induced metric) is a geodesic of N.*

PROOF.   The proposition follows from $\nabla_{\alpha'}\alpha' = \bar{\nabla}_{\alpha'}\alpha' - (\bar{\nabla}_{\alpha'}\alpha')^{\nu}$, and $(\bar{\nabla}_{\alpha'}\alpha')^{\nu} = II(\alpha', \alpha')$.   ∎

A Riemannian manifold with constant sectional curvature is called a *space form*. We have seen in Example 1.2.5 that R^n with the standard metric has constant sectional curvature 0. In the following we will describe complete simply connected space forms with nonzero curvature.

Let $g = dx_1^2 + \ldots + dx_{n+k}^2$ be the standard metric on R^{n+k}, and $\hat{\nabla}$ the Levi-Civita connection of g. Then we have seen in Example 1.2.5 that

$$\hat{\nabla}(u) = du,$$

if we identify $C^\infty(TR^{n+k})$ with the space of smooth maps from R^{n+k} to R^{n+k}. Let M^n be a submanifold of (R^{n+k}, g), and $X : M \to R^{n+k}$ the inclusion map. Let e_A and ω_A be as in section 2.1. First note that the differential $dX_p : TM_p \to TN_p$ of the map X at $p \in M$ is the inclusion i of TM_p in TN_p. Under the natural isomorphism $L(TM, TN) \simeq T^*M \otimes TN$, i corresponds to $\sum_i \omega_i(p) \otimes e_i(p)$. Hence we have

$$dX = \sum_i \omega_i \otimes e_i. \tag{2.2.1}$$

2.2.3. Example. Let S^n denote the unit sphere of R^{n+1}. Note that the inclusion map $X : S^n \to R^{n+1}$ is also the outward unit normal field on S^n, i.e., we may choose $e_{n+1} = X$. The exterior derivative of e_{n+1} gives

$$de_{n+1} = \sum \omega_{n+1,i} \otimes e_i.$$

Using (2.2.1), we have

$$\omega_{i,n+1} = -\omega_i,$$

So it follows from the Gauss equation (2.1.15) that S^n has constant sectional curvature 1. This induced metric of S^n is called the standard metric.

2.2.4. Example. Let $R^{n,1}$ denote the Lorentz space (N, g), i.e., $N = R^{n+1}$ and g is the non-degenerate metric $dx_1^2 + \ldots + dx_n^2 - dx_{n+1}^2$ of index 1. So TN is an $O(n, 1)-$ bundle, and results similar to those of section 1.1 and 2.1 can be derived. Let ∇ denote the unique connection TN, that is torsion free and compatible with g. Let

$$M = \{x \in R^{n,1} \mid g(x, x) = -1\},$$

and $X : M \to R^{n,1}$ denote the inclusion map. Then the induced metric on M is positive definite, and X is a unit normal field on M, i.e., $g(x, v) = 0$, for all $v \in TM_x$. Let $e_{n+1} = X$ and $e_1, \ldots, e_{n+1}$ a local frame field on $R^{n,1}$ such that

$$g(e_i, e_j) = \epsilon_i \delta_{ij}, \quad \text{where } \epsilon_1 = \ldots = \epsilon_n = -\epsilon_{n+1} = 1.$$

So $e_1(x), \ldots, e_n(x)$ are tangent to M for $x \in M$. Let $\omega^1, \ldots, \omega^{n+1}$ be the dual coframe, i.e., $\omega^A(e_B) = \delta_{AB}$. Let ω_A^B be the connection 1-form corresponding to $\bar\nabla$, i.e.,

$$\bar\nabla e_A = de_A = \sum_B \omega_A^B \otimes e_B.$$

By (1.1.6), we have

$$\epsilon_A \omega_A^B + \omega_B^A \epsilon_B = 0, \quad \text{and}$$

$$\bar{\nabla}\omega^A = -\sum_B \omega^A_B \otimes \omega^B.$$

Set

$$\omega_A = \epsilon_A \omega^A.$$

Since $e_{n+1} = X$, we have

$$de_{n+1} = \sum_i \omega^i_{n+1} \otimes e_i = dX = \sum_i \omega^i \otimes e_i.$$

So $\omega^i_{n+1} = \omega_i$. By the Gauss equation we have

$$\Omega^j_i = -\omega^{n+1}_i \wedge \omega^j_{n+1} = -\omega^{n+1}_i \wedge \omega^{n+1}_j$$
$$= -\omega^i \wedge \omega^j = -\omega_i \wedge \omega^j.$$

So M has constant sectional curvature -1. From now on we will let H^n denote M with the induced metric from $R^{n,1}$. H^n is also called the hyperbolic n-space.

It is well-known ([KN]) that every simply connected space form of sectional curvature c is isometric to R^n, S^n, H^n if $c = 0, 1$, or -1 respectively. We will let $N^n(c)$ denote these complete, simply connected Riemannian n-manifold with constant sectional curvature c.

2.2.5. Definition. An immersed hypersurface M^n of the simply connected space form $N^{n+1}(c)$ is called *totally umbilic* if $II(x) = f(x)I(x)$ for some smooth function $f : M \to R$.

In the following we will give examples of totally umbilic hypersurface of space forms.

2.2.6. Example. An affine n-plane E of R^{n+k} is totally geodesic. For we can choose e_α to be a constant orthonormal normal frame on E. Then $de_\alpha \equiv 0$. So we have $II \equiv 0$. Let $S^n(x_0, r)$ be the sphere of radius r centered at x_0 in R^{n+1}. Then $e_{n+1}(x) = (x - x_0)/r$ is a unit normal vector field on $S^n(x_0, r)$, and $de_{n+1} = (1/r)\sum_i \omega_i \otimes e_i$. So we have $\omega_{i,n+1} = -(1/r)\omega_i$ and $II = -(1/r)I$, i.e., $S^n(x_0, r)$ is totally umbilic, and has constant sectional curvature $\frac{1}{r^2}$.

2.2.7. Example. Let V be an affine hyperplane of R^{n+2}, v_0 a unit normal vector of V, $\cos\theta$ the distance from the origin to V, and $M = S^{n+1} \cap V$. Then $e_{n+1} = -\cot\theta X + \csc\theta v_0$ is a unit normal field to M in S^{n+1}. Taking the exterior derivative of e_{n+1}, we obtain

$$de_{n+1} = -\cot\theta dX = -\cot\theta \sum \omega_i \otimes e_i,$$

i.e., $\omega_{i,n+1} = \cot\theta \omega_i$ and $II = \cot\theta I$. So M is totally umbilic in S^{n+1} with sectional curvature equal to $1 + \cot^2\theta = \csc^2\theta$, and M is t.g. in S^{n+1}, if $\cos\theta = 0$ (or equivalently V is a linear hyperplane).

2.2.8. Example. Let v_0 be a non-zero vector of the Lorentz space $R^{n+1,1}$, and

$$M = \{x \in R^{n+1,1} \mid \langle x, x \rangle = -1, \langle x, v_0 \rangle = a\}.$$

Then

$$0 = \langle dX, X \rangle = \sum_i \langle e_i, X \rangle \omega_i,$$

$$0 = \langle dX, v_0 \rangle = \sum_i \langle e_i, v_0 \rangle \omega_i.$$

So $\langle X, e_i \rangle = \langle v_0, e_i \rangle = 0$, which implies that

$$v_0 = -aX + be_{n+1}, \tag{2.2.2}$$

for some b. Note that

$$\langle v_0, v_0 \rangle = -a^2 + b^2.$$

Taking the differential of (2.2.2), we have $\sum_i (a\omega_i + b\omega_{i,n+1})e_i = 0$. So

$$a\omega_i + b\omega_{i,n+1} = 0. \tag{2.2.3}$$

(i) If $\langle v_0, v_0 \rangle = 1$, then $-a^2 + b^2 = 1$ and we may assume that $a = \sinh t_0$ and $b = \cosh t_0$. So (2.2.3) implies that $\omega_{i,n+1} = -\tanh t_0 \, \omega_i$, i.e., $II = -\tanh t_0 \, I$, i.e., M is totally umbilic with sectional curvature $-1 + \tanh^2 t_0 = -\operatorname{sech}^2 t_0$.

(ii) If $\langle v_0, v_0 \rangle = 0$, then $-a^2 + b^2 = 0$, $\omega_{i,n+1} = \omega_i$. So $II = I$, and M is totally umbilic with sectional curvature 0.

(iii) If $\langle v_0, v_0 \rangle = -1$, then $-a^2 + b^2 = -1$ and we may assume that $a = \cosh t_0$, $b = \sinh t_0$. Then we have $\omega_{i,n+1} = -\coth t_0$, which implies that $II = -\coth t_0 I$, i.e., M is totally umbilic with sectional curvature $-\operatorname{csch}^2 t_0$.

2.2.9. Proposition. *Suppose* $X : M^n \to R^{n+1}$ *is an immersed, totally umbilic connected hypersurface, and* $n > 1$. *Then*

(i) $II = cI$ *for some constant* c

(ii) $X(M)$ *is either contained in a hyperplane, or is contained in a standard hypersphere of* R^{n+1}.

PROOF. Let e_A, ω_i and ω_{AB} as before. By assumption we have

$$\omega_{i\alpha} = f(x)\omega_i. \tag{2.2.4}$$

Taking the exterior derivative of (2.2.4), and using (2.1.4) and (2.1.13), we obtain

$$dw_{i\alpha} = df \wedge \omega_i + f \sum_j \omega_{ij} \wedge \omega_j$$

$$= \sum_j f_j \omega_j \wedge \omega_i + f \sum_j \omega_{ij} \wedge \omega_j$$

$$= \sum_j \omega_{ij} \wedge \omega_{j\alpha} = f \sum_j \omega_{ij} \wedge \omega_j.$$

So $\sum_j f_j \omega_j \wedge \omega_i = 0$, which implies that $f_j = 0$ for all $j \neq i$. Since $n > 1$, $df = 0$, i.e., $f = c$ a constant.

If $c = 0$, then $\omega_{i\alpha} = 0$. So $de_\alpha = 0$, e_α is a constant vector v_0, and

$$d\langle X, v_0 \rangle = \sum_i \langle e_i, v_0 \rangle \omega_i = 0,$$

i.e., $X(M)$ is contained in a hyperplane. If $c \neq 0$, then $\omega_{i\alpha} = c\omega_i$ and

$$d\left(X + \frac{e_\alpha}{c}\right) = \sum_i \left(\omega_i e_i - \frac{1}{c}\omega_{i\alpha}\right) e_i = 0.$$

So $X + e_\alpha/c$ is equal to a constant vector $x_0 \in R^{n+1}$, which implies that $\|X - x_0\|^2 = (1/c)^2$.   ∎

The concept of totally umbilic was generalized to submanifolds in [NR] as follows:

2.2.10. Definition. An immersed submanifold M^n of the simply connected space form $N^{n+k}(c)$ is called *totally umbilic* if $II = \xi I$, where ξ is a parallel normal field on M.

2.2.11. Proposition. *Let $X : M^n \rightarrow R^{n+k}$ be a connected, immersed totally umbilic submanifold, i.e., $II = \xi I$, where ξ is a parallel normal field on M. Then either*

(i) $\xi = 0$ and M is contained in an affine n-plane of R^{n+k}, or

(ii) $X + (\xi/a)$ is a constant vector x_0, where $a = \|\xi\|$; and M is contained in a standard $n-$sphere of R^{n+k}.

PROOF. If $\xi = 0$, then $\omega_{i\alpha} = 0$ for all i, α. The Ricci equation (2.1.14) gives $d\omega_{\alpha\beta} = \omega_{\alpha\gamma} \wedge \omega_{\gamma\beta}$, which implies that the normal connection is flat. It follows from Proposition 1.1.5 that there exists a parallel orthonormal normal frame e_α^*. So we may assume that e_α are parallel, i.e., $\omega_{\alpha\beta} = 0$. This implies that $de_\alpha = 0$, i.e., the e_α are constant vectors. Then

$$d\langle X, e_\alpha \rangle = \langle dX, e_\alpha \rangle = 0,$$

so the $\langle X, e_\alpha \rangle$ are constant c_α, and M is contained in the n–plane defined by $\langle X, e_\alpha \rangle = c_\alpha$.

If $\xi \neq 0$, then $a = \|\xi\|$ is a constant, and we may assume that $\xi = a e_{n+1}$, $\nabla^\nu e_{n+1} = 0$, so

$$\omega_{i,n+1} = a\omega_i, \quad \omega_{i\alpha} = 0, \quad \omega_{n+1,\alpha} = 0, \qquad (2.2.5)$$

for all $\alpha > (n+1)$. Then

$$d\left(X + \frac{e_{n+1}}{a}\right) = \sum_i \left(\omega_i - \frac{1}{a}\omega_{i,n+1}\right) e_i = 0,$$

so $X + (e_{n+1}/a)$ is a constant vector x_0. Using (2.2.5), we have

$$d(e_1 \wedge \ldots \wedge e_{n+1}) = \sum_{i,\alpha > n+1} e_1 \wedge \ldots \omega_{i\alpha} e_\alpha \wedge e_{i+1} \ldots \wedge e_{n+1}$$

$$+ \sum_{\alpha > n+1} e_1 \wedge \ldots \wedge e_n \wedge \omega_{n+1,\alpha} e_\alpha = 0.$$

Hence the span of $e_1(x), \ldots, e_{n+1}(x)$ is a fixed $(n+1)$–dimensional linear subspace V of R^{n+k} for all $x \in M$. But $X = x_0 - e_{n+1}/a$, so M is contained in the intersection of the affine $(n+1)$–plane $x_0 + V$ and the hypersphere of R^{n+k} of center x_0 and radius $1/a$. ∎

Exercises.

1. Prove the analogue of Proposition 2.2.9 for totally umbilic hypersurfaces of S^{n+1} and H^{n+1}.

2. Prove the analogue of Proposition 2.2.11 for totally umbilic submanifolds of S^{n+1} and H^{n+1}.

2.3 Fundamental theorem for submanifolds of space forms.

Given a submanifold M^n of a complete, simply connected space form, we have associated to M an orthogonal bundle (the normal bundle $\nu(M)$) with a compatible connection, and also the first and second fundamental forms of M. Together these satisfy the Gauss, Codazzi and Ricci equations. In the following, we will show that these data determine the submanifold up to isometries of of the space form.

2.3.1. Theorem. *Suppose* (M^n, g) *is a Riemannian manifold,* ξ *is a smooth rank* k *orthogonal vector bundle over* M *with a compatible connection* ∇^1, *and* $A : \xi \to S^2 T^* M$ *is a vector bundle morphism. Let* $e_1, \ldots, e_n$ *be a local orthonormal frame field on* TM, $\omega_1, \ldots, \omega_n$ *its dual coframe, and* ω_{ij} *the corresponding Levi-Civita connection 1-form, i.e.,* ω_{ij} *is determined by the structure equations*

$$d\omega_i = \sum_j \omega_{ij} \wedge \omega_j, \qquad \omega_{ij} + \omega_{ji} = 0. \qquad (2.3.1)$$

Let $e_{n+1}, \ldots, e_{n+k}$ *be an orthonormal local frame field of* ξ, *and* $\omega_{\alpha\beta}$ *is the* $o(k)$*-valued 1-form corresponds to* ∇^1. *Let* $\omega_{i\alpha}$ *be the 1-forms determined by the vector bundle morphism* A:

$$A(e_\alpha) = \sum_i \omega_{i\alpha} \otimes \omega_i.$$

Set $\omega_{\alpha i} = -\omega_{i\alpha}$, *and suppose* ω_{AB} *satisfy the Gauss, Codazzi and Ricci equations:*

$$d\omega_{ij} = \sum_k \omega_{ik} \wedge \omega_{kj} + \sum_\alpha \omega_{i\alpha} \wedge \omega_{\alpha j}, \qquad (2.3.2)$$

$$d\omega_{i\alpha} = \sum_k \omega_{ik} \wedge \omega_{k\alpha} + \sum_\beta \omega_{i\beta} \wedge \omega_{\beta\alpha}, \qquad (2.3.3)$$

$$d\omega_{\alpha\beta} = \sum_\gamma \omega_{\alpha\gamma} \wedge \omega_{\gamma\beta} + \sum_i \omega_{\alpha i} \wedge \omega_{i\beta}. \qquad (2.3.4)$$

Then given $x_0 \in M$, $p_0 \in R^{n+k}$, *and an orthonormal basis* $v_1, \ldots, v_{n+k}$ *of* R^{n+k}, *for small enough connected neighborhoods* U *of* x_0 *in* M *there is a unique immersion* $f : U \to R^{n+k}$ *and vector bundle isomorphism* $\eta : \xi \to \nu(M)$ *such that* $f(x_0) = p_0$ *and* $v_1, \ldots, v_n$ *are tangent to* $f(U)$ *at* p_0, g *is the first fundamental form,* $A(\eta(e_\alpha))$ *are the shape operators of the immersion, and* ∇^1 *corresponds to the induced normal connection under the isomorphism* η.

PROOF.   It follows from the definition of ω_{AB} that $\varpi = (\omega_{AB})$ is an $o(n+k)$-valued 1-form on M. Then (2.3.2)-(2.3.4) imply that ϖ satisfies Maurer-Cartan equation:

$$d\varpi = \varpi \wedge \varpi,$$

which is the integrability condition for the first order system

$$d\varphi = \varpi\varphi.$$

So there exist a small neighborhood U of x_0 in M and maps $e_A : U \to \mathbf{R}^{n+k}$ such that

$$de_A = \sum_B \omega_{AB} \otimes e_B,$$

where $e_A(x_0) = v_A$ and $\{e_A(x)\}$ is orthonormal for all $x \in U$. To solve the system

$$dX = \sum_i \omega_i \otimes e_i,$$

we prove the right hand side is a closed 1-form as follows:

$$\begin{aligned}
d\left(\sum_i \omega_i \otimes e_i\right) &= \sum_i d\omega_i \otimes e_i - \omega_i \wedge \sum_A \omega_{iA} \otimes e_A \\
&= \sum_i \left(d\omega_i - \sum_j \omega_{ij} \wedge \omega_j\right) \otimes e_i - \sum_{i,j,\alpha} (\omega_i \wedge \omega_{i\alpha}) \otimes e_\alpha, \\
&= \sum_i \left(d\omega_i - \sum_j \omega_{ij} \wedge \omega_j\right) \otimes e_i - \sum_{i,j,\alpha} h_{i\alpha j} \omega_i \wedge \omega_j \otimes e_\alpha,
\end{aligned}$$

$$(2.3.5)$$

the structure equations (2.3.1) implies the first term of (2.3.5) is zero and $h_{i\alpha j} = h_{j\alpha i}$ implies the second term is zero. ∎

2.3.2. Corollary. *Let $\varphi_0 : (M, g) \to \mathbf{R}^{n+k}$ and $\varphi_1 : (M, g) \to \mathbf{R}^{n+k}$ be immersions. Suppose that they have the same first, second fundamental forms and the normal connections. Then there is a unique orthogonal transformation B and a vector $v_0 \in \mathbf{R}^{n+k}$ such that $\varphi_0(x) = B(\varphi_1(x)) + v_0$.*

The group G_m of isometries of $\mathbf{R}^m$ is the semi-direct product of the orthogonal group $O(m)$ and the translation group $\mathbf{R}^m$; $gT_v g^{-1} = T_{g(v)}$, where $g \in O(m)$ and T_v is the translation defined by v. So its Lie algebra $\mathcal{G}_m$ can be identified as the Lie subalgebra of $gl(m+1)$ consisting of matrices of the form

$$\begin{pmatrix} A & v \\ 0 & 0 \end{pmatrix},$$

where $A \in o(m)$, and v is an $m \times 1$ matrix.

Let M, ω_i, ω_{AB} be as in Theorm 2.3.1. Let τ denote the following $gl(n+k+1)$–valued 1-form on M:

$$\tau = \begin{pmatrix} \varpi & \theta \\ 0 & 0 \end{pmatrix},$$

where $\varpi = (\omega_{AB})$ is an $o(n+k)$–valued 1-form, and θ is an $(n+k) \times 1$-valued 1-form $(\omega_1, \ldots, \omega_n, 0, \ldots, 0)^t$.

Then τ is a $\mathcal{G}_{n+k}$–valued 1-form on M. The Gauss, Codazzi and Ricci equations are equivalent to the Maurer-Cartan equations

$$d\tau = \tau \wedge \tau.$$

Hence there exists a unique map $F : U \to GL(n+k+1)$ such that $dF = \tau F$, the mth row of $F(x_0)$ is $(v_m, 0)$ for $m \leq (n+k)$, and the $(n+k+1)^{st}$ row of $F(x_0)$ is $(p_0, 1)$. Then the $(n+k+1)^{st}$ row is of the form $(X, 1)$, and X is the immersion of M into R^{n+k}.

A similar argument will give the fundamental theorem for submanifolds of the sphere and the hyperbolic space. For S^{n+k}, we have $F : U \to O(n+k+1)$, and the $(n+k+1)^{st}$ row of F gives the immersion of M into S^{n+k}. For H^{n+k}, we have $F : U \to O(n+k, 1)$, and the $(n+k+1)^{st}$ row of F gives the immersion of M into H^{n+k}.

2.3.3. Theorem.   *Given $(M, g), \xi, \nabla^1, A, \omega_i, \omega_{AB}$ as in Theorem 2.3.1. Let c denote the integer $0, 1$ or -1. Set*

$$\tau_c = \begin{pmatrix} \varpi & \theta \\ -c\theta^t & 0 \end{pmatrix},$$

where $\varpi = (\omega_{AB})$ is an $o(n+k)$ valued 1-form, and θ is the $(n+k) \times 1$ valued 1-form $(\omega_1, \ldots, \omega_n, 0, \ldots, 0)^t$ on M. Then

(i) τ_c is a $\mathcal{G}_{n+k}, o(n+k+1)$, or $o(n+k,1)$–valued 1-form on M for $c = 0, 1$ or -1 respectively.

(ii) If τ_c satisfies the Maurer-Cartan equations

$$d\tau_c = \tau_c \wedge \tau_c,$$

then

(1) the system

$$dF = \tau_c F \tag{2.3.6}$$

for the $GL(n+k+1)$–valued map F is solvable,

(2) if F is a solution for (2.3.6) and X denotes the $(n+k+1)^{st}$ row, then $X : M \to N^{n+k}(c)$ is an isometric immersion such that g, ξ, ∇^1, A are the first fundamental form, normal bundle, induced normal connection, and the shape operators respectively for the immersion X.

(3) The data g, ξ, ∇^1, A determine the isometric immersions of M into $N^{n+k}(c)$ uniquely up to isometries of $N^{n+k}(c)$.

Exercises.

1. Show that the group of isometries of (S^n, g) is $O(n+1)$, where g is the standard metric of S^n.
2. Show that the group of isometries of the hyperbolic space (H^n, g) is $O(n, 1)$.
3. Prove Theorem 2.3.3 for S^{n+k} and H^{n+k}.
4. Show that the $n-1$ smooth functions $k_1(s), \ldots, k_{n-1}(s)$ obtained in Ex. 3 of section 2.1 determine the curve uniquely up to rigid motions (this is the classical fundamental theorem for curves in R^n).

Chapter 3.

Weingarten Surfaces in three dimensional space forms.

In this chapter we will consider smooth, *oriented* surfaces M in three-dimensional simply-connected space forms $N^3(c)$. Such an M is called a *Weingarten surface* if its two principal curvatures λ_1, λ_2 satisfy a non-trivial functional relation, e.g., surfaces with constant mean curvature or constant Gaussian curvature. We will use the Gauss and Codazzi equations for surfaces to derive some basic properties of Weingarten surfaces.

Let $X : M \to N^3(c)$ be an immersed surface. Using the same notation as in section 2.1, we have

$$dX = \omega_1 \otimes e_1 + \omega_2 \otimes e_2, \tag{3.0.1}$$

$$d\omega_1 = \omega_{12} \wedge \omega_2, \qquad d\omega_2 = \omega_1 \wedge \omega_{12}, \tag{3.0.2}$$

and the Gauss equation (2.1.12), Codazzi equations (2.1.13) become:

$$d\omega_{12} = -K\omega_1 \wedge \omega_2 = -\omega_{13} \wedge \omega_{23} = -(\lambda_1 \lambda_2 + c)\omega_1 \wedge \omega_2, \tag{3.0.3}$$

$$d\omega_{13} = \omega_{12} \wedge \omega_{23}, \quad d\omega_{23} = \omega_{13} \wedge \omega_{12}. \tag{3.0.4}$$

The mean curvature and the Gaussian curvature are given by

$$H = \lambda_1 + \lambda_2, \quad K = c + \lambda_1 \lambda_2.$$

A point $p \in M$ is called an umbilic point if $II_p = \lambda I_p$, i.e., the two principal curvatures at p are equal. The eigendirections of the shape operator of M at a non-umbilic point are called the principal directions. Local coordinates (x, y) on M are called *line of curvature coordinates* if the vector fields $\frac{\partial}{\partial x}$ and $\frac{\partial}{\partial y}$ are principal directions. If $p \in M$ is not an umbilic point then there is a neighborhood U of p consisting of only non-umbilic points, and the frame field given by the unit eigenvectors of the shape operator is smooth and orthonormal. So it follows form Ex. 1 of section 1.4 that there exist line of curvature coordinates near p. A tangent vector $v \in TM_p$ is called *asymptotic* if $II(v, v) = 0$, and a coordinate system (x, y) is called *asymptotic* if $\frac{\partial}{\partial x}$ and $\frac{\partial}{\partial y}$ are asymptotic.

A local coordinate system on a Riemannian surface is called *isothermal* if the metric tensor is of the form $f^2(dx^2 + dy^2)$. It is well-known that on a Riemannian 2-manifold there always exists isothermal coordinates locally ([Ch2]). If (x, y) and (u, v) are two isothermal coordinate systems on M, then the coordinate change from $z = x + iy$ to $w = u + iv$ is a complex analytic function. Hence every two dimensional Riemannian manifold has a natural complex structure given by the metric.

3.1 Constant mean curvature surfaces in $N^3(c)$.

In this section we derive a special coordinate system for surfaces of $N^3(c)$ with constant mean curvature, and obtain some immediate consequences.

3.1.1. Theorem.   *Let M be an immersed surface in $N^3(c)$ with constant mean curvature H. Suppose $p_0 \in M$ is not an umbilic point. Then there is a local coordinate system (u, v) defined on a neighborhood U of p_0, which is both isothermal and a line of curvature coordinate system for M. In fact, if $\lambda_1 > \lambda_2$ denote the two principal curvatures of M then on U the two fundamental forms are:*

$$I = \frac{2}{(\lambda_1 - \lambda_2)}(du^2 + dv^2),$$

$$II = \frac{2}{(\lambda_1 - \lambda_2)}(\lambda_1 du^2 + \lambda_2 dv^2).$$

PROOF. We will prove this theorem for $H = 0$, and the proof for H being a non-zero constant is similar. We may assume that (x, y) is a line of curvature coordinate system for M near p_0, i.e.,

$$\omega_1 = A(x, y)dx, \quad \omega_2 = B(x, y)dy,$$

$$\omega_{13} = \lambda\omega_1 = \lambda A dx, \quad \omega_{23} = -\lambda\omega_2 = -\lambda B dy, \qquad (3.1.1)$$

where λ and $-\lambda$ are the principal curvatures. We may also assume that $\lambda > 0$. By Example 1.2.4 we have

$$\omega_{12} = \frac{-A_y}{B}dx + \frac{B_x}{A}dy. \qquad (3.1.2)$$

Substituting (3.1.1) and (3.1.2) to the Codazzi equations (3.0.4) we obtain

$$\lambda_y A + 2\lambda A_y = 0, \quad \lambda_x B + 2\lambda B_x = 0.$$

This implies that
$$(A\sqrt{\lambda})_y = 0, \quad (B\sqrt{\lambda})_x = 0.$$

So $A\sqrt{\lambda}$ is a function $a(x)$ of x alone, and $B\sqrt{\lambda}$ is a function $b(y)$ of y alone. Let (u, v) be the coordinate system defined by

$$du = a(x)dx, \quad dv = b(y)dy.$$

Then we have

$$I = A^2 dx^2 + B^2 dy^2 = \frac{1}{\lambda}(du^2 + dv^2),$$

$$II = \lambda(A^2 dx^2 - B^2 dy^2) = du^2 - dv^2. \quad \blacksquare$$

3.1.2. Proposition.   *Let U be an open subset of R^2 with metric $ds^2 = f^2(dx^2 + dy^2)$, and $u : U \to R$ a smooth function. Then*
(i) with respect to the dual frame $\omega_1 = f\,dx$ and $\omega_2 = f\,dy$, we have

$$\omega_{12} = -(\log f)_y dx + (\log f)_x dy, \tag{3.1.3}$$

(ii) if $u : U \to R$ is a smooth function then

$$\Delta u = \frac{u_{xx} + u_{yy}}{f^2}, \tag{3.1.4}$$

where Δ is the Laplacian with respect to ds^2,
(iii) the Gaussian curvature K of ds^2 is

$$K = -\Delta(\log f) = -\frac{(\log f)_{xx} + (\log f)_{yy}}{f^2}. \tag{3.1.5}$$

PROOF.   (i) follows from Example 1.2.4. To see (ii), note that

$$du = u_x dx + u_y dy = u_1 \omega_1 + u_2 \omega_2,$$

so

$$u_1 = u_x/f, \quad u_2 = u_y/f.$$

Set $\nabla^2 u = \sum u_{ij}\omega_i \otimes \omega_j$, then by (1.3.6)

$$du_1 + u_2 \omega_{21} = \sum_i u_{1i}\omega_i, \tag{3.1.6}$$

$$du_2 + u_1 \omega_{12} = \sum_i u_{2i}\omega_i. \tag{3.1.7}$$

Comparing coefficients of dx in (3.1.6) and dy in (3.1.7), we obtain

$$u_{11}f = (u_x/f)_x + (u_y f_y/f^2),$$

$$u_{22}f = (u_y/f)_y - (u_x f_x/f^2),$$

which implies that

$$(\Delta u)f = f_{11} + f_{22} = (u_{xx} + u_{yy})/f^2.$$

Since $d\omega_{12} = -K\omega_1 \wedge \omega_2$, (iii) follows. ∎

As a consequence of the Gauss equation (3.0.3), Theorem 3.1.1 and Proposition 3.1.2 we have

3.1.3. Theorem. *Let M be an immersed surface in $N^3(c)$ with constant mean curvature H. Let K be the Gaussian curvature, and Δ the Laplacian with respect to the induced metric on M. Then K satisfies the following equation:*

$$\Delta \, \log(H^2 - 4K + 4c) = 4K.$$

3.1.4. Theorem. *If M is an immersed surface of $N^3(c)$ with constant mean curvature H, then the traceless part of the second fundamental form of M, i.e., $II - \frac{H}{2}I$, is the real part of a holomorphic quadratic differential. In fact, if $z = x_1 + ix_2$ is an isothermal coordinate on M and $II - \frac{H}{2}I = \sum b_{ij}dx_idx_j$. Then*
 (i) $\alpha = b_{11} - ib_{12}$ is analytic,
 (ii) $II - \frac{H}{2}I = \, Re(\alpha(z)dz^2)$.

PROOF. We may assume that $\omega_1 = fdx_1$, $\omega_2 = fdx_2$, and $\omega_{i3} = \sum h_{ij}\omega_j$. Then we have

$$\omega_{12} = -(\log f)_y dx + (\log f)_x dy,$$

$$b_{11} = -b_{22} = (h_{11} - \frac{H}{2})f^2,$$

$$b_{12} = h_{12}f^2.$$

Using (1.3.7), and the fact that $h_{11} - h_{22} = 2h_{11} - H$, the covariant derivative of II is given as follows:

$$dh_{11} + 2h_{12}\omega_{21} = \sum h_{11k}\omega_k, \tag{3.1.8}$$

$$dh_{12} + (2h_{11} - H)\omega_{12} = \sum h_{12k}\omega_k. \tag{3.1.9}$$

Equating the coefficient of dx in (3.1.8) and the coefficient of dy in (3.1.9), we obtain

$$(h_{11})_x + 2h_{12}\frac{f_y}{f} = h_{111}f,$$

$$(h_{12})_y + (2h_{11} - H)\frac{f_x}{f} = h_{122}f.$$

Since H is constant and ∇ commutes with contractions, we have $h_{11k} + h_{22k} = 0$. Thus $h_{111} = -h_{221}$, which is equal to $-h_{122}$ by Proposition 2.1.3. So

$$(h_{11})_x + 2h_{12}\frac{f_y}{f} = -(h_{12})_y - (2h_{11} - H)\frac{f_x}{f}.$$

It then follows from a direct computation that

$$(b_{11})_x = (h_{11})_x f^2 + 2ff_x(h_{11} - H/2)$$
$$= -(h_{12})_y f^2 - 2h_{12}ff_y = -(b_{12})_y.$$

Similarly, by equating the coefficient of dy in (3.1.8) and the coefficient of dx in (3.1.9), we can prove that

$$(b_{11})_y = (b_{12})_x.$$

These are Cauchy-Riemann equations for α, so α is an analytic function. ∎

Since the only holomorphic differential on S^2 is zero ([Ho]), $II - \frac{H}{2}I = 0$ for any immersed sphere in $N^3(c)$ with constant mean curvature H, i.e., they are totally umbilic. Hence we have

3.1.5. Corollary ([Ho]). *If S^2 is immersed in R^3 with non-zero constant mean curvature H, then S^2 is a standard sphere embedded in R^3.*

3.1.6. Corollary ([Al],[Cb]). *If S^2 is minimally immersed in S^3, then S^2 is an equator (i.e., totally geodesic)*

3.1.7. Corollary. *If S^2 is immersed in S^3 with non-zero constant mean curvature H, then S^2 is a standard sphere, which is the intersection of S^3 and an affine hyperplane of R^4.*

Next we discuss the immersions of closed surfaces with genus greater than zero in $N^3(c)$. Given a minimal surface M in $N^3(c)$, we have associated to it a holomorphic quadratic differential Q, and locally we can find isothermal coordinate system (x, y) such that $Q = \alpha(z)dz^2$ for some analytic function $\alpha = b_{11} - ib_{12}$, and

$$I = e^{2u}(dx^2 + dy^2), \quad II = Re(\alpha(z)dz^2). \qquad (3.1.10)$$

Then

$$\omega_{12} = -u_y dx + u_x dy,$$

$$b_{11} = h_{11}e^{2u}, \quad b_{12} = h_{12}e^{2u}.$$

So

$$\det(h_{ij}) = -(b_{11}^2 + b_{12}^2)e^{-4u} = -e^{-4u}\,|\alpha|^2,$$

and the Gaussian curvature is

$$K = \det(h_{ij}) + c. \tag{3.1.11}$$

The Gauss equation (3.0.3) gives

$$u_{xx} + u_{yy} = e^{-2u}\,|\alpha|^2 - ce^{2u}, \tag{3.1.12}$$

and the Codazzi equations are exactly the Cauchy Riemann equations for α. It follows from the Fundamental Theorem 2.3.3 for surfaces in $N^3(c)$, that the following propositions are valid.

3.1.8. Proposition. *Let U be an open subset of the complex plane C, α an analytic function on U, and u a smooth function, which satisfies equation (3.1.12). Then there is a minimal immersion defined on an open subset of U such that its two fundamental forms are given by (3.1.10).*

3.1.9. Proposition ([Lw1]). *Suppose $X : M^2 \to N^3(c)$ is a minimal immersion with fundamental forms I, II, and Q is the associated holomorphic quadratic differential. Then there is a family of minimal immersions X_θ whose fundamental forms are:*

$$I_\theta = I, \quad II_\theta = Re(e^{i\theta}Q),$$

where θ is a constant.

Let M be a closed complex surface (i.e., a Riemann surface) of genus g. Then it is well-known that there is a metric ds^2 on M, whose induced complex structure is the given one, and that has constant Gaussian curvature 1, 0, or -1, for $g = 0$, $g = 1$, or $g \geq 1$ respectively.

Now we assume that (M, ds^2) is a closed surface of genus $g \geq 1$ with constant Gaussian curvature k, and Q is a holomorphic quadratic differential on M. Suppose z is a local isothermal coordinate system for M, $ds^2 = f^2|dz|^2$ and $Q = \alpha(z)dz^2$. Then $\|Q\|^2 = |\alpha|^2 f^{-4}$ is a well-defined smooth function on M (i.e., independent of the choice of z), and

$$k = -\triangle \, \log f, \tag{3.1.13}$$

where $\triangle$ is the Laplacian of ds^2. If M can be minimally immersed in S^3 such that the induced metric is conformal to ds^2, and Q is the quadratic differential associated to the immersion, then there exists a smooth function φ on M such that the induced metric is

$$I = e^{2\varphi}ds^2 = f^2 e^{2\varphi}(dx^2 + dy^2),$$

and

$$K = -e^{-4\varphi}\|Q\|^2 + c.$$

So the conformal equation (1.3.11) implies that φ satisfies the following equation:

$$1 + \Delta\varphi = -e^{2\varphi} + \|Q\|^2 e^{-2\varphi}, \tag{3.1.14}$$

for $g > 1$, or

$$\Delta\varphi = -e^{2\varphi} + \|Q\|^2 e^{-2\varphi}, \tag{3.1.15}$$

for $g = 1$, where Δ is the Laplacian for the metric ds^2. These equations are the same as the Gauss equation.

If $g = 1$, then M is a torus, so we may assume that $M \simeq R^2/\Lambda$, where Λ is the integer lattice generated by $(1,0)$, and $(r\cos\theta, r\sin\theta)$, $ds^2 = |dz|^2$, and $\|Q\|^2$ is a constant a. Then equation (3.1.15) become

$$\Delta\varphi = -e^{2\varphi} + ae^{-2\varphi}. \tag{3.1.16}$$

Let $b = \frac{1}{4}\log a$, and $u = \varphi - b$. Then (3.1.16) becomes

$$\Delta u = -2\sqrt{a}\ \sinh(2u).$$

So one natural question that arises from this discussion is: For what values of r and θ is there a doubly periodic smooth solution for

$$u_{xx} + u_{yy} = a\ \sinh u, \tag{3.1.17}$$

with periods $(1,0)$, and $(r\cos\theta, r\sin\theta)$?

If $g > 1$, then there are two open problems that arise naturally from the above discussion:

(i) Fix one complex structure on a closed surface M with genus $g > 1$, and determine the set of quadratic differentials Q on M such that (3.1.14) admits smooth solutions on M.

(ii) Fix a smooth closed surface M with genus $g > 1$ and determine the possible complex structures on M such that the set in (i) is not empty.

However the understanding of the equation (3.1.14) on closed surfaces is only a small step toward the classification of closed minimal surfaces of S^3, because a solution of these equations on a closed surface need not give a closed minimal surface of S^3. In the following we will discuss where the difficulties lie. Suppose u is a doubly periodic solution for (3.1.17), i.e., u is a solution on a torus. Then the coefficients τ of the first order system of partial differential equations

$$dF = \tau F, \tag{3.1.18}$$

as in the fundamental theorem 2.2.5 for surfaces in S^3, are doubly periodic. But the solution F need *not* to be doubly periodic, i.e., such u need not give an immersed minimal torus of S^3. For example, if we assume that u depends only on x, then (3.1.17) reduces to an ordinary differential equation, $u'' = a \sinh u$, which always has periodic solution. But it was proved by Hsiang and Lawson in [HL] that there are only countably many immersed minimal tori in S^3, that admit an S^1-action. If the closed surface M has genus greater than one, then for a given solution u of (3.1.14), the local solution of the corresponding system (3.1.18) may not close up to a solution on M (the period problem is more complicated than for the torus case).

Let (M, ds^2) be a closed surface with constant curvature k, and $d\tilde{s}^2 = e^{2\varphi}ds^2$. Suppose $(M, d\tilde{s}^2)$ is isometrically immersed in $N^3(c)$ with constant mean curvature H, and Q is the associated holomorphic quadratic differential. Then we have

$$e^{-4\varphi}\|Q\|^2 = -\det(h_{ij}) + H^2/4,$$

and φ satisfies the conformal equation (1.3.11):

$$-k + \triangle\varphi = \|Q\|^2 e^{-2\varphi} - (H^2/4 + c)e^{2\varphi}, \tag{3.1.19}$$

where $\triangle$ is the Laplacian for ds^2. Moreover (3.1.19) is the Gauss equation for the immersion. Note that if $X : M \to R^3$ is an immersion with mean curvature $H \neq 0$ and a is a non-zero constant, then aX is an immersion with mean curvature H/a and the induced metric on M via aX is conformal to that of X. So for the study of constant mean curvature surfaces of R^3, we may assume that $H = 2$. Then (3.1.19) is the same as the above equations for minimal surfaces of S^3. It is known that the only *embedded* closed surface (no assumption on the genus) with constant mean curvature in R^3 is the standard sphere (for a proof see [Ho]), and Hopf conjectured that there is no immersed closed surface of genus bigger than 0 in R^3 with non-zero constant mean curvature. Recently Wente found counter examples for this conjecture, he constructed many immersed tori of R^3 with constant mean curvature ([We]).

Exercises.

1. Suppose (M, g) is a Riemannian surface, and (x, y), (u, v) are local isothermal coordinates for g defined on U_1 and U_2 respectively. Then

the coordinate change from $z = x + iy$ to $w = u + iv$ on $U_1 \cap U_2$ is a complex analytic function.

3.2 Surfaces of R^3 with constant Gaussian curvature.

In the classical surface theory, a *congruence of lines* is an immersion $f : U \to Gr$, where U is an open subset of R^2 and Gr is the Grassman manifold of all lines in R^3 (which need not pass through the origin). We may assume that $f(u, v)$ is the line passes through $p(u, v)$ and parallel to the unit vector $\xi(u, v)$ in R^3. Let $t(u, v)$ be a smooth function. Then a necessary and sufficient condition for

$$X(u, v) = p(u, v) + t(u, v)\xi(u, v)$$

to be an immersed surface of R^3 such that $\xi(u, v)$ is tangent to the surface at $X(u, v)$ is

$$\det(\xi, X_u, X_v) = 0.$$

This gives the following quadratic equation in t:

$$\det(\xi, \ p_u + t \ \xi_u, \ p_v + t \ \xi_v) = 0,$$

which generically has two distinct roots. So given a congruence of lines there exist two surfaces M and M^* such that the lines of the congruence are the common tangent lines of M and M^*. They are called *focal surfaces* of the congruence. There results a mapping $\ell : M \to M^*$ such that the congruence is given by the line joining $P \in M$ to $\ell(P) \in M^*$. This simple construction plays an important role in the theory of surface transformations.

We rephrase this in more current terminology:

3.2.1. Definition. A *line congruence* between two surfaces M and M^* in R^3 is a diffeomorphism $\ell : M \to M^*$ such that for each $P \in M$, the line joining P and $P^* = \ell(P)$ is a common tangent line for M and M^*. The line congruence ℓ is called *pseudo-spherical* (p.s.), or a *Bäcklund transformation*, if

(i) $\|\overrightarrow{PP^*}\| = r$, a constant independent of P.

(ii) The angle between the normals ν_P and ν_{P^*} at P and P^* is a constant θ independent of P.

The following theorems were proved over a hundred years ago:

3.2.2. Bäcklund Theorem.   *Suppose $\ell : M \to M^*$ is a p.s. congruence in R^3 with distance r and angle $\theta \neq 0$. Then both M and M^* have constant negative Gaussian curvature equal to* $-\frac{\sin^2 \theta}{r^2}$.

PROOF.   There exists a local orthonormal frame field e_1, e_2, e_3 on M such that $\overrightarrow{PP^*} = re_1$, and e_3 is normal to M. Let

$$\begin{aligned}
e_1^* &= -e_1, \\
e_2^* &= \cos\theta\, e_2 + \sin\theta\, e_3, \\
e_3^* &= -\sin\theta\, e_2 + \cos\theta\, e_3.
\end{aligned} \tag{3.2.1}$$

Then $\{e_1^*, e_2^*\}$ is an orthonormal frame field for TM^*. If locally M is given by the immersion $X : U \to \mathbf{R}^3$, then M^* is given by

$$X^* = X + r\, e_1. \tag{3.2.2}$$

Taking the exterior derivative of (3.2.2), we get

$$\begin{aligned}
dX^* &= dX + r\,de_1 \\
&= \omega_1 e_1 + \omega_2 e_2 + r(\omega_{12}e_2 + \omega_{13}e_3) \\
&= \omega_1 e_1 + (\omega_2 + r\omega_{12})e_2 + r\omega_{13}e_3.
\end{aligned} \tag{3.2.3}$$

On the other hand, letting ω_1^*, ω_2^* be the dual coframe of e_1^*, e_2^*, we have

$$\begin{aligned}
dX^* &= \omega_1^* e_1^* + \omega_2^* e_2^*, \quad \text{using}(3.2.1) \\
&= -\omega_1^* e_1 + \omega_2^*(\cos\theta\, e_2 + \sin\theta\, e_3).
\end{aligned} \tag{3.2.4}$$

Comparing coefficients of e_1, e_2, e_3 in (3.2.3) and (3.2.4), we get

$$\begin{aligned}
\omega_1^* &= -\omega_1, \\
\cos\theta\, \omega_2^* &= \omega_2 + r\omega_{12}, \\
\sin\theta\, \omega_2^* &= r\omega_{13}.
\end{aligned} \tag{3.2.5}$$

This gives

$$\omega_2 + r\omega_{12} = r\cot\theta\, \omega_{13}. \tag{3.2.6}$$

In order to compute the curvature, we compute the following 1-forms:

$$\begin{aligned}
\omega_{13}^* &= \langle de_1^*, e_3^* \rangle \\
&= -\langle de_1, -\sin\theta\, e_2 + \cos\theta\, e_3 \rangle \\
&= \sin\theta\, \omega_{12} - \cos\theta\, \omega_{13}, \quad \text{using (3.2.6)} \\
&= -\frac{\sin\theta}{r}\omega_2, \\
\omega_{23}^* &= \langle de_2^*, e_3^* \rangle \\
&= \langle \cos\theta\, de_2 + \sin\theta\, de_3, -\sin\theta\, e_2 + \cos\theta\, e_3 \rangle \\
&= \omega_{23}.
\end{aligned} \tag{3.2.7}$$

By the Gauss equation (3.0.3), we have

$$\Omega_{12}^* = \omega_{13}^* \wedge \omega_{23}^*, \text{ using } (3.2.7)$$

$$= -\frac{\sin\theta}{r}\omega_2 \wedge \omega_{23}$$

$$= \frac{\sin\theta}{r}h_{12}\omega_1 \wedge \omega_2 = \frac{\sin\theta}{r}\omega_1 \wedge \omega_{13}$$

$$= -(\frac{\sin\theta}{r})^2\omega_1^* \wedge \omega_2^*,$$

i.e., M^* has constant curvature $-(\frac{\sin\theta}{r})^2$. By symmetry, M also has Gaussian curvature $-(\frac{\sin\theta}{r})^2$. ∎

3.2.3. Integrability Theorem. *Let M be an immersed surface of R^3 with constant Gaussian curvature -1, $p_0 \in M$, v_0 a unit vector in TM_{p_0}, and r, θ constants such that $r = \sin\theta$. Then there exist a neighborhood U of M at p_0, an immersed surface M^*, and a p.s. congruence $\ell : U \to M^*$ such that the vector joining p_0 and $p_0^* = \ell(p_0)$ is equal to rv_0 and θ is the angle between the normal planes at p_0 and p_0^*.*

PROOF. A unit tangent vector field e_1 on M determines a local orthonormal frame field e_1, e_2, e_3 such that e_3 is normal to M. In order to find the p.s. congruence, it suffices to find a unit vector field e_1 such that the corresponding frame field satisfies the differential system (3.2.6), i.e.,

$$\tau = \omega_2 + \sin\theta\,\omega_{12} - \cos\theta\,\omega_{13} = 0. \tag{3.2.8}$$

Since the curvature of M is equal to -1, the Gauss equation (3.0.3) implies that

$$d\omega_{12} = \omega_1 \wedge \omega_2, \quad \omega_{13} \wedge \omega_{23} = -\omega_1 \wedge \omega_2. \tag{3.2.9}$$

Using (3.2.8) and (3.2.9), we compute directly:

$$d\tau = \omega_{21} \wedge \omega_1 + \sin\theta\,\omega_1 \wedge \omega_2 - \cos\theta\,\omega_{12} \wedge \omega_{23}$$

$$= -\omega_{12} \wedge (\omega_1 + \cos\theta\,\omega_{23}) + \sin\theta\,\omega_1 \wedge \omega_2,$$

$$\equiv \frac{1}{\sin\theta}(-\cos\theta\,\omega_{13} + \omega_2) \wedge (\omega_1 + \cos\theta\,\omega_{23}) + \sin\theta\,\omega_1 \wedge \omega_2, \text{ mod } \tau,$$

$$= \frac{1}{\sin\theta}(-1 + \cos^2\theta + \cos\theta\,h_{12} - \cos\theta\,h_{21})\omega_1 \wedge \omega_2 + \sin\theta\,\omega_1 \wedge \omega_2,$$

which is 0, because $h_{12} = h_{21}$. Then the result follows from the Frobenius theorem. ∎

The proof of the following theorem is left as an exercise.

3.2.4. Bianchi's Permutability Theorem.   *Let $\ell_1 : M_0 \to M_1$ and $\ell_2 : M_0 \to M_2$ be p.s. congruences in R^3 with angles θ_1, θ_2 and distance $\sin\theta_1, \sin\theta_2$ respectively. If $\sin\theta_1 \neq \sin\theta_2$, then there exist a unique hyperbolic surface M_3 in R^3 and two p.s. congruences $\ell_1^* : M_1 \to M_3$ and $\ell_2^* : M_2 \to M_3$ with angles θ_2, θ_1 respectively, such that $\ell_1^*(\ell_1(p)) = \ell_2^*(\ell_2(p))$ for all $p \in M_0$. Moreover M_3 is obtained by an algebraic method.*

Next we will discuss some special coordinates for surfaces immersed in R^3 with constant Gaussian curvature -1, and their relations to the Bäcklund transformations.

3.2.5. Theorem.   *Suppose M is an immersed surface of R^3 with constant Gaussian curvature $K \equiv -1$. Then there exists a local coordinate system (x, y) such that*

$$I = \cos^2\varphi \, dx^2 + \sin^2\varphi \, dy^2, \qquad (3.2.10)$$

$$II = \sin\varphi\cos\varphi(dx^2 - dy^2), \qquad (3.2.11)$$

and $u = 2\varphi$ satisfies the Sine-Gordon equation (SGE):

$$u_{xx} - u_{yy} = \sin u. \qquad (3.2.12)$$

This coordinate system is called the Tchebyshef curvature coordinate system.

PROOF.   Since $K = -1$, there is no umbilic point on M. So we may assume (p, q) are line of curvature coordinates and $\lambda_1 = \tan\varphi$, $\lambda_2 = -\cot\varphi$, i.e.,

$$\omega_1 = A(p, q)\, dp, \quad \omega_2 = B(p, q)\, dq,$$

$$\omega_{13} = \tan\varphi\,\omega_1 = \tan\varphi A\, dp, \quad \omega_{23} = -\cot\varphi\,\omega_2 = -\cot\varphi B\, dq.$$

By Example 1.2.4, we have

$$\omega_{12} = \frac{-A_q}{B}\, dp + \frac{B_p}{A}\, dq.$$

Substituting the above 1-forms in the Codazzi equations (3.0.4) we obtain

$$A_q \cos\varphi + A\varphi_q \sin\varphi = 0, \quad B_p \sin\varphi - B\varphi_p \cos\varphi = 0,$$

which implies that $\frac{A}{\cos\varphi}$ is a function $a(p)$ of p alone and $\frac{B}{\sin\varphi}$ is a function $b(q)$ of q alone. Then the new coordinate system (x, y), defined by $dx = a(p)\, dp$, $dy = b(q)\, dq$, gives the fundamental forms as in the theorem.

With respect to the coordinates (x, y) we have

$$\omega_{12} = \varphi_y \, dx + \varphi_x \, dy,$$

and the Gauss equation (3.0.3) becomes

$$\varphi_{xx} - \varphi_{yy} = \sin \varphi \cos \varphi,$$

i.e., $u = 2\varphi$ is a solution for the Sine-Gordon equation. $\blacksquare$

Note that the coordinates (s, t), where

$$x = s + t \quad y = s - t,$$

are asymptotic coordinates, the angle u between the asymptotic curves, i.e., the s-curves and t-curves, is equal to 2φ, and

$$I = ds^2 + 2 \cos u \, ds \, dt + dt^2, \tag{3.2.13}$$

$$II = 2 \sin u \, ds \, dt. \tag{3.2.14}$$

(s, t) are called the *Tchebyshef* coordinates. The Sine-Gordon equation becomes

$$u_{st} = \sin u. \tag{3.2.15}$$

3.2.6. Hilbert Theorem.   *There is no isometric immersion of the simply connected hyperbolic 2-space H^2 into R^3.*

PROOF.   Suppose H^2 can be isometrically immersed in R^3. Because $\lambda_1 \lambda_2 = -1$, there is no umbilic points on H^2, and the principal directions gives a global orthonormal tangent frame field for H^2. It follows from the fact that H^2 is simply connected that the line of curvature coordinates (x, y) in Theorem 3.2.5 is defined for all $(x, y) \in R^2$, and so is the Tchebyshef coordinates (s, t). They are global coordinate systems for H^2. Then using (3.2.10) and (3.2.12), the area of the immersed surface can be computed as follows:

$$\int_{R^2} \omega_1 \wedge \omega_2 = \int_{R^2} \sin \varphi \cos \varphi \, dx \wedge dy$$

$$= -\int_{R^2} \sin(2\varphi) \, ds \wedge dt = -\int_{R^2} 2\varphi_{st} \, ds \wedge dt$$

$$= -\lim_{a \to \infty} \int_{D_a} 2\varphi_{st} \, ds \wedge dt = -\lim_{a \to \infty} \int_{\partial D_a} -\varphi_s \, ds + \varphi_t \, dt,$$

where D_a is the square in the (s,t) plane with $P(-a,-a)$, $Q(a,-a)$, $R(a,a)$ and $S(-a,a)$ as vertices, and ∂D_a is its boundary. The last line integral can be easily seen to be

$$2(\varphi(Q) + \varphi(S) - \varphi(P) - \varphi(R)).$$

Since $I = \cos^2 \varphi dx^2 + \sin^2 \varphi dy^2$ is the metric on H^2, $\sin \varphi$ and $\cos \varphi$ never vanish. Hence we may assume that the range of φ is contained in the interval $(0, \pi/2)$, which implies that the area of the immersed surface is less than 4π. On the other hand, the metric on H^2 can also be written as $(dx^2 + dy^2)/y^2$ for $y > 0$ and the area of H^2 is

$$\int_{-\infty}^{\infty} \int_{0}^{\infty} \frac{1}{y} \, dy \, dx,$$

which is infinite, a contradiction. ∎

It follows from the fundamental theorem of surfaces in R^3 that there is a bijective correspondence between the local solutions u of the Sine-Gordon equation (3.2.12) whose range is contained in the interval $(0, \pi)$ and the immersed surfaces of R^3 with constant Gaussian curvature -1. In fact, using the same proof as for the Fundamental Theorem, we obtain bijection between the global solutions u of the Sine-Gordon equation (3.2.12) and the smooth maps $X : R^2 \to R^3$ which satisfy the following conditions:

(i) rank $X \geq 1$ everywhere,

(ii) if X is of rank 2 in an open set U of R^2, then $X|U$ is an immersion with Gaussian curvature -1.

Theorem 3.2.3 and 3.2.4 give methods of generating new surfaces of R^3 with curvature -1 from a given one. So given a solution u of the SGE (3.2.12), we can use these theorems to obtain a new solution of the SGE by the following three steps:

(1) Use the fundamental theorem of surfaces to construct a hyperbolic surface M of R^3 with (3.2.10) and (3.2.11) as its fundamental forms with $\varphi = u/2$.

(2) Solve the first order system (3.2.9) of partial differential equations on M to get a family of new hyperbolic surfaces M_θ in R^3.

(3) On each M_θ, find the Tchebyshef coordinate system, which gives a new solution u_θ for the SGE.

However, the first and third steps in this process may not be easier than solving SGE. Fortunately, the following theorem shows that these steps are not necessary.

3.2.7. Theorem.   Let $\ell : M \to M^*$ be a p.s. congruence with angle θ and distance $\sin \theta$. Then the Tchebyshef curvature coordinates of M and M^* correspond under ℓ.

PROOF.   Let (x, y) be the line of curvature coordinates of M as in Theorem 3.2.5, and φ the angle associated to M, i.e.,

$$I = \cos^2 \varphi \, dx^2 + \sin^2 \varphi \, dy^2, \quad II = \cos \varphi \sin \varphi (dx^2 - dy^2).$$

Let $v_1 = \frac{1}{\cos \varphi} \frac{\partial}{\partial x}$, $v_2 = \frac{1}{\sin \varphi} \frac{\partial}{\partial y}$ (the principal directions), τ_1, τ_2 the dual coframe, and τ_{AB} the corresponding connection 1-forms. Then we have

$$\tau_1 = \cos \varphi \, dx, \quad \tau_2 = \sin \varphi \, dy,$$

$$\tau_{12} = \varphi_y \, dx + \varphi_x \, dy,$$

$$\tau_{13} = \tan \varphi \, \tau_1 = \sin \varphi \, dx, \quad \tau_{23} = -\cot \varphi \, \tau_2 = -\cos \varphi \, dy.$$

Use the same notation as in the proof of Theorem 3.2.2, and suppose

$$e_1 = \cos \alpha v_1 + \sin \alpha v_2, \quad e_2 = -\sin \alpha v_1 + \cos \alpha v_2, \qquad (3.2.16)$$

where e_1 is the congruence direction. We will show that the angle associated to M^* is α. It is easily seen that

$$\omega_1 = \cos \alpha \cos \varphi \, dx + \sin \alpha \sin \varphi \, dy,$$

$$\omega_2 = -\sin \alpha \cos \varphi \, dx + \cos \alpha \sin \varphi \, dy,$$

$$\omega_{13} = \langle de_1, e_3 \rangle = \cos \alpha \sin \varphi \, dx - \sin \alpha \cos \varphi \, dy,$$

$$\omega_{23} = \langle de_2, e_3 \rangle = -\sin \alpha \sin \varphi \, dx - \cos \alpha \cos \varphi \, dy.$$

Using (3.2.5), the first fundamental forms of M^* can be computed directly as follows:

$$\begin{aligned}
I^* &= (\omega_1^*)^2 + (\omega_2^*)^2 \\
&= (\omega_1)^2 + (\omega_{13})^2 \\
&= (\cos \alpha \cos \varphi \, dx + \sin \alpha \sin \varphi \, dy)^2 + (\cos \alpha \sin \varphi \, dx - \sin \alpha \cos \varphi \, dy)^2 \\
&= \cos^2 \alpha \, dx^2 + \sin^2 \alpha \, dy^2.
\end{aligned}$$

Similarly,

$$\begin{aligned}
II^* &= \omega_1^* \omega_{13}^* + \omega_2^* \omega_{23}^* \\
&= \omega_1 \omega_2 + \omega_{13} \omega_{23} \\
&= \cos \alpha \sin \alpha (-dx^2 + dy^2). \quad \blacksquare
\end{aligned}$$

Using the same notation as in the proof of Theorem 3.2.7, we have $\tau_{12} = \varphi_y dx + \varphi_x dy$, and $\omega_{12} = \tau_{12} + d\alpha$. Comparing coefficients of dx, dy in (3.2.8), we get the Bäcklund transformation for the SGE (3.2.12):

$$\begin{cases} \alpha_x + \varphi_y = -\cot \theta \, \cos \varphi \, \sin \alpha + \csc \theta \, \sin \alpha \, \cos \varphi, \\ \alpha_y + \varphi_x = -\cot \theta \, \sin \varphi \, \cos \alpha - \csc \theta \, \cos \alpha \, \sin \varphi. \end{cases} \qquad (3.2.17)$$

The integrability theorem 3.2.3 implies that (3.2.17) is solvable, if φ is a solution for (3.2.12). And Theorem 3.2.7 implies that the solution α for (3.2.17) is also a solution for (3.2.12).

The classical Bäcklund theory for the SGE played an important role in the study of soliton theory (see [Lb]). Both the geometric and analytic aspects of this theory were generalized in [18:39], [Te1] for hyperbolic n-manifolds in $\boldsymbol{R}^{2n-1}$.

É. Cartan proved that a small piece of the simply connected hyperbolic space $\boldsymbol{H}^n$ can be isometrically embedded in $\boldsymbol{R}^{2n-1}$, and it cannot be locally isometrically embedded in $\boldsymbol{R}^{2n-2}$ ([Ca1,2], [Mo]). It is still not known whether the Hilbert theorem 3.2.6 is valid for $n > 2$, i.e., whether or not $\boldsymbol{H}^n$ can be isometrically immersed in $\boldsymbol{R}^{2n-1}$?

Exercises.

1. Let M be an immersed surface in $\boldsymbol{R}^3$. Two tangent vectors u and v of M at x are conjugate if $II(u,v) = 0$. Two curves α and β on M are conjugate if $\alpha'(t)$ and $\beta'(t)$ are conjugate vectors for all t. Let $\ell : M \to M^*$ be a line congruence in $\boldsymbol{R}^3$, e_1 and e_1^* denote the common tangent direction on M and M^* respectively. Then the integral curves of e_1^* and $d\ell(e_1)$ are conjugate curves on M^*.

2. Prove Theorem 3.2.4.

3. Let M_i be as in Theorem 3.2.4, and φ_i the angle associated to M_i. Show that

$$\tan \frac{\varphi_3 - \varphi_0}{2} = \frac{\cos \theta_2 - \cos \theta_1}{\cos(\theta_1 - \theta_2) - 1} \tan \frac{\varphi_2 - \varphi_1}{2}.$$

3.3   Immersed flat tori in S^3.

Suppose M is an immersed surface in S^3 with $K = 0$. Since $K = 1 + \det(h_{ij})$, we have $\det(h_{ij}) = -1$. So using a proof similar to that for Theorem 3.2.5, we obtain the following local results for immersed flat surfaces in S^3.

3.3.1.   Theorem.   *Let M be an immersed surface in S^3 with Gaussian curvature 0. Then locally there exist line of curvature coordinates (x,y) such that*

$$I = \cos^2 \varphi \, dx^2 + \sin^2 \varphi \, dy^2, \tag{3.3.1}$$

$$II = \sin \varphi \cos \varphi (dx^2 - dy^2), \tag{3.3.2}$$

where φ satisfies the linear wave equation:

$$\varphi_{xx} - \varphi_{yy} = 0. \tag{3.3.3}$$

Let $u = 2\varphi$, $x = s + t$, and $y = s - t$. Then we have

3.3.2. Corollary.   *Let M be an immersed surface in S^3 with Gaussian curvature 0. Then locally there exist asymptotic coordinates (s, t) (Tchebyshef coordinates) such that*

$$I = ds^2 + 2\cos u\, ds dt + dt^2, \tag{3.3.4}$$

$$II = 2\sin u\, ds dt, \tag{3.3.5}$$

where u is the angle between the asymptotic curves and

$$u_{st} = 0, \tag{3.3.6}$$

Suppose M is an immersed surface of S^3 with $K = 0$. Let e_A be the frame field such that

$$e_1 = \frac{1}{\cos \varphi} \frac{\partial}{\partial x}, \quad e_2 = \frac{1}{\sin \varphi} \frac{\partial}{\partial y}, \quad e_4 = X,$$

and e_3 normal to M in S^3. Using the same notation as in section 2.1, we have

$$\omega_1 = \cos \varphi\, dx, \quad \omega_2 = \sin \varphi\, dy, \quad \omega_{12} = \varphi_y\, dx + \varphi_x\, dy,$$
$$\omega_{13} = \sin \varphi\, dx, \quad \omega_{23} = -\cos \varphi\, dy,$$
$$\omega_{14} = -\cos \varphi\, dx, \quad \omega_{24} = -\sin \varphi\, dy, \quad \omega_{34} = 0.$$

Then

$$dg = \Theta g,$$

where g is the $O(4)$–valued map whose i^{th} row is e_i, and $\Theta = (\omega_{AB})$.

Conversely, given a solution φ of (3.3.3), let I, II be given as in Theorem 3.3.1. Then (3.3.3) implies that the Gaussian curvature of the metric I is 0. Moreover, the Gauss and Codazzi equations are satisfied. So by the fundamental theorem of surfaces in S^3 (Theorem 2.2.3), there exists an immersed local surface in S^3 with zero curvature. In fact (see section 2.1), the system for $g : R^2 \to O(4)$:

$$dg = \Theta g, \tag{3.3.7}$$

is solvable, and the fourth row of g gives an immersed surface into S^4 with I, II as fundamental forms.

Similarly, we can also use the Tchebyshef coordinates and the following frame to write down the immersion equation. Let v_1, v_2, v_3, v_4 be the local orthonormal frame field such that $v_1 = \frac{\partial}{\partial s}$, and $v_3 = e_3, v_4 = e_4$. So

$$v_1 = \cos \varphi \, e_1 + \sin \varphi \, e_2, \quad v_2 = -\sin \varphi \, e_1 + \cos \varphi \, e_2.$$

Let τ_i be the dual of v_i, $\tau_{AB} = \langle dv_A, v_B \rangle$, and $u = 2\varphi$. Then we have

$$\tau_1 = ds + \cos u \, dt, \quad \tau_2 = -\sin u \, dt, \quad \tau_{12} = u_s ds,$$
$$\tau_{13} = \sin u \, dt, \quad \tau_{23} = -ds + \cos u \, dt,$$
$$\tau_{14} = -\tau_1, \quad \tau_{24} = -\tau_2,$$

where $u = 2\varphi$. The corresponding $o(4)$–valued 1-form as in the fundamental theorem of surfaces in S^3 is $\tau = P \, ds + Q \, dt$, where

$$P = \begin{pmatrix} 0 & u_s & 0 & -1 \\ -u_s & 0 & -1 & 0 \\ 0 & 1 & 0 & 0 \\ 1 & 0 & 0 & 0 \end{pmatrix},$$

$$Q = \begin{pmatrix} 0 & 0 & \sin u & -\cos u \\ 0 & 0 & \cos u & \sin u \\ -\sin u & -\cos u & 0 & 0 \\ \cos u & -\sin u & 0 & 0 \end{pmatrix}.$$

If $u_{st} = 0$, then the following system for $g : R^2 \to O(4)$:

$$dg = \tau g, \tag{3.3.8}$$

is solvable, and the fourth row of g gives an immersed surface into S^4 with (3.3.4) (3.3.5) as fundamental forms and $K = 0$. Note that (3.3.8) can be rewritten as

$$\begin{cases} g_s = Pg, \\ g_t = Qg. \end{cases} \tag{3.3.9}$$

Every solution of (3.3.6) is of the form $\xi(s) + \eta(t)$, and in the following we will show that (3.3.9) reduces to two ordinary differential equations.

Identifying R^4 with the 2–dimensional complex plane C^2 via the map

$$F(x_1, \ldots, x_4) = (x_1 + ix_2, ix_3 + x_4),$$

we have

$$P = \begin{pmatrix} i\xi' & -1 \\ 1 & 0 \end{pmatrix}.$$

The first equation of (3.3.9) gives a system of ODE:

$$z' = i\xi'z - w, \quad w' = z, \tag{3.3.10}$$

which is equivalent to the second order equation for $z : R \rightarrow C$:

$$z'' + i\xi' z' + z = 0. \tag{3.3.11}$$

Identifying R^4 with the C^2 via the map

$$F(x_1, \ldots, x_4) = (x_1 + ix_2, x_3 + ix_4),$$

we have

$$Q = \begin{pmatrix} 0 & ie^{-iu} \\ ie^{iu} & 0 \end{pmatrix}.$$

And the second equation in (3.3.9) gives a system of ODE:

$$z' = ie^{-iu}w, \quad w' = ie^{iu}z, \tag{3.3.12}$$

which is equivalent to the second order equation for $z : R \rightarrow C$:

$$z'' + i\eta'z' + z = 0. \tag{3.3.13}$$

So the study of the flat tori in S^3 reduces to the study of the above ODE.

In the following we describe some examples given by Lawson: Let $S^3 = \{(z, w) \in C^2 \mid |z|^2 + |w|^2 = 1\}$. then $CP^1 \simeq S^2$ is obtained by identifying $(z, w) \in S^3$ with $e^{i\theta}(z, w)$, and the quotient map $\pi : S^3 \rightarrow S^2$ is the Hopf fibration. If $\gamma = (x, y, z) : S^1 \rightarrow S^2$ is an immersed closed curve on S^2, then $\pi^{-1}(\gamma)$ is an immersed flat torus of S^3. In fact, $X(\sigma, \theta) = e^{i\theta}(x(\sigma), y(\sigma) + iz(\sigma))$ gives a parametrization for the torus. It follows from direct computation that this torus has curvature zero, the θ−curves are asymptotic, the Tchebyshef coordinates (s, t) are given by $t = \sigma$ and $s = \theta + \alpha(\sigma)$ for some function α, and the corresponding angle u as in the above Corollary depends only on t. These s−curves are great circles, but the other family of asymptotics (the t−curves) in general need not be closed curves. It is not known whether these examples are the only flat tori in S^3.

3.4 Bonnet transformations.

Let M be an immersed surface in $N^3(c)$, and e_3 its unit normal vector. The *parallel set* M_t of constant distance t to M is defined to be $\{\exp_x(te_3(x)) \mid x \in M\}$. Note that

$$\exp_x(tv) = \begin{cases} x + tv, & \text{if } c = 0; \\ \cos t\, x + \sin t\, v, & \text{if } c = 1; \\ \cosh t\, x + \sinh t\, v, & \text{if } c = -1. \end{cases}$$

If M_t is an immersed surface, then we call it a *parallel surface*. The classical Bonnet transformation is a transformation from a surface in $\mathbf{R}^3$ to one of its parallel sets. Bonnet's Theorem can be stated as follows:

3.4.1. Theorem.   *Let* $X : M^2 \to \mathbf{R}^3$ *be an immersed surface, e_3 its unit normal field, and H, K the mean curvature and Gaussian curvature of M.*

(i) If $H = a \neq 0$, and K never vanishes, then the parallel set $M_{1/a}$ (defined by the map $X^* = X + \frac{1}{a}e_3$) is an immersed surface with constant Gaussian curvature a^2.

(ii) If K is a positive constant a^2 and suppose that M has no umbilic points, then its parallel set $M_{1/a}$ is an immersed surface with mean curvature $-a$.

This theorem is a special case of the following simple result :

3.4.2. Theorem.   *Let $X : M^2 \to N^3(c)$ be an immersed surface, e_3 its unit normal field, and A the shape operator of M. Then the parallel set M^* of constant distance t to M defined by*

$$X^* = aX + be_3 \tag{3.4.1}$$

is an immersion if and only if $(aI - bA)$ is non-degenerate on M, where $(a, b) = (1, 0)$ for $c = 0$, $(\cos t, \sin t)$ for $c = 1$, and $(\cosh t, \sinh t)$ for $c = -1$. Moreover, $e_3^* = -cbX + ae_3$ is a unit normal field of M^*, and the corresponding shape operator is

$$A^* = (cb + aA)(a - bA)^{-1}. \tag{3.4.2}$$

PROOF.   We will consider only the case $c = 0$. The other cases are similar. Let e_A be an adapted local frame for the immersed surface M in $\mathbf{R}^3$ as in section 2.3. Taking the differential of (3.4.1), we get

$$dX^* = dX + tde_3,$$
$$= \sum \omega_i e_i - t \sum h_{ij}\omega_i e_j, \tag{3.4.3}$$
$$= \sum (\delta_{ij} - t\,h_{ij})\omega_i e_j = I - tA$$

Hence X^* is an immersion if and only if $(I - tA)$ is non-degenerate. It also follows from (3.4.3) that e_A is an adapted frame for M^*, and the dual coframe is $\omega_i^* = \sum_j (\delta_{ij} - h_{ij})\omega_j$. Moreover, $\omega_{i3}^* = \langle de_i^*, e_3^* \rangle = \omega_{i3}$, so we have

$$A^* = A(I - tA)^{-1}. \quad \blacksquare \qquad\qquad (3.4.4)$$

3.4.3. Corollary. *If M^2 is an immersed Weingarten surface in $N^3(c)$ then so is each of its regular parallel surfaces. Conversely, if one of the parallel surface of M is Weingarten then M is Weingarten.*

Let λ_1, λ_2 be the principal curvatures for the immersed surface M in $N^3(c)$, and λ_1^*, λ_2^* the principal curvatures for the parallel surface M^*. Then (3.4.4) becomes

$$\lambda_i^* = (cb + a\lambda_i)/(a - b\lambda_i).$$

As consequences of Theorem 3.4.2, we have

3.4.4. Corollary. *Suppose $X : M^2 \to S^3$ has constant Gaussian curvature $K = (1 + r^2) > 1$, and $t = \tan^{-1}(1/r)$. Then*

$$X^* = \cos t\, X + \sin t\, e_3$$

is a branched immersion with constant mean curvature $(1 - r^2)/r$.

3.4.5. Corollary. *Suppose $X : M^2 \to S^3$ has constant mean curvature $H = r$, and $t = \cot^{-1}(r/2)$. Then*

$$X^* = \cos t\, X + \sin t\, e_3$$

is a branched immersion with constant mean curvature $-r$.

We note that when $r = 0$ the above corollary says that the unit normal of a minimal surface M in S^3 gives a branched minimal immersion of M in S^3. This was proved by Lawson [Lw1], who called the new minimal surface the *polar variety* of M.

3.4.6. Corollary. *Suppose $X : M^2 \to H^3$ has constant Gaussian curvature $K = (-1 + r^2) > 2$, and $t = \tanh^{-1}(1/r)$. Then*

$$X^* = \cosh t\, X + \sinh t\, e_3$$

is a branched immersion with constant mean curvature $(1 + r^2)/r$.

3.4.7. Corollary.   *Suppose* $X : M^2 \to H^3$ *has constant mean curvature* $H = r, r > 2,$ *and* $t = \tanh^{-1}(2/r)$. *Then*

$$X^* = \cosh t\, X + \sinh t\, e_3$$

is a branched immersion with constant mean curvature $-r$.

Exercises.

1. Prove an analogue of Theorem 3.4.2 for immersed hypersurfaces in $N^{n+1}(c)$.
2. Suppose M^3 is an immersed, orientable, minimal hypersurface of S^4 and the Gauss-Kronecker (i.e., the determinant of the shape operator) never vanishes on M. Use the above exercise to show that $\pm e_4 : M^3 \to S^4$ is an immersion, and the induced metric on M has constant scalar curvature 6 ([De]).

Chapter 4.
Focal Points.

One important method for obtaining information on the topology of an immersed submanifold M^n of R^{n+k} is applying Morse theory to the Euclidean distance functions of M. This is closely related to the focal structure of the submanifold. In this chapter, we give the definition of focal points and calculate the gradient and the Hessian of the height and Euclidean distance functions in terms of the geometry of the submanifolds.

4.1 Height and Euclidean distance functions.

In the following we will assume that M^n is an immersed submanifold of R^{n+k}, and X is the immersion. For $v \in R^{n+k}$, we let v^{T_x} and v^{ν_x} denote the orthogonal projection of v onto TM_x and $\nu(M)_x$ respectively.

4.1.1. Proposition. *Let a denote a non-zero fixed vector of R^{n+k}. and $h_a : M \to R$ denote the restriction of the height function of R^{n+k} to M, i.e., $h_a(x) = \langle x, a \rangle$. Then we have*
(i) $\nabla h_a(x) = a^{T_x}$, by identifying T^*M with TM,
(ii) $\nabla^2 h_a(X) = \langle II(x), a \rangle$, which is equal to $A_{a^{\nu_x}}$ if we identify $\otimes^2 T^*M$ with $L(TM, TM)$,
(iii) $\triangle h_a = \langle H, a \rangle$, where H is the mean curvature vector of M.

PROOF. Since $dh_a = \langle dX, a \rangle = \sum \omega_i \langle e_i, a \rangle = \sum (h_a)_i \omega_i$, we have $(h_a)_i = \langle e_i, a \rangle$. So $\nabla h_a = \sum_i \langle e_i, a \rangle \omega_i$. If we identify T^*M with TM via the metric, then $\nabla h_a = \sum_i \langle e_i, a \rangle e_i = a^{T_x}$. Using (1.3.6), we have

$$\sum_j (h_a)_{ij} \omega_j = d(\langle e_i, a \rangle) + \sum_m \langle e_m, a \rangle \omega_{mi}.$$

Since

$$de_i = \sum_j \omega_{ij} e_j + \sum_\alpha \omega_{i\alpha} e_\alpha,$$

we have

$$(h_a)_{ij} = h_{i\alpha j} \langle e_\alpha, a \rangle,$$

which proves (ii), and (iii) follows from the definition of the Laplacian. ∎

4.1.2. Corollary. *With the same assumptions as in Proposition 4.1.1,*
(i) A point $x_0 \in M$ is a critical point of h_a if and only if $a \in \nu(M)_{x_0}$.
(ii) The index of h_a at the critical point x_0 is the sum of the dimension of the negative eigenspace of A_a.

4.1.3. Corollary. *Let $X = (u_1, \ldots, u_{n+k}) : M \to R^{n+k}$ be an immersion. Then*

$$\triangle X = H,$$

where $\triangle$ is the Laplacian on smooth functions on M given by the induced metric, and $\triangle X = (\triangle u_1, \ldots, \triangle u_{n+k})$.

4.1.4. Corollary. *A closed (i.e., compact without boundary) n-manifold can not be minimally immersed in R^{n+k}.*

PROOF. It follows from Stoke's theorem that if M is closed and $f : M \to R$ is a smooth function satisfying $\triangle f = 0$, then f is a constant (cf. Exercise 6(iv) of section 1.3). If M is minimal, then $\triangle h_a = 0$, so X is constant, contradicting that X is an immersion. ■

A similar argument as for 4.1.1. gives

4.1.5. Proposition. *Let a denote a fixed vector of R^{n+k}, and $f_a : M \to R$ the restriction of the square of the Euclidean distance function of R^{n+k} to M, i.e., $f_a(x) = \|x - a\|^2$. Then we have*
(i) $\nabla f_a(x) = 2(x - a)^{Tx}$, if we identify T^*M with TM.
(ii) $\frac{1}{2} \nabla^2 f_a(x) = I(x) + \langle II(x), (x - a) \rangle$, and by identifying $\otimes^2 T^*M$ with $L(TM, TM)$, we have $\nabla^2 f_a(x) = Id - A_{(a-x)^{\nu x}}$,
(iii) $\triangle f_a(x) = n - \langle H, (a - x) \rangle$, where H is the mean curvature vector of M.

In Part II, Chapter 9, we define the Hessian of a smooth function f at a critical point x_0. Given two smooth vector fields X and Y, $X(Yf)(x_0)$ depends only on the value of X, Y at x_0, so it defines a bilinear form $\text{Hess}(f, x_0)$ on TM_{x_0}. Moreover, because $XY - YX = [X, Y]$ is a tangent vector field and $df_{x_0} = 0$, $\text{Hess}(f, x_o)$ is a symmetric bilinear form.

4.1.6. Corollary. *With the same assumption as in Proposition 4.1.5,*
(i) a point $x_0 \in M$ is a critical point of f_a if and only if $(a - x_0) \in \nu(M)_{x_0}$.
(ii) If x_0 is a critical point of f, then $\text{Hess}(f, x_0) = \nabla^2 f(x_0)$.
(iii) The index of f_a at the critical point x_0 is the sum of the dimension of the eigenspace E_λ of A_a corresponding to the eigenvalue $\lambda > 1$.

The critical points of h_a and f_a are closely related to the singular points of the normal maps and the endpoint maps of M, which are defined as follows:

4.1.7. Definition. The *normal map* $N : \nu(M) \to R^{n+k}$ and the *endpoint map* $Y : \nu(M) \to R^{n+k}$ of an immersed submanifold M of R^{n+k} are defined respectively by $N(v) = v$, and $Y(v) = x + v$, for $v \in \nu(M)_x$.

4.1.8. Proposition. Let M be an immersed submanifold of R^{n+k}, and N, Y the normal map and the endpoint map of M respectively. Suppose $v \in \nu(M)_{x_0}$, and e_α is an orthonormal frame field of $\nu(M)$ defined on a neighborhood U of x_0, which is parallel at x_0 (i.e., $\nabla^\nu e_\alpha(x_0) = 0$ for all α). Then using the trivialization $\nu(M)|U \simeq U \times R^k$ via the frame field e_α, we have

(i) $dN_v(u, z) = (-A_v(u), z)$,
(ii) $dY_v(u, z) = (I - A_v(u), z)$.

PROOF. Let X denote the immersion of M into R^{n+k}. Then $N = \sum_\alpha z_\alpha e_\alpha$, and $Y = X + \sum_\alpha z_\alpha e_\alpha$. So

$$dN = \sum_{\alpha,i} z_\alpha \omega_{\alpha i} \otimes e_i + \sum_{\alpha,\beta} z_\alpha \omega_{\alpha \beta} \otimes e_\beta + \sum_\alpha dz_\alpha \otimes e_\alpha,$$

$$dY = dX + dN = \sum_i \omega_i \otimes e_i + dN.$$

Then the proposition follows from the fact that e_α is parallel at x_0, i.e., $\omega_{\alpha\beta}(x_0) = 0$. ∎

4.1.9. Corollary. With the same assumption as in Proposition 4.1.8. Then for $v \in \nu(M)_x$ we have

(i) v is a singular point of the normal map N (i.e., the rank of dN_v is less than $(n+k)$) if and only if A_v is singular; in fact the dimension of $\mathrm{Ker}(dN_v)$ and $\mathrm{Ker}\, A_v$ are equal.

(ii) v is a singular point of the end point map Y if and only if $I - A_v$ is singular; in fact the dimension of $\mathrm{Ker}(dY_v)$ and $\mathrm{Ker}(I - A_v)$ are equal.

Let $X : M \to S^{n+k} \subset R^{n+k+1}$ be an immersion. We may choose a local orthonormal frame $e_0, e_1, \ldots, e_{n+k}$ such that $e_1, \ldots, e_n$ are tangent to M, $e_0 = X$, and $e_{n+1}, \ldots, e_{n+k}$ are normal to M in S^{n+k}. Then we have

$$de_0 = dX = \sum_i \omega_i \otimes e_i,$$

so $\omega_{0i} = \omega_i$, and $\omega_{0\alpha} = 0$. Since

$$de_i = \sum_j \omega_{ij} \otimes e_j + \sum_\alpha \omega_{i\alpha} \otimes e_\alpha + \omega_{i0} \otimes e_0,$$

we have:

4.1.10. Proposition.   Let $X : M \to S^{n+k}$ be an immersion, and $a \in S^{n+k}$. Then

(i) $\nabla h_a(x) = a^{Tx}$, by identifying T^*M with TM,

(ii) $\nabla^2 h_a = -h_a I + \langle II, a \rangle$, if we identify $\otimes^2 T^*M$ with $L(TM, TM)$, then $\nabla^2 h_a(x) = -h_a I + A_{a^\nu x}$,

(iii) $\triangle h_a = -n h_a + \langle H, a \rangle$, where H is the mean curvature vector of M in S^{n+k}.

(iv) $\triangle X = -nX + H$.

4.1.11. Corollary.   Let $X = (u_1, \ldots, u_{n+k+1}) : M^n \to S^{n+k}$ be an isometric immersion. Then M is minimal in S^{n+k} if and only if $\triangle u_i = -n u_i$ for all i, where $\triangle$ is the Laplacian with respect to the metric on M.

Let $X : M^n \to H^{n+k} \subseteq R^{n+k,1}$ be an isometric immersion, and e_A as above. Since

$$\omega_{0i} = \omega_{i0} = \omega_i \quad,$$

we have

4.1.12. Proposition.   Let $X : M \to H^{n+k} \subset R^{n+k,1}$ be an immersion, and $a \in H^{n+k}$. Then

(i) $\nabla h_a(x) = a^{Tx}$, by identifying T^*M with TM,

(ii) $\nabla^2 h_a = h_a I + \langle II, a \rangle$, and if we identify $\otimes^2 T^*M$ with $L(TM, TM)$, then $\nabla^2 h_a(x) = h_a I + A_{a^\nu x}$,

(iii) $\triangle h_a = n h_a + \langle H, a \rangle$, where H is the mean curvature vector of M in H^{n+k}.

(iv) $\triangle X = nX + H$.

4.1.13. Corollary.   There are no immersed closed minimal submanifolds in the hyperbolic space H^n.

If M is immersed in S^{n+k}, then $f_a = 1 + \|a\|^2 - 2h_a$. If M is immersed in H^{n+k}, then $f_a = -1 + \|a\|^2 - 2h_a$. It follows that for immersed submanifolds of S^{n+k} or H^{n+k}, f_a and $-h_a$ differ only by a constant.

Exercises.

1. Let $f : M \to R$ be a smooth function on the Riemannian manifold M, and p a critical point of f. Show that $\nabla^2 f(p) = \text{Hess}(f)_p$.

4.2 The focal points of submanifolds of R^n.

Let $a \in R^{n+k}$, and define $f_a : M \to R$ by $f_a(x) = \|x - a\|^2$ as in section 4.1. It follows from Proposition 4.1.6 that q is a critical point of f_a if and only if $(a - q) \in \nu(M)_q$, and the Hessian of f_a at a critical point q is $I - A_{(a-q)}$. Note that $I - A_{(a-q)}$ is also the tangential part of $dY_{(q,a-q)}$, where Y is the endpoint map. This leads us to the study of focal points ([Mi1]).

4.2.1. Definition. Let $X : M^n \to R^{n+k}$ be an immersion. A point $a = Y(x,e)$ in the image of the endpoint map Y of M, is called a *non-focal point* of M with respect to x if $dY_{(x,e)}$ is an isomorphism. If $m = \dim(\operatorname{Ker} dY_{(x,e)}) > 0$, then a is called a *focal point* of multiplicity m with respect to x. The *focal set* Γ of M in R^{n+k} is the set of all focal points of M.

Note that a is a focal point of M if and only if a is a critical value of the endpoint map Y, and the focal set Γ of M is the set of all critical values of Y. It follows from Proposition 4.1.8 that

$$\Gamma = \{x + e \mid x \in M, \ e \in \nu(M)_x, \text{ and } \det(I - A_e) = 0\}.$$

4.2.2. Example. Let M^n be an immersed hypersurface in R^{n+1}, and $\lambda_1, \ldots, \lambda_n$ the principal curvatures of M with respect to the unit normal field e_α. Using Proposition 4.1.8, we have $dY_{(x,te_\alpha)} = I - A_{te_\alpha} = I - tA_{e_\alpha}$. So (x, te_α) is a singular point of Y if and only if

$$\det(dY_{(x,te_\alpha)}) = \prod_i (1 - t\lambda_i) = 0.$$

Therefore $\Gamma \cap (x + \nu(M)_x)$ is equal to the finite set $\{x + \frac{1}{\lambda_i(x)} e_\alpha(x) \mid \lambda_i \neq 0\}$. For example if M^n is the sphere of radius r and centered at a_0 in R^{n+1}, then $\Gamma = \{a_0\}$; and if $M = S^1 \times R \subseteq R^3$, a right cylinder based on the unit circle, then $\Gamma = 0 \times R$.

4.2.3. Example. Let M^n be an immersed submanifold of R^{n+k}, and $\{e_\alpha\}$ a local orthonormal normal frame field. Then it follows from Proposition 4.1.8 that

$$\det(dY_{(x,e)}) = \det(I - \sum z_\alpha A_{e_\alpha}), \tag{4.2.1}$$

where $e = \sum_\alpha z_\alpha e_\alpha$, and A_{e_α} is the shape operator in the normal direction e_α. Note that (4.2.1) is a degree k polynomial with real coefficients, and in general it can not be decomposed as a product of degree one polynomials. Hence the focal set Γ of M can be rather complicated.

4.2.4. Example.   Let M^n be an immersed submanifold of R^{n+k} with flat normal bundle. It follows from Proposition 2.1.2 that $\{A_e | e \in \nu(M)_x\}$ is a family of commuting self-adjoint operators on TM_x. So there exist a common eigendecomposition $TM_x = \bigoplus_{i=1}^p E_i$ and p linear functionals α_i on $\nu(M)_x$ such that $A_e | E_i = \alpha_i(e) id_{E_i}$. Since $\nu(M)_x^*$ can be identified as $\nu(M)_x$, there exist $v_i \in \nu(N)_x$ such that $\alpha_i(e) = \langle e, v_i \rangle$. So we have

$$A_e | E_i = \langle e, v_i \rangle id_{E_i},$$

$$\det(dY_e) = \det(I - A_e) = \prod_{i=1}^p (1 - \langle v_i, e \rangle)^{m_i}.$$

So $\Gamma \cap \nu_x$ is the union of p hyperplanes ℓ_i in ν_x, where ν_x is the affine normal plane $x + \nu(M)_x$. We call the normal vectors v_i the *curvature normals* and ℓ_i the *focal hyperplanes* at x. In general, the focal hyperplanes at x do not have common intersection points. But if M is contained in a sphere centered at a, then $a \in \nu_x$ and is a focal point of M with respect to x with multiplicities n for all $x \in M$. Moreover, if $k = 2$, M is contained in S^{n+1}, and $\lambda_1, \ldots, \lambda_n$ are the principal curvatures of M as a hypersurface of S^{n+1}, then let e_{n+1} be the normal of M in S^{n+1}, and $e_{n+2}(x) = x$, we have $\lambda_{i,n+1} = \lambda_i$, $\lambda_{i,n+2} = -1$, and ℓ_i is the line that passes through the origin with slope $1/\lambda_i$.

4.2.5. Proposition.   *If M^n is an immersed submanifold of codimension k in S^{n+k} with flat normal bundle, then, as an immersed submanifold of codimension $k + 1$ in R^{n+k+1}, M^n also has flat normal bundle.*

PROOF.   Let $X : M \to S^{n+k}$ be the immersion, and $\{e_A\}$ be an adapted local orthonormal frame for M such that $\{e_\alpha\}$ is parallel with respect to the induced normal connection of M, i.e., $\omega_{\alpha\beta} = 0$. Set $e_0 = X$. Then $\{e_{n+1}, \ldots, e_{n+k}, e_0\}$ is an orthonormal frame field for the normal bundle $\nu(M)$ in R^{n+k+1}. Since $dX = \sum \omega_i e_i$,

$$\omega_{\alpha 0} = 0.$$

This proves that $\{e_{n+1}, \ldots, e_{n+k}, e_0\}$ is a parallel frame field for $\nu(M)$.   ∎

Since a hypersurface always has flat normal bundle, any hypersurface of S^{n+1} is a codimension 2 submanifold of R^{n+2} with flat normal bundle. Proposition 4.2.5 also implies that the study of submanifolds of sphere with flat normal bundles is included in the study of submanifolds of Euclidean space with flat normal bundles.

4.2.6. Theorem.   *Let M^n be an immersed submanifold of R^{n+k}, $q \in M$, $e \in \nu(M)_q$, and $a = Y(q, e) = q + e$. Then*

(i) q is a critical point of f_a,

(ii) q is a non-degenerate critical point of f_a if and only if a is a non-focal point of M,

(iii) q is a degenerate critical point of f_a with nullity m if and only if a is a focal point of M with multiplicity m with respect to q,

(iv) Index(f_a, q) is equal to the number of focal points of M with respect to q on the line segment joining q to a, each counted with its multiplicities.

PROOF. Suppose A_e has eigenvalues $\lambda_1, \ldots, \lambda_r$ with multiplicities m_i, and eigenspace E_i. Since $\text{Hess}(f_a, q) = \nabla^2 f_a(q) = I - A_e$, the negative space of the Hessian is equal to $\bigoplus \{E_i | \lambda_i > 1\}$. If $\lambda_i > 1$, then $0 < 1/\lambda_i < 1$ and $\det(I - A_{e/\lambda_i}) = 0$, which implies that $q + (e/\lambda_i)$ is a focal point with respect to q with multiplicity m_i. ∎

Chapter 5.

Transformation Groups.

The theory of Lie groups of transformations of finite dimensional manifolds is a complex, rich, and beautiful one, with many applications to different branches of mathematics. For a systematic introduction to this subject we refer the reader to [Br] and [Dv]. Because our interest is in the Riemannian geometry of Hilbert manifolds, we will concentrate on *isometric* actions on such manifolds. In studying the action of a Lie group G on a finite dimensional manifold M, it is well known that without the assumption that the group G is compact or, more generally, that the action is proper (cf. definition below) all sorts of comparatively pathological behavior can occur. For example, orbits need not be regularly embedded closed submanifolds, the action may not admit slices, and invariant Riemannian metrics need not exist. In fact, in the finite dimensional case, properness is both necessary and sufficient for G to be a closed subgroup of the group of isometries of M with respect to some Riemannian metric. In infinite dimensions properness is no longer necessary for the latter, but it is sufficient when coupled with one other condition. This other condition on an action, defined below as "Fredholm", is automatically satisfied in finite dimensions. As we shall see, much of the richness of the classical theory of compact transformation groups carries over to proper, Fredholm actions on Hilbert manifolds.

5.1 G-manifolds.

A Hilbert manifold M is a differentiable manifold locally modeled on a separable Hilbert space $(V, \langle \ , \ \rangle)$. The foundational work on Hilbert (and Banach) manifolds was carried out in the 1960's. The standard theorems of differential calculus (e.g., the inverse function theorem and the local existence and uniqueness theorem for ordinary differential equations) remain valid ([La]), and in [Sm2] Smale showed that one of the basic tools of finite dimensional differential topology, Sard's Theorem, could be recovered in infinite dimensions if one restricted the morphisms to be smooth Fredholm maps.

A Riemannian metric on M is a smooth section g of $S^2(T^*M)$ such that $g(x)$ is an inner product for TM_x equivalent to the inner product $\langle \ , \ \rangle$ on V for all $x \in M$. Such an (M, g) is called a Riemannian Hilbert manifold. For fixed vector fields X and Z the right hand side of (1.2.2) defines a continuous linear functional of TM_x. Since TM_x^* is isomorphic to TM_x, (1.2.2) defines a unique element $(\nabla_Z X)(x)$ in TM_x, and the argument for a unique compatible, torsion free connection for g is valid for infinite dimensional Riemannian manifolds, so geodesics and the exponential map $\exp : TM \to M$ can be defined just as in finite dimensions. A diffeomorphism $\varphi : M \to M$ is an isometry if $d\varphi_x : TM_x \to TM_{\varphi(x)}$ is a linear isometry for all $x \in M$.

5.1.1. Definition.   Let M and N be Hilbert manifolds. A smooth map $\varphi : M \to N$ is called an *immersion* if $d\varphi_x$ is injective and $d\varphi_x(TM_x)$ is a closed linear subspace of $TN_{\varphi(x)}$ for all $x \in M$.

If the dimension of N is finite, then $d\varphi_x(TM_x)$ is always a closed linear subspace of $TN_{\varphi(x)}$. So this definition agrees with the finite dimensional case.

5.1.2. Definition.   A Hilbert Lie group G is a Hilbert manifold with a group structure such that the map $(g_1, g_2) \mapsto g_1 g_2^{-1}$ from $G \times G \to G$ is smooth.

In this chapter we will always assume that manifolds are Hilbert manifolds and that Lie groups are Hilbert Lie groups. They can be either of finite or infinite dimension.

Let G be a Lie group, and M a smooth manifold. A smooth G-action on M is a smooth map $\rho : G \times M \to M$ such that

$$ ex = x, \quad (g_1 g_2)x = g_1(g_2 x), $$

for all $x \in M$ and $g_1, g_2 \in G$. Here e is the identity element of G and $gx = \rho(g, x)$. This defines a group homomorphism, again denoted by ρ, from G to the group $\mathrm{Diff}(M)$ of diffeomorphisms of M; namely $\rho(g)(x) = gx$. Given a fixed such G-action, we say that G acts on M, or that M is a G-manifold.

5.1.3. Definition.   A G-manifold M with action ρ is
   (i) *linear*, if M is a vector space V and $\rho(G) \subseteq GL(V)$, i.e., ρ is a linear representation of G,
   (2) *affine*, if M is an affine space V and $\rho(G)$ is a subgroup of the affine group of V,
   (3) *orthogonal*, if M is a Hilbert space V and $\rho(G)$ is a subgroup of the group of linear isometries of V, i.e., ρ is an orthogonal representation of G,
   (4) *Riemannian* or *isometric*, if M is a Riemannian manifold and $\rho(G)$ is included in the group of isometries of M.

5.1.4. Examples.

(1) The natural orthogonal action of $SO(n)$ on R^n given by taking ρ to be the inclusion of $SO(n)$ into $\mathrm{Diff}(R^n)$.

(2) The Adjoint action of G on G, defined by by $Ad(g)(h) = ghg^{-1}$.

(3)The adjoint action of G on its Lie algebra $\mathcal{G}$ given by $g \to d(Ad(g))_e$, the differential of $Ad(g)$ at the identity e. If G is compact and semi-simple then the Killing form b is negative definite, and the adjoint action is orthogonal with respect to the inner product $-b$.

(4) $SO(n)$ acts on the linear space S of trace zero symmetric $n \times n$ matrices by conjugation, i.e., $g \cdot x = gxg^{-1}$. This action is orthogonal with respect to the inner product $\langle x, y \rangle = \mathrm{tr}(xy)$.

(5) $SU(n)$ acts on the linear space $\mathcal{M}$ of Hermitian $n \times n$ trace zero matrices by conjugation. This action is orthogonal with respect to the inner product $\langle x, y \rangle = \mathrm{tr}(x\bar{y})$.

The differential of the group homomorphism $\rho : G \to \mathrm{Diff}(M)$ at the identity e gives a Lie algebra homomorphism from the Lie algebra $\mathcal{G}$ to the Lie algebra $C^\infty(TM)$ of $\mathrm{Diff}(M)$. We will denote the vector field $d\rho_e(\xi)$ by ξ again, and identify $\mathcal{G}$ as a Lie subalgebra of $C^\infty(TM)$. In fact, if g_t is the one-parameter subgroup in G generated by ξ then $\xi(x) = \frac{d}{dt}\big|_{t=0}\, g_t x$. If M is a Riemannian G-manifold, then the vector field ξ is a Killing vector field.

5.1.5. Definition. If M is a G-manifold and $x \in M$ then Gx, the *G-orbit through* x, and G_x, *the isotropy subgroup at* x are defined respectively by:

$$Gx = \{gx \mid g \in G\},$$

$$G_x = \{g \in G \mid gx = x\}.$$

The *orbit map* $\omega_x : G \to M$ is the map $g \mapsto gx$. It is constant on G_x cosets and hence defines a map $\varpi_x : G/G_x \to M$ that is clearly injective, with image Gx. Since G/G_x has a smooth quotient manifold structure, this means that we can (and will) regard each orbit as a smooth manifold by carrying over the differentiable structure from G/G_x. Since the action is smooth the orbits are even smoothly "immersed" in M, but it is important to note that without additional assumptions the orbits will *not* be regularly embedded in M, i.e., the manifold topology that Gx inherits from G/G_x will not in general be the topology induced from M. Moreover Gx will not in general be closed in M, and even the tangent space of Gx at x need not be closed in TM_x. The assumptions of properness and Fredholm, defined below, are required to avoid these unpleasant possibilities.

To prepare for the definition of Fredholm actions we recall the definition of a Fredholm map between Hilbert manifolds. If V, W are Hilbert spaces, then a bounded linear map $T : V \to W$ is *Fredholm* if $\mathrm{Ker}\, T$ and $\mathrm{Coker}\, T$ are of finite dimension. It is then a well-known, easy consequence of the closed graph theorem that $T(V)$ is closed in W. If M and N are Hilbert manifolds, then a differentiable map $f : M \to N$ is *Fredholm* if df_x is Fredholm for all x in M.

5.1.6. Definition.   The G-action on M is called *Fredholm* if for each $x \in M$ the orbit map $\omega_x : G \to M$ is Fredholm. In this case we also say that M is a *Fredholm G-manifold*.

5.1.7. Remark.   Clearly any smooth map between finite dimensional manifolds is Fredholm, so if G is a finite dimensional Lie group and M is a finite dimensional G-manifold then the action of G on M is automatically Fredholm.

5.1.8. Proposition.   *If M is any G manifold, then*
(i) $G_{gx} = gG_x g^{-1}$,
(ii) *if* $Gx \cap Gy \neq \emptyset$, *then* $Gx = Gy$,
(iii) $T(Gx)_x = \{\xi(x) \mid \xi \in \mathcal{G}\}$.
(iv) *If the action is Fredholm then each isotropy group G_x has finite dimension and each orbit Gx has finite codimension in M.*

Let M/G denote the set of all orbits, and $\pi : M \to M/G$ the orbit map defined by $x \mapsto Gx$. The set M/G equipped with the quotient topology is called the *orbit space* of the G-manifold M and will also be denoted by $\tilde{M}$. The conjugacy class of a closed subgroup H of G will be denoted by (H) and is called a G-*isotropy type*. If Gx is any orbit of a G-manifold M, then the set of isotropy groups $G_{gx} = gG_x g^{-1}$ at points of Gx is an isotropy type, called the isotropy type of the orbit, and two orbits (of possibly different G-manifolds) are said to be of the same type if they have the same isotropy types.

5.1.9. Definition.   Let M and N be G-manifolds. A mapping $F : M \to N$ is *equivariant* if $F(gx) = gF(x)$ for all $(g, x) \in G \times M$. A function $f : M \to R$ is invariant if $f(gx) = f(x)$ for all $(g, x) \in G \times M$.

If $F : M \to N$ is equivariant, then it is easily seen that $F(Gx) = G(F(x))$, and $G_x \subseteq G_{F(x)}$ with equality if and only if F maps Gx one-to-one onto $G(F(x))$. It follows that two orbits have the same type if and only if they are equivariantly diffeomorphic.

5.1.10. Definition.   Let M be a G-manifold. An orbit Gx is a *principal orbit* if there is a neighborhood U of x such that for all $y \in U$ there exists a G-equivariant map from Gx to Gy (or equivalently there exists $g \in G$ such that $G_x \subseteq gG_y g^{-1}$). (G_x) is a principal isotropy type of M if Gx is a principal orbit.

A point x is called a *regular point* if Gx is a principal orbit, and x is called a *singular point* if Gx is not a principal orbit. The set of all regular points, and the set of all singular points of M will be denoted by M_r and M_s respectively.

5.1.11. Definition.   Let M be a G-manifold. A submanifold S of M is called a *slice* at x if there is a G-invariant open neighborhood U of Gx and a smooth equivariant retraction $r : U \to Gx$, such that $S = r^{-1}(x)$.

5.1.12. Proposition.   *If M is a G-manifold and S is a slice at x, then*
(i) $x \in S$ and $G_x S \subseteq S$,
(ii) $gS \cap S \neq \emptyset$ implies that $g \in G_x$,
(iii) $GS = \{gs|\ (g,s) \in G \times S\}$ is open in M.

PROOF.   Let $r : U \to Gx$ be an equivariant retraction and $S = r^{-1}(x)$. Then $G_y \subseteq G_x$ for all $y \in S$, hence $r|Gy$ is a submersion. This implies that x is a regular value of r, so S is a submanifold of M. If $y \in S$ and $gy \in S$, then $r(gy) = x = gr(y) = gx$, i.e., $g \in G_x$. If $g \in G_x$ and $s \in S$, then $r(gs) = gr(s) = gx = x$. So we have $G_x S \subseteq S$.   ∎

5.1.13. Corollary.   *If S is a slice at x, then*
(1) S is a G_x-manifold,
(2) if $y \in S$, then $G_y \subseteq G_x$,
(3) if Gx is a principal orbit and G_x is compact, then $G_y = G_x$ for all $y \in S$, i.e., all nearby orbits of Gx are principal of the same type.
(4) two G_x-orbits $G_x s_1$ and $G_x s_2$ of S are of the same type if and only if the two G-orbits Gs_1 and Gs_2 of M are of the same type,
(5) $S/G_x = GS/G$, which is an open neighborhood of the orbit space M/G near Gx.

PROOF.   (1) and (2) follow from the definition of slice. If $y \in S$ then G_y is a closed subgroup of G_x, hence if G_x is compact so is G_y. If Gx is principal then, by definition, for y near x we also have that G_x is conjugate to a subgroup of G_y. But if two compact Lie groups are each isomorphic to a subgroup of the other then they clearly have the same dimension and the same number of components. It then follows that for y in S we must have $G_y = G_x$. Let $K = G_x$ and $s \in S$. Using the condition (ii) of the Proposition we see that $K_s = G_s$, and (4) and (5) follow.   ∎

Exercises.

1. What are the orbits of the actions in (1), (4) and (5) of Example 5.1.4 ?

2. Find all orbit types of the actions in (1), (4) and (5) of Example 5.1.4.
3. Describe the orbit space of the actions in (1), (4) and (5) of Example 5.1.4.
4. Let S be the $SO(n)$-space in Example 5.1.4 (4), and Σ the set of all trace zero $n \times n$ real diagonal matrices. Show that:
   (i) Σ meets every orbit of S,
   (ii) if $x \in \Sigma$, then Gx is perpendicular to Σ,
   (iii) let $\Sigma^0 = \{\mathrm{diag}(x_1,\ldots,x_n) \mid x_1,\ldots,x_n$ are distinct$\}$, and S a connected component of Σ^0. Then S is a slice at x for all $x \in S$.
5. Let $\mathcal{M}$ be the $SU(n)$-space in Example 5.1.4 (5), and Σ the set of all trace zero $n \times n$ real diagonal matrices. Show that
   (i) Σ meets every orbit of $\mathcal{M}$,
   (ii) if $x \in \Sigma$, then Gx is orthogonal to Σ,
   (iii) let $\Sigma^0 = \{\mathrm{diag}(x_1,\ldots,x_n) \mid x_1,\ldots,x_n$ are distinct$\}$, and S a connected component of Σ^0, then S is a slice at x for all $x \in S$.
6. Describe the orbit space of the action of Example 5.1.4 (3) for $G = SU(n)$ and $G = SO(n)$.

5.2 Proper actions.

5.2.1. Definition.   A G-action on M is called *proper* if $g_n x_n \to y$ and $x_n \to x$ imply that g_n has a convergent subsequence.

5.2.2. Remark.   Either of the following two conditions is necessary and sufficient for a G-action on M to be proper:
   (i) the map from $G \times M$ to $M \times M$ defined by $(g, x) \mapsto (gx, x)$ is proper,
   (ii) given compact subsets K and L of M, the set $\{g \in G \mid gK \cap L \neq \emptyset\}$ is compact.

5.2.3. Remark.   If G is compact then clearly any G-action is proper. Also, if G acts properly on M, then all the isotropy subgroups G_x are compact.

Next we discuss the relation between proper actions and Riemannian actions.

5.2.4.   Proposition.   *Let M be a finite dimensional Riemannian G-manifold. If G is closed in the group of all isometries of M then the action of G on M is proper.*

PROOF.   Suppose $g_n x_n \to y$ and $x_n \to x$ in M. Since M is of finite dimension, there exist compact neighborhoods K and L of x and y such that

$x_n \in K$ and $g_n K \subseteq L$. Because the $g_n : M \to M$ are isometries, $\{g_n\}$ is equicontinuous, and it then follows from Ascoli's theorem that a subsequence of $\{g_n\}$ converges uniformly to some isometry g of M. Thus if G is closed in the group of isometries of M, g_n has a convergent subsequence in G.   ∎

The above proposition is not true for infinite dimensional Riemannian G-manifolds. A simple counterexample is the standard orthogonal action on an infinite dimensional Hilbert space V (the isotropy subgroup at the origin is the group $O(V)$, which is not compact). However, if M is a proper Fredholm (PF) G-manifold, then there exists a G-invariant metric on M, i.e., G acts on M isometrically with respect to this metric. In order to prove this fact, we need the following two theorems. (Although these two theorems were proved in [Pa1] for proper actions on finite dimensional G-manifolds, they generalize without difficulty to infinite dimensional PF G-manifolds):

5.2.5. Theorem.   *If M is a PF G-manifold and $\{U_\alpha\}$ is a locally finite open cover consisting of G-invariant open sets, then there exists a smooth partition of unity $\{f_\alpha\}$ subordinate to $\{U_\alpha\}$ such that each f_α is G-invariant.*

Such $\{f_\alpha\}$ is called a *G-invariant partition of unity*. Roughly speaking, it is a partition of unity subordinate to the open cover $\{\tilde{U}_\alpha\}$ of the orbit space $\tilde{M}$.

5.2.6. Theorem.   *If M is a PF G-manifold, then given any $x \in M$ there exists a slice at x.*

5.2.7. Theorem.   *If M is a PF G-manifold, then there exists a G-invariant metric on M, i.e., the G-action on M is isometric with respect to this metric.*

PROOF.   Using Theorem 5.2.6, given any $x \in M$ there exists a slice S_x at x. Then $\{U_x = GS_x \mid x \in M\}$ is a G-invariant open cover of M. So we may assume that there exists a locally finite G-invariant open cover $\{U_\alpha\}$ such that $U_\alpha = GS_\alpha$ and S_α is the slice at x_α. Let $\{f_\alpha\}$ be a G-invariant partition of unity subordinate to $\{U_\alpha\}$.

Since G_{x_α} is compact, by the averaging method we can obtain an orthogonal structure b_α on $TM|S_\alpha$, which is G_{x_α}-invariant. Extend b_α to $TM|U_\alpha$ by requiring that $b_\alpha(gs)(dg_s(u_1), dg_s(u_2)) = b_\alpha(s)(u_1, u_2)$ for $g \in G$ and $s \in S_\alpha$. This is well-defined because b_α is G_{x_α}-invariant. Then $b = \sum f_\alpha b_\alpha$ is a G-invariant metric on M.   ∎

As a consequence of Proposition 5.2.4 and Theorem 5.2.7 we see that a finite dimensional G-manifold M is proper if and only if there exists a Riemannian metric on M such that G is a closed subgroup of $\mathrm{Iso}(M)$.

5.3 Coxeter groups.

In this sections we will review some of the standard results concerning Coxeter groups. For details see [BG] and [Bu].

Coxeter groups can be defined either algebraically, in terms of generators and relations, or else geometrically. We will use the geometric definition. In the following we will use the term hyperplane to mean a translate ℓ of a linear subspace of codimension one in some R^k, and we let R_ℓ denote the reflection in the hyperplane ℓ. Given a constant vector $v \in R^k$, we let $T_v : R^k \to R^k$ denote the translation given by v, i.e., $T_v(x) = x + v$. Recall that any isometry φ of R^k is of the form $\varphi(x) = g(x) + v_0$ (i.e., the composition of T_{v_0} and g) for some $g \in O(k)$ and $v_0 \in R^k$.

5.3.1. Definition. Let $\{\ell_i \mid i \in I\}$ be a family of hyperplanes in R^k. A subgroup W of $\mathrm{Iso}(R^k)$ generated by reflections $\{R_{\ell_i} \mid i \in I\}$ is a *Coxeter group* if the topology induced on W from $\mathrm{Iso}(R^n)$ is discrete and the W-action on R^k is proper. An infinite Coxeter group is also called an *affine Weyl group*.

5.3.2. Definition. Let W be a subgroup of $\mathrm{Iso}(R^k)$ generated by reflections. A hyperplane ℓ of R^k is called a *reflection hyperplane* of W if the reflection R_ℓ is an element of W. A unit normal vector to a reflection hyperplane of W is called a *root* of W.

5.3.3. Definition. A family $\mathfrak{H}$ of hyperplanes in R^k is *locally finite* if given any $x \in R^k$ there exists a neighborhood U of x such that $\{\ell \mid \ell \cap U \neq \emptyset, \ell \in \mathfrak{H}\}$ is finite.

5.3.4. Definition. Let $\mathfrak{H} = \{\ell_i \mid i \in I\}$ be a family of hyperplanes in R^k, and v_i a unit vector normal to ℓ_i. The rank of $\mathfrak{H}$ is defined to be the maximal number of independent vectors in $\{v_i \mid i \in I\}$. If W is the Coxeter group generated by $\{R_\ell \mid \ell \in \mathfrak{H}\}$, then the rank of W is defined to be the rank of $\mathfrak{H}$.

5.3.5. Proposition. *Suppose $\mathfrak{H}$ is a locally finite family of hyperplanes in R^k with rank $m < k$. Then there exists an m-dimensional plane E in R^k such that the subgroup of $\mathrm{Iso}(R^k)$ generated by $\{R_\ell \mid \ell \in \mathfrak{H}\}$ is isomorphic to the subgroup of $\mathrm{Iso}(E)$ generated by reflections of E in the hyperplanes $\{\ell \cap E \mid \ell \in \mathfrak{H}\}$, the isomorphism being given by $g \mapsto g|E$.*

Thus, without loss of generality, we may assume that a rank k Coxeter group is a subgroup of $\mathrm{Iso}(R^k)$.

5.3.6. Theorem [Te5]. *Let W the subgroup of $\mathrm{Iso}(R^k)$ generated by a set of reflections $\{R_i \mid i \in I\}$, and let $\mathfrak{H}$ denote the set of all reflection* hyperplanes of W. Then W is a Coxeter group if and only if $\mathfrak{H}$ is locally finite.

5.3.7. Corollary.   Let W be a subgroup of $\mathrm{Iso}(\boldsymbol{R}^k)$ generated by a set of reflections $\{R_i \mid i \in I\}$, and $\mathfrak{H}$ the set of reflection hyperplanes of W. Suppose that $\mathfrak{H}$ is locally finite and $\mathrm{rank}(\mathfrak{H}) = k$. Then

(i) W is a Coxeter group of rank k,

(ii) W permutes the hyperplanes in $\mathfrak{H}$,

(iii) if $\mathfrak{H}$ has finitely many hyperplanes, then W is a finite group and $\bigcap\{\ell \mid \ell \in \mathfrak{H}\} = \{x_o\}$ is a point,

(iv) if $\mathfrak{H}$ has infinitely many hyperplanes, then W is an infinite group.

5.3.8. Theorem.   Let W be a rank k Coxeter group on $\boldsymbol{R}^k$, and $\mathfrak{H}$ the set of reflection hyperplanes of W. Let U be a connected component of $\boldsymbol{R}^k \setminus \bigcup\{\ell_i \mid i \in I\}$, and $\bar{U}$ the closure of U. Then

(i) $\bar{U}$ is a fundamental domain of W, i.e., each W-orbit meets $\bar{U}$ at exactly one point, and $\bar{U}$ is called a Weyl chamber of W,

(ii) $\bar{U}$ is a simplicial cone if W is finite, and $\bar{U}$ is a simplex if W is infinite.

5.3.9. Theorem.   Let W be a rank k finite Coxeter group on $\boldsymbol{R}^k$, and $\bar{U}$ a Weyl chamber of W. Then there are k reflection hyperplanes $\ell_1, \ldots, \ell_k$ of W such that

(i) $\bigcap\{\ell_i \mid 1 \le i \le k\} = \{x_0\}$, is one point,

(ii) the boundary of $\bar{U}$ is contained in $\bigcup\{\ell_i \mid 1 \le i \le k\}$,

(iii) W is generated by reflections $\{R_{\ell_i} \mid 1 \le i \le k\}$,

(iv) there exist unit vectors v_i normal to ℓ_i such that

$$\bar{U} = \{x \in \boldsymbol{R}^k \mid \langle x, v_i \rangle \ge 0 \text{ for all } 1 \le i \le k\},$$

and $\{v_1, \ldots, v_k\}$ is called a simple root system of W.

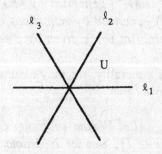

5.3.10. Theorem. *Let W be a rank k infinite Coxeter group on R^k, and $\bar{U}$ a Weyl chamber of W. Then there are $k+1$ reflection hyperplanes $\ell_1, \ldots, \ell_{k+1}$ of W such that*

(i) $\bigcap \{\ell_i \mid 1 \leq i \leq k+1\} = \emptyset$,

(ii) the boundary of $\bar{U}$ is contained in $\bigcup \{\ell_i \mid 1 \leq i \leq k+1\}$,

(iii) W is generated by reflections $\{R_{\ell_i} \mid 1 \leq i \leq k+1\}$,

(iv) there exist unit vectors v_i normal to ℓ_i such that

$$\bar{U} = \{x \in R^k \mid \langle x, v_i \rangle \geq 0 \text{ for all } 1 \leq i \leq k+1\},$$

and $\{v_1, \ldots, v_{k+1}\}$ is called a simple root system for W,

(v) if $Q = \{v \in R^k \mid T_v \in W\}$ is the subgroup of translations in W, then W is the semi-direct product of W_p and Q, where p is a vertex of $\bar{U}$ and W_p is the isotropy subgroup of W at p.

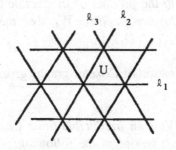

5.3.11. Definition. A rank k Coxeter group W on R^k is called *crystallographic*, if there is a rank k integer lattice Γ which is invariant under W. A finite crystallographic group is also called a *Weyl group*.

5.3.12. Theorem. *Let W be a Coxeter group generated by reflections in affine hyperplanes $\{\ell_i \mid i \in I_0\}$. Then W is crystallographic if and only if the angles between any ℓ_i and ℓ_j is π/p, for some $p \in \{1, 2, 3, 4, 6\}$, or equivalently, if and only if the order m_{ij} of $R_{\ell_i} \circ R_{\ell_j}$, $i \neq j$ is either infinite or is equal to $2, 3, 4$ or 6.*

Note that if, for $i = 1, 2$, W_i is a Coxeter group on R^{k_i}, then $W_1 \times W_2$ is a Coxeter group on $R^{k_1 + k_2}$.

5.3.13. Definition. A Coxeter group W on R^k is *irreducible* if it cannot be written as a product two Coxeter groups.

5.3.14. Theorem. *Every Coxeter group can be written as the direct product of finitely many irreducible Coxeter groups.*

Let W be a finite Coxeter group of rank k, $\{v_1, \ldots, v_k\}$ a system of simple roots, R_i the reflection along v_i, and m_{ij} the order of $R_i \circ R_j$. The *Coxeter graph* associated to W is a graph with k vertices with the ith and j^{th} vertices joined by a line (called a branch) with a mark m_{ij} if $m_{ij} > 2$ and is not joined by a branch if $m_{ij} = 2$. As a matter of convenience we shall usually suppress the label on any branch for which $m_{ij} = 3$. The Dynkin diagram is a Coxeter graph with the further restriction that $m_{ij} = 2, 3, 4, 6$ or ∞, in which branches marked with 4 are replaced by double branches and branches marked with 6 are replaced by triple branches. Similarly, we associate to an infinite Coxeter group of rank k a graph of $k + 1$ vertices.

5.3.15. Theorem.

(1) A Coxeter group is irreducible if and only if its Coxeter graph is connected.

(2) If the Coxeter graph of W_1 and W_2 are the same then W_1 is isomorphic to W_2.

(3) If W is isomorphic to the product of irreducible Coxeter groups $W_1 \times \ldots \times W_r$ and D_i is the Coxeter graph for W_i, then the Coxeter graph of W is the disjoint union of $D_1, \ldots, D_r$.

Therefore the classification of Coxeter graphs gives the classifications of Coxeter groups.

5.3.16. Theorem.   *If W is an irreducible finite Coxeter group of rank k, then its Coxeter graph must be one of the following:*

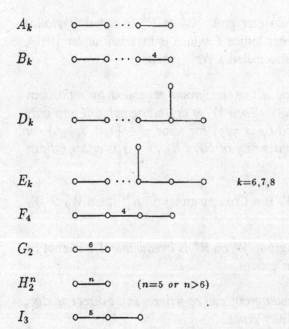

A_k

B_k

D_k

E_k $k = 6, 7, 8$

F_4

G_2

H_2^n $(n = 5 \text{ or } n > 6)$

I_3

I_4 o—^{5}—o———o———o

5.3.17. Corollary. *If W is a rank k finite Weyl group, then its Dynkin diagram must be one of the following:*

A_k o———o $\cdots$ o———o

B_k o———o $\cdots$ o=====o

D_k o———o $\cdots$ o———o———o (with branch)

E_k o———o $\cdots$ o———o———o $k=6,7,8$

F_4 o———o=====o———o

G_2 o=====o

5.3.18. Theorem. *If W is an irreducible infinite Coxeter group of rank k, then its Dynkin diagram must be one of the following:*

$\tilde{A}_1$ o—$^\infty$—o

$\tilde{A}_k$ o———o $\cdots$ o———o (with apex)

$\tilde{B}_2$ o=====o=====o

$\tilde{B}_k$ o———o———o $\cdots$ o=====o

$\tilde{C}_k$ o=====o———o $\cdots$ o———o=====o

$\tilde{D}_k$ o———o———o $\cdots$ o———o———o

$\tilde{E}_6$ o———o———o———o———o (with branch)

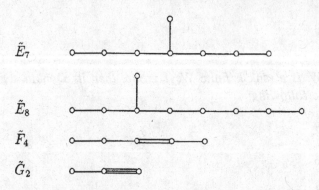

$\tilde{E}_7$

$\tilde{E}_8$

$\tilde{F}_4$

$\tilde{G}_2$

5.3.19. Chevalley Theorem. *Let W be a finite Coxeter group of rank k on R^k. Then there exist k W-invariant polynomials $u_1, \ldots, u_k$ such that the ring of W-invariant polynomials on R^k is the polynomial ring $R[u_1, \ldots, u_k]$.*

Exercises.

1. Classify rank 1 and 2 Coxeter groups directly by analytic geometry and standard group theory.
2. Suppose W is a rank 3 finite Coxeter group on R^3.
   (i) Show that W leaves S^2 invariant,
   (ii) Describe the fundamental domain of W on S^2 for $W = A_3$, B_3.

5.4 Riemannian G-manifolds.

Let M be a Riemannian Hilbert manifold. Recall that a smooth curve α is a geodesic if $\nabla_{\alpha'} \alpha' = 0$. Let $\exp_p : TM_p \to M$ denote the exponential map at p. That is, $\exp_p(v) = \alpha(1)$, where α is the unique geodesic with $\alpha(0) = p$ and $\alpha'(0) = v$. Then $\exp_p(0) = p$ and $d(\exp_p)_0 = id$. It follows that for $r > 0$ sufficiently small the restriction φ of $\exp_p$ to the ball $B_r(0)$ of radius r about the origin of TM_p is a diffeomorphism of $B_r(0)$ onto a neighborhood of p in M. Then φ is called a geodesic coordinate system for M at p. The supremum of all such r is called the injectivity radius of M at p. If $\varphi : M \to M$ is an isometry and σ is a geodesic, then $\varphi(\sigma)$ is also a geodesic. In particular we have

5.4.1. Proposition. *If M is a Riemannian Hilbert manifold and $\varphi : M \to M$ is an isometry, then*

$$\varphi(\exp_p(tv)) = \exp_{\varphi(p)}(t\,d\varphi_p(v)),$$

for $p \in M$, and $v \in TM_p$. In particular, if $\varphi(p_0) = p_0$ then in geodesic coordinates near p_0, φ is linear.

5.4.2. Corollary.   If M is a Riemannian Hilbert manifold and $\varphi : M \to M$ is an isometry, then the fixed point set of φ:

$$F = \{x \in M \mid \varphi(x) = x\}$$

is a totally geodesic submanifold of M.

PROOF.   This follows from the fact that TF_x is the eigenspace of the linear map $d\varphi_x$ with respect to the eigenvalue 1. ∎

In section 5.2 we used the existence of slices for PF G-manifolds to prove the existence of G-invariant metrics. We will now see that conversely the existence of slices for PF Riemannian actions is easy.

Let N be an embedded closed submanifold of a Riemannian manifold M. For $r > 0$ we let $S_r(x) = \{\exp_x(u) \mid x \in N, \ u \in \nu(N)_x, \ \|u\| < r\}$, and $\nu_r(N) = \{u \in \nu(N)_x \mid x \in N, \ \|u\| < r\}$. If $\exp$ maps $\nu_r(N)$ diffeomorphically onto the open subset $U_r = \exp(\nu_r(N))$, then U_r is called a *tubular neighborhood* of N. Suppose M is a PF Riemannian G-manifold and $N = Gp$. Then there exists an $r > 0$ such that $\exp_p$ is diffeomorphic on the r-ball B_r of TM_p and $\exp_p(B_r) \cap N$ has only one component (or, equivalently, $d_M(p, N \backslash \exp_p(B_r)) \geq r$). Then $U_{r/2}$ is a tubular neighborhood of $N = Gp$ in M.

5.4.3. Proposition.   Let M be a Riemannian PF G-manifold. Let $r > 0$ be small enough that $U_r = \exp(\nu_r(Gx))$ is a tubular neighborhood of Gx in M. Let S_x denote $\exp_x(\nu_r(Gx)_x)$. Then
(1) $S_{gx} = gS_x$,
(2) S_x is a slice at x, which will be called the normal slice at x.

PROOF.   (1) is a consequence of Proposition 5.4.1. Since $\nu_r(Gx)$ is a tubular neighborhood, S_x and S_y are disjoint if $x \neq y$. So if $gS_x \cap S_x \neq \emptyset$, then $S_{gx} = S_x$ and $gx = x$.

Let M be a G-manifold. The differential of the action G_x defines a linear representation ι of G_x on TM_x called the *isotropy representation* at x. Now suppose that M is a Riemannian G-manifold. Then ι is an orthogonal representation, and the tangent space $T(Gx)_x$ to the orbit of x is an invariant linear subspace. So the orthogonal complement $\nu(Gx)_x$, i.e., the normal plane of Gx in M at x, is also an invariant linear subspace, and the restriction of the isotropy representation of G_x to $\nu(Gx)_x$ is called the *slice representation* at x.

5.4.4. Example. Let $M = G$ be a compact Lie group with a bi-invariant metric. Let $G \times G$ act on G by $(g_1, g_2) \cdot g = g_1 g g_2^{-1}$. Then M is a Riemannian $G \times G$-manifold (in fact a symmetric space), G_e is the diagonal subgroup $\{(g, g) | \ g \in G\}$, and the isotropy representation of $G_e \simeq G$ on $TG_e = \mathcal{G}$ is just the adjoint action as in Example 5.1.4 (3).

5.4.5. Example. Let $M = G/K$ be a compact symmetric space, and $\mathcal{G} = \mathcal{K} + \mathcal{P}$ is the orthogonal decomposition with respect to $-b$, where b is the Killing form on $\mathcal{G}$. Then $TM_{eK} = \mathcal{P}$ and $G_{eK} = K$. Let ad denote the adjoint representation of G on $\mathcal{G}$. Then $ad(K)(\mathcal{P}) \subseteq \mathcal{P}$. So it gives a representation of K on $\mathcal{P}$, which is the isotropy representation of M at eK. For example, $M = (G \times G)/G$ gives Example 5.4.4.

5.4.6. Remark. The set of all isotropy representations for non-compact symmetric spaces is the same as the set of all isotropy representations for compact symmetric spaces.

5.4.7. Proposition. *Let M be a Riemannian PF G-manifold, and $x \in M$. Then Gx is a principal orbit if and only if the slice representation at x is trivial.*

PROOF. Let S denote the normal slice at x. Then $G_y \subseteq G_x$ for all $y \in S$. So Gx is a principal orbit if and only $G_y = G_x$ for all $y \in S$, i.e., G_x fixes S. Then the result follows from Proposition 5.4.1. ∎

5.4.8. Corollary. *Let M be a Riemannian G-manifold, x a regular point, and S_x the normal slice at x as in Proposition 5.4.3. Then $G_y = G_x$ for all $y \in S_x$.*

5.4.9. Corollary. *Let M be a Riemannian G-manifold, Gx a principal orbit, and $v \in \nu(Gx)_x$. Then $\hat{v}(gx) = dg_x(v)$ is a well-defined smooth normal vector field of Gx in M.*

PROOF. If $gx = hx$, then $g^{-1}h \in G_x$. By Proposition 5.4.7, $d(g^{-1}h)_x(v) = v$, which implies that $dg_x(v) = dh_x(v)$. ∎

5.4.10. Definition. Let M be a Riemannian G-manifold, and N an orbit of M. A section u of $\nu(N)$ is called an *equivariant normal field* if $dg_x(u(x)) = u(gx)$ for all $g \in G$ and $x \in N$.

5.4.11. Corollary. *Let M be a Riemannian G-manifold, Gx a principal orbit, and $\{v_\alpha\}$ an orthonormal basis for $\nu(Gx)_x$. Let $\hat{v}_\alpha$ be the equivariant normal field defined by v_α as in Corollary 5.4.9. Then $\{\hat{v}_\alpha\}$ is a global* smooth orthonormal frame field on Gx. In particular, the normal bundle of Gx in M is trivial.

5.4.12. Proposition.   Let M be a Riemannian G-manifold, N an orbit in M, and v an equivariant normal field on N. Then
   (1) $A_{v(gx)} = dg_x \circ A_{v(x)} \circ dg_x^{-1}$ for all $x \in N$, where A_v is the shape operator of N with respect to the normal vector v,
   (2) the principal curvatures of N along v are constant,
   (3) $\{\exp(v(x))|\ x \in N\}$ is again a G-orbit.

   PROOF.   Since $dg_x(TN_x) = TN_{gx}$ and g is an isometry, (1) follows. (2) is a consequence of (1). Since $v(gx) = dg_x(v(x))$, (3) follows from Proposition 5.4.1. ∎

5.4.13. Corollary.   Let $N^n(c)$ be the simply connected space form with constant sectional curvature c, G a subgroup of $\mathrm{Iso}(N^n(c))$, M a G-orbit, and v an equivariant normal field on M. Then $\{Y(v(x)) \mid x \in M\}$ is again a G-orbit, where Y is the endpoint map of M in $N^n(c)$.

We will now consider the orbit types of PF actions.

5.4.14. Proposition.   If M is a PF G-manifold, then there exists a principal orbit type.

   PROOF.   By Remark 5.2.3 all of the isotropy subgroups of M are compact. It follows that there exists an isotropy subgroup, G_x, having minimal dimension and, for that dimension, the smallest number of components. By Theorem 5.2.6, there exists a slice S at x. Then GS is an open subset, and $G_s \subseteq G_x$ for all $s \in S$. By the choice of x it follows that in fact $G_s = G_x$ for all $s \in S$. But then $G_{gs} = gG_sg^{-1} = gG_xg^{-1}$, so (G_x) is a principal orbit type. ∎

5.4.15. Theorem.   If M is a PF G-manifold, then the set M_r of regular points is open and dense.

   PROOF.   Openness follows from the existence of slice. To prove denseness, we proceed as follows: Let U be an open subset of M, $x \in U$, and S a slice at x. Choose $y \in GS \cap U$ so that G_y has smallest dimension and, for that dimension, the smallest number of components. Let S_0 be a slice at y, and $z \in GS_0 \cap U \cap GS$. It follows form Corollary 5.1.13 (2) that there exists $g \in G$ such that $G_z \subseteq gG_yg^{-1}$. Since the dimension of G_y is less than or equal to the dimension of G_z, we conclude that G_z and G_y in fact have the same dimension, and then since the number of components of G_y is less than or equal to the number of components of G_z, $G_z = gG_yg^{-1}$. This proves that Gy is a principal orbit. ∎

5.4.16. Theorem.   *If M is a PF G-manifold then given a point $p \in M$ there exists a G-invariant open neighborhood U containing p such that U has only finitely many G-orbit types.*

PROOF.   By Theorem 5.2.7 we may assume that M is a PF Riemannian G-manifold. Let S be the normal slice at p. Then S is of finite dimension, and G_p is a compact group acting isometrically on S so, by 5.1.13(4), it will suffice to prove this theorem for Riemannian G-manifolds of finite dimension n. We prove this by induction. For $n = 0$ the theorem is trivial. Suppose it is true for all proper G-manifolds of dimension less than n and let M be a proper Riemannian G-manifold of dimension n, $p \in M$, and S the normal slice at p. By 5.1.13(4) again, it will suffice to prove that locally S has only finitely many orbit types. If $\dim(S) < n$, then this follows from the induction hypothesis, so assume that $\dim(S) = n$. Then by Proposition 5.4.1 the G_p-action ρ on S is an orthogonal action on $TM_p = R^n$ with respect to geodesic coordinates. Now G_p leaves S^{n-1} invariant and, by the induction hypothesis, locally S^{n-1} has only finitely many orbit types. But then because S^{n-1} is compact, it has finitely many orbit types altogether. Now note that, in a linear representation, the isotropy group (and hence the type of an orbit) is constant on any line through the origin, except at the origin itself. So ρ has at most one more orbit type on S than on S^{n-1}, and hence only finitely many orbit types. ∎

5.4.17. Theorem.   *If M is a PF G-manifold, then the set $\tilde{M}_s = M_s/G$ of singular orbits does not locally disconnect the orbit space $\tilde{M} = M/G$.*

PROOF.   Using the slice representation as in the previous theorem, it suffices to prove this theorem for linear orthogonal G-action on R^n. We proceed by induction. If $n = 1$, then we may assume that $G = O(1) = Z_2$. It is easily seen that R/G is the half line $\{x\mid x \geq 0\}$ with 0 as the only singular orbit. So $\{0\}$ does not locally disconnect R/G. Suppose $G \subseteq O(n)$. Applying the induction hypothesis to the slice representation of S^{n-1}, we conclude that the set of singular orbits of S^{n-1} does not locally disconnect S^{n-1}/G. But R^n/G is the cone over S^{n-1}/G. So the set of singular orbits of R^n does not locally disconnect R^n/G. ∎

5.4.18. Corollary.   *If M is a connected PF G-manifold, then*
(1) M/G is connected,
(2) M has a unique principal orbit type.

5.4.19. Corollary.   *Suppose M is a connected PF G-manifold. Let*
$m = \inf\{\dim(G_x) \mid x \in M\}$, *and k the smallest number of components of*
all the dimension m isotropy subgroups. Then an orbit Gx_0 is principal if
and only if G_{x_0} has dimension m and k components.

5.5 Riemannian submersions.

A smooth map $\pi : E \to B$ is a *submersion* if B is a finite dimension
manifold and the rank of $d\pi_x$ is equal to the dimension of B. Then $V =$
$\ker(d\pi)$ is a smooth subbundle of TE called the tangent bundle along the
fiber (or the vertical subbundle). In case E and B are Riemannian manifolds
we define the horizontal subbundle $\mathcal{H}$ of TE to be the orthogonal complement
$V^{\perp}$ of the vertical bundle.

5.5.1. Definition.   Let E and B be Riemannian manifolds. A submersion
$\pi : E \to B$ is called a *Riemannian submersion* if $d\pi_x$ maps $\mathcal{H}_x$ isometrically
onto $TB_{\pi(x)}$ for all $x \in E$.

The theory of Riemannian submersions, first systematically studied by
O'Neil [On], plays an important role in the study of isometric actions, as we
will see in the following.

5.5.2. Remark.   Let M be a PF Riemannian G-manifold. Suppose M has
a single orbit type (H), and $H = G_x$. Then the orbit map $p : M \to \tilde{M}$ is
a smooth fiber bundle. If S is a slice at x, then we get a local trivialization
of p on the neighborhood GS of the orbit Gx using the diffeomorphism
$G/H \times S \approx GS$ defined by $(gH, s) \mapsto gs$. There is a unique metric on
$\tilde{M}$ such that p is a Riemannian submersion. To see this, we define the
inner product on $T\tilde{M}_{p(x)}$ by requiring that $dp_x : \nu(Gx)_x \to T\tilde{M}_{p(x)}$ is an
isometry. Since $dg_x(T(Gx)_x) = T(Gx)_{gx}$ and dg_x is an isometry, dg_x maps
the inner product space $\nu(Gx)_x$ isometrically onto $\nu(Gx)_{gx}$. This shows
that the metric on $\tilde{M}$ is well-defined, and it is easily seen to be smooth.
Actually, in this case M is a smooth fiber bundle in a completely different
but important way. First, it is clear that M is partitioned into the closed,
totally geodesic submanifolds $F(gHg^{-1})$, where the latter denotes the fixed
point set of the subgroup gHg^{-1}. Clearly $F(g_1 H g_1^{-1}) = F(g_2 H g_2^{-1})$ if and

only if $g_1 N(H) = g_2 N(H)$, where $N(H)$ denotes the normalizer of H in G. Thus we get a smooth map $\Pi : M \to G/N(H)$ having the $F(gHg^{-1})$ as fibers. Note that $N(H)$ acts on $F(H)$, and it is easily seen that the fibration $\Pi : M \to G/N(H)$ is the bundle with fiber $F(H)$ associated to the principal $N(H)$-bundle $G \to G/N(H)$.

What is most important about this *second* realization of M as the total space of a differentiable fiber bundle is that it points the way to generalize the first when M has more than one orbit type. In this case let (H) be a fixed orbit type of M, say $H = G_x$. Then $F = F(H)$ is again a closed, totally geodesic submanifold of M, and $F^* = F^*(H) = \{x \in M \mid G_x = H\}$ is an open submanifold of F. Just as above, we see that $M_{(H)}$ is a smooth fiber bundle over $G/N(H)$ with fiber F^*. In particular each orbit type $M_{(H)}$ is a smooth G-invariant submanifold of M. But of course $M_{(H)}$ has a single orbit type, so as above its orbit space $\tilde{M}_{(H)}$ has a natural differentiable structure making the orbit map $p : M_{(H)} \to \tilde{M}_{(H)}$ a smooth fiber bundle, and a smooth Riemannian structure making p a Riemannian submersion. Now we have already seen that the decompositions of M and $\tilde{M}$ into the orbit types $M_{(H)}$ and $\tilde{M}_{(H)}$ are locally finite. In fact they have all the best properties one can hope for in such a situation. To be technical, they are stratifications of M and $\tilde{M}$ respectively and, by what we have just noted, the orbit map $p : M \to \tilde{M}$ is a stratified Riemannian submersion.

5.5.3. Definition.   Let $\pi : E \to B$ be a Riemannian submersion, V the vertical subbundle, and $\mathcal{H}$ the horizontal subbundle. Then a vector field ξ on E is

(1) *vertical*, if $\xi(x)$ is in V_x for all $x \in E$,

(2) *horizontal*, if $\xi(x)$ is in $\mathcal{H}_x$ for all $x \in E$,

(3) *projectable*, if there exists a vector field η on B such that $d\pi(\xi) = \eta$,

(4) *basic*, if it is both horizontal and projectable.

5.5.4. Proposition.   *Let $\pi : E \to B$ be a Riemannian submersion.*

(1) If τ is a smooth curve on B then given $p_0 \in \pi^{-1}(\tau(t_0))$ there exists a unique smooth curve $\tilde{\tau}$ on E such that $\tilde{\tau}'(t)$ is horizontal for all t, $\pi(\tilde{\tau}) = \tau$, and $\tilde{\tau}(t_0) = p_0$. $\tilde{\tau}$ is called the horizontal lifting of τ at p_0.

(2) If η is a vector field on B, then there exists a unique basic field $\tilde{\eta}$ on E such that $d\pi(\tilde{\eta}) = \eta$, which is called the horizontal lift of η. In fact, this gives a one to one correspondence between $C^\infty(TB)$ and the space of basic vector fields on E.

5.5.5. Proposition.   *If X is vertical and Y is projectable then $[X, Y]$ is vertical.*

PROOF.   This follows from the fact that

$$d\pi([X, Y]) = [d\pi(X), d\pi(Y)]. \quad \blacksquare$$

5.5.6. Proposition.   *Let $\pi : E \to B$ be a Riemannian submersion, τ a geodesic in B, and $\tilde{\tau}$ its horizontal lifting in E. Let $L(\alpha)$ denote the arc length of the smooth curve α, and $E_b = \pi^{-1}(b)$. Then*

(1) $L(\tilde{\tau}) = L(\tau)$,

(2) $\tilde{\tau}$ perpendicular to the fiber $E_{\tau(t)}$ for all t,

(3) if τ is a minimizing geodesic joining p to q in B, then $L(\tilde{\tau}) = d(E_p, E_q)$, the distance between the fibers E_p and E_q,

(4) $\tilde{\tau}$ is a geodesic of E.

PROOF.   (1) and (2) are obvious. (4) is a consequence of (3). It remains to prove (3). Suppose τ is a minimizing geodesic joining p and q in B. If α is a smooth curve in E joining a point in E_p to a point in E_q, then $\pi \circ \alpha$ is a curve on B joining p, q. So $L(\pi \circ \alpha) \geq L(\tau)$. Let $\alpha' = u + v$, where u is the horizontal component and v is the vertical component of α'. Since $\|d\pi(u)\| = \|u\|$ and $d\pi(v) = 0$, $\|d\pi(\alpha')\| \leq \|\alpha'\|$. So we have

$$L(\alpha) \geq L(\pi(\alpha)) \geq L(\tau) = L(\tilde{\tau}),$$

which implies that $\tilde{\tau}$ is a geodesic, $d(E_p, E_q) = L(\tau)$. $\blacksquare$

5.5.7. Corollary.   *Let $\pi : E \to B$ be a Riemannian submersion. If σ is a geodesic in E such that $\sigma'(t_0)$ is horizontal then $\sigma'(t)$ is horizontal for all t (or equivalently, if a geodesic σ of E is perpendicular to $E_{\sigma(t_0)}$ then it is perpendicular to all fibers $E_{\sigma(t)}$).*

PROOF.   Let $p_0 = \sigma(t_0)$, τ the geodesic of B such that $\tau(t_0) = \pi(p_0)$ and $\tau'(t_0) = d\pi(\sigma'(t_0))$. Let $\tilde{\tau}$ be the horizontal lifting of τ at p_0. Then both σ and $\tilde{\tau}$ are geodesics of E passing through p_0 with the same tangent vector at p_0. So $\sigma = \tilde{\tau}$. $\blacksquare$

5.5.8. Corollary.   *Let $\pi : E \to B$ be a Riemannian submersion, and $\mathcal{H}$ the horizontal subbundle (or distribution).*

(1) If $\mathcal{H}$ is integrable then the leaves are totally geodesic.

(2) If $\mathcal{H}$ is integrable and S is a leaf of $\mathcal{H}$ then $\pi|S$ is a local isometry.

5.5.9. Remark.   If $F = \pi^{-1}(b)$ is a fiber of π then $\mathcal{H}|F$ is just the normal bundle of F in E. There exists a canonical global parallelism on the normal bundle $\nu(F)$: a section $\tilde{v}$ of $\nu(F)$ is called π-*parallel* if $d\pi(\tilde{v}(x))$ is a fixed vector $v \in TB_b$ independent of x in F. Clearly $\tilde{v} \mapsto v$ is a bijective correspondence between π-parallel fields and TB_b. There is another parallelism on $\nu(F)$ given by the induced normal connection ∇^ν as in the submanifold geometry, i.e., a normal field ξ is parallel if $\nabla^\nu \xi = 0$. It is important to note that in general the π-parallelism in $\nu(F)$ is *not* the same as the parallel translation defined by the normal connection ∇^ν. (The latter is in general not flat, while the former is always both flat and without holonomy.) Nevertheless we shall see later that if $\mathcal{H}$ is integrable then these two parallelisms *do* coincide.

5.5.10. Remark. Let M be a Riemannian G-manifold, (H) the principal orbit type, and $\pi : M_{(H)} \to \tilde{M}_{(H)}$ the Riemannian submersion given by the orbit map. Then a normal field ξ of a principal orbit Gx is G-equivariant if and only if ξ is π-parallel.

5.5.11. Definition. A Riemannian submersion $\pi : E \to B$ is called *integrable* if the horizontal distribution $\mathcal{H}$ is integrable.

We will first discuss the local theory of Riemannian submersions. Let $\pi : E \to B$ be a Riemannian submersion. Then there is a local orthonormal frame field e_A on E such that $e_1, \ldots, e_n$ are vertical and $e_{n+1}, \ldots, e_{n+k}$ are basic. Then $\{e_\alpha^* = d\pi(e_\alpha)\}$ is a local orthonormal frame field on B. We use the same index convention as in section 2.1, i.e.,

$$1 \leq i, j, k \leq n, \ n+1 \leq \alpha, \beta, \gamma \leq n+k, \ 1 \leq A, B, C \leq n+k.$$

Let ω_A and ω_α^* be the dual coframe, and ω_{AB}, $\omega_{\alpha\beta}^*$ the Levi-Civita connections on E and B respectively. Then $\pi^*(\omega_\alpha^*) = \omega_\alpha$. Assume that

$$\omega_{i\alpha} = \sum_\beta a_{i\alpha\beta}\omega_\beta + \sum_j r_{i\alpha j}\omega_j \ , \tag{5.5.1}$$

$$\omega_{\alpha\beta} = \pi^*(\omega_{\alpha\beta}^*) + \sum_i b_{\alpha\beta i}\omega_i \ . \tag{5.5.2}$$

Note that

$$d\omega_\alpha = d(\pi^*\omega_\alpha^*) = \pi^*(d\omega_\alpha^*)$$

$$= \pi^*\left(\sum_\beta \omega_{\alpha\beta}^* \wedge \omega_\beta^*\right)$$

$$= \sum_\beta \pi^*(\omega_{\alpha\beta}^*) \wedge \omega_\beta,$$

which does not have $\omega_i \wedge \omega_\beta$ and $\omega_i \wedge \omega_j$ terms. But the structure equation gives

$$d\omega_\alpha = \sum_j \omega_{\alpha\beta} \wedge \omega_\beta + \sum_i \omega_{\alpha i} \wedge \omega_i. \tag{5.5.3}$$

So the coefficients of $\omega_i \wedge \omega_\beta$ and $\omega_i \omega_j$ in (5.5.3) are zero, i.e.,

$$b_{\alpha\beta i} = a_{i\alpha\beta}, \qquad (5.5.4)$$

$$r_{i\alpha j} = r_{j\alpha i}. \qquad (5.5.5)$$

Note that the restriction of $\omega_{i\alpha}$ and $\omega_{\alpha\beta}$ to the fiber F are the second fundamental forms and the normal connection of F in E. In fact, $\sum r_{i\alpha j}\omega_i \otimes \omega_j \otimes e_\alpha$ is the second fundamental form of F and $\omega_{\alpha\beta} = \sum_i b_{\alpha\beta i}\omega_i = \sum_i a_{i\alpha\beta}\omega_i$ is the induced normal connection of the normal bundle $\nu(F)$ in E.

Next we describe in our notation the two fundamental tensors A and T associated to Riemannian submersions by O'Neil in [On]. Let u^h and u^v denote the horizontal and vertical components of $u \in TE_p$. Then it is easy to check that

$$T(X, Y) = (\nabla_{X^v} Y^v)^h + (\nabla_{X^v} Y^h)^v,$$

$$A(X, Y) = (\nabla_{X^h} Y^h)^v + (\nabla_{X^h} Y^v)^h,$$

define two tensor fields on E. Using (5.5.1) and (5.5.2), these two tensors are

$$T = \sum r_{j\alpha i}(\omega_i \otimes \omega_j \otimes e_\alpha - \omega_i \otimes \omega_\alpha \otimes e_j),$$

$$A = \sum a_{j\beta\alpha}(\omega_\alpha \otimes \omega_i \otimes e_\beta - \omega_\alpha \otimes \omega_\beta \otimes e_i).$$

If $\mathcal{H}$ is integrable then, by Corollary 5.5.8, each leaf S of $\mathcal{H}$ is totally geodesic and $e_\alpha | S$ is a local frame field on S. Thus the second fundamental form on S is zero, i.e., $\nabla_{e_\alpha} e_j$ is vertical, or $a_{i\alpha\beta} = 0$. Note that $e_i | F$ form a tangent frame field for the fiber F, and $e_\alpha | F$ is a normal vector field of F. By Proposition 5.5.5, $[e_j, e_\alpha] = \nabla_{e_j} e_\alpha - \nabla_{e_\alpha} e_j$ is vertical, so we have $\nabla_{e_j} e_\alpha$ is vertical, i.e., $e_\alpha | F$ is parallel with respect to the induced normal connection of F in E.

Conversely, suppose $e_\alpha | F$ is parallel for every fiber F of π, i.e., $\nabla_{e_i} e_\alpha$ is vertical, or $\omega_{\alpha\beta}(e_i) = 0$. By (5.5.1) (5.5.2) and (5.5.4), we have

$$0 = \omega_{\alpha\beta}(e_i) = b_{\alpha\beta i} = a_{i\alpha\beta} = \omega_{i\alpha}(e_\beta).$$

The torsion equation implies that

$$[e_\alpha, e_\beta] = \nabla_{e_\alpha} e_\beta - \nabla_{e_\beta} e_\alpha = \sum(\omega_{\beta A}(e_\alpha) - \omega_{\alpha A}(e_\beta))e_A.$$

Hence $[e_\alpha, e_\beta]$ is horizontal, i.e., $\mathcal{H}$ is integrable. So we have proved:

5.5.12. Theorem. *Let $\pi : E \to B$ be a Riemannian submersion. Then the following statements are equivalent:*

(i) π is integrable,

(ii) every π-parallel normal field on the fiber $F = \pi^{-1}(b)$ is parallel with respect to the induced normal connection of F in E,

(iii) the O'Neil tensor A is zero.

5.6 Sections.

Henceforth M will denote a connected, complete Riemannian G-manifold, and M_r is the set of regular points of M. As noted above, we have a Riemannian submersion $\pi : M_r \to \tilde{M}_r$. We assume all the previous notational conventions. In particular we identify the Lie algebra $\mathcal{G}$ of G with the Killing fields on M generating the action of G.

5.6.1. Proposition.   *If $\xi \in \mathcal{G}$ and σ is a geodesic on M, then the quantity $\langle \sigma'(t), \xi(\sigma(t)) \rangle$ is a constant independent of t.*

PROOF.   If ξ is a Killing field and $\nabla \xi = \sum \xi_{ij} e_i \otimes \omega_j$, then $\xi_{ij} + \xi_{ji} = 0$. So $\langle \nabla_{\sigma'} \xi, \sigma' \rangle = 0$. Since σ is a geodesic, $\nabla_{\sigma'} \sigma' = 0$, which implies that

$$\frac{d}{dt} \langle \xi(\sigma), \sigma' \rangle = \langle \nabla_{\sigma'} \xi, \sigma' \rangle + \langle \xi(\sigma), \nabla_{\sigma'} \sigma' \rangle = 0. \quad \blacksquare$$

It will be convenient to introduce for each regular point x the set $T(x)$, defined as the image of $\nu(Gx)_x$ under the exponential map of M, and also $T_r(x) = T(x) \cap M_r$ for the set of regular points of $T(x)$. Note that $T(x)$ may have singularities.

5.6.2. Proposition.   *For each regular point x of M:*

(1) $gT(x) = T(gx)$ and $gT_r(x) = T_r(gx)$,

(2) for $v \in \nu(Gx)_x$ the geodesic $\sigma(t) = \exp_x(tv)$ is orthogonal to each orbit it meets,

(3) if G is compact then $T(x)$ meets every orbit of M.

PROOF.   (1) follows from Proposition 5.4.1, and (2) follows from Proposition 5.6.1. Finally suppose G is compact and given any $y \in M$, since Gy is compact, we can choose $g \in G$ so that gy minimizes the distance from x to Gy. Let $\sigma(t) = \exp(tv_0)$ be a minimizing geodesic from $x = \sigma(0)$ to $gy = \sigma(s)$. Then σ is perpendicular to Gy. By (2), σ is also orthogonal to Gx. In particular $v_0 = \sigma'(0) \in \nu(Gx)_x$ so the arbitrary orbit Gy meets $T(x) = \exp(\nu(Gx)_x)$ at $\exp(sv_0) = gy$. $\quad \blacksquare$

Let x be a regular point and S a normal slice at x. If S is orthogonal to each orbit it meets then so is gS. This implies that the Riemannian submersion $\pi : M_r \to \tilde{M}_r$ is integrable. Since for most Riemannian G-manifold M the submersion $\pi : M_r \to \tilde{M}_r$ is not integrable, a normal slice is in general *not* orthogonal to each orbit it meets.

5.6.3. Example. Let S^1 act on $R^2 \times R^2$ by $e^{it}(z_1, z_2) = (e^{it}z_1, e^{it}z_2)$. Then $p = (1, 0)$ is a regular point. It is easy to check that $y = (1, 1) \in T(p)$ and $T(p)$ is not orthogonal to the orbit Sy.

5.6.4. Definition. A connected, closed, regularly embedded smooth submanifold Σ of M is called a *section* for M if it meets all orbits orthogonally.

The conditions on Σ are, more precisely, that $G\Sigma = M$ and that for each $x \in \Sigma$, $T\Sigma_x \subseteq \nu(Gx)_x$. But since $T(Gx)_x$ is just the set of $\xi(x)$ where $\xi \in \mathcal{G}$, this second condition has the more explicit form

(*) For each $x \in \Sigma$ and $\xi \in \mathcal{G}$, $\xi(x)$ is orthogonal to $T\Sigma_x$.

In the following we will discuss some basic properties for G-manifolds that admit sections. For more detail, we refer the reader to [PT2].

It is trivial that if Σ is a section for M then so is $g\Sigma$ for each $g \in G$. Since $G\Sigma = M$, it follows that if one section Σ exists then in fact there is a section through each point of M, and we shall say that M *admits sections*.

5.6.5. Example. All the examples in 5.1.4 admit sections. In fact, for (1), $\{ru \mid r \in R\}$ is a section, where u is any unit vector in R^n; for (2) a maximal torus is a section; for (3) a maximal abelian (Cartan) subalgebra is a section; for (4) and (5), the space of all trace zero real diagonal matrices is a section.

5.6.6. Definition. The *principal horizontal distribution* of a Riemannian G-manifold M is the horizontal distribution of the Riemannian submersion on the principal stratum $\pi : M_r \to \tilde{M}_r$.

If Σ is a section of M then the set $\Sigma_r = \Sigma \cap M_r$ of regular points of Σ is an integral submanifold of the principal horizontal distribution $\mathcal{H}$ of the G-action. Since $\tilde{M}_r$ is always connected, it follows from Corollary 5.5.8, Remark 5.5.10 and Theorem 5.5.12 that we have:

5.6.7. Theorem. *If M admits sections, and Σ is a section, then*
 (1) *the principal horizontal distribution $\mathcal{H}$ is integrable;*
 (2) *each connected component of $\Sigma_r = \Sigma \cap M_r$ is a leaf of $\mathcal{H}$;*
 (3) *if F is the leaf of $\mathcal{H}$ through a regular point x then $\pi|F$ is a covering isometry onto $\tilde{M}_r$;*
 (4) Σ *is totally geodesic;*

(5) there is a unique section through each regular point x of M, namely $T(x) = \exp(\nu(Gx)_x)$.

(6) an equivariant normal field on a principal orbit is parallel with respect to the induced normal connection.

5.6.8. Remark.   One might naively hope that, conversely to Theorem 5.6.7(1), if $\mathcal{H}$ is integrable then M admits sections. To give a counterexample take $M = S^1 \times S^1$ and let $G = S^1 \times \{e\}$ acting by translation. Let ξ denote the vector field on M generating the action of G and let η denote an element of the Lie algebra of $S^1 \times S^1$ generating a nonclosed one parameter group γ. If we choose the invariant Riemannian structure for M making ξ and η orthonormal then a section for M would have to be a coset of γ, which is impossible since γ is not closed in M. This also gives a counter example to the weaker conjecture that if a compact G-manifold M has codimension 1 principal orbits then any normal geodesic to the principal orbit is a section. It *is* probably true that if $\mathcal{H}$ is integrable, then a leaf of $\mathcal{H}$ can be extended to be a complete immersed totally geodesic submanifold of M, which meets every orbit orthogonally. However we can prove this only in the real analytic case.

5.6.9. Proposition.   *Suppose G is a compact Lie group, and M a Riemannian G-manifold. Let x_0 be a regular point of M, and $T = \exp(\nu(Gx_0)_{x_0})$. If $\mathcal{H}$ is integrable and T is a closed properly embedded submanifold of M, then T is a section.*

PROOF.   By Proposition 5.6.2(3), it suffices to show that T is orthogonal to Gx for all $x \in T$. Let F denote the leaf of $\mathcal{H}$ through x_0. By Corollary 5.5.8, F is totally geodesic. So F is open in T and T is orthogonal to Gy for all $y \in F$. Now suppose $x \in T \setminus F$. Since $\exp_x : TT_x \to T$ is regular almost everywhere, there is an open neighborhood U of the unit sphere of TT_x such that for all $v \in U$ there is an $r > 0$ such that $\sigma_v(r) = \exp_x(rv)$ is in F. Then by Proposition 5.6.2(2) $\sigma'_v(0) = v$ is normal to Gx.   ∎

It is known that any connected totally geodesic submanifold of a simply connected, complete symmetric space can be extended uniquely to one that is complete and properly embedded (cf. [KN] Chapter 9, Theorem 4.3). So we have

5.6.10. Corollary.   *Let $M = G/K$ be a simply connected complete symmetric space, and H a subgroup of G. Then the action of H on M admits sections if and only if the principal horizontal distribution of this action is integrable. In particular if the principal H-orbit is of codimension one then the H-action on M has a section.*

It follows from Theorem 5.5.12 that

5.6.11. Theorem. *The following statements are equivalent for a Riemannian G-manifold M:*

(1) the principal horizontal distribution $\mathcal{H}$ is integrable,

(2) every G-equivariant (i.e., π-parallel) normal vector field on a principal orbit is parallel with respect to the induced normal connection for the normal bundle $\nu(Gx)$ in M,

(3) for each regular point x of M, if S is the normal slice at x then for all $\xi \in \mathcal{G}$ and $s \in S$, $\xi(s)$ is normal to S.

5.6.12. Proposition. *Let V be an orthogonal representation of G, x a regular point of V, and Σ the linear subspace of V orthogonal to the orbit Gx at x. Then the following are equivalent:*

(i) V admits sections,

(ii) Σ is a section for V,

(iii) for each v in Σ and ξ in $\mathcal{G}$, $\xi(v)$ is normal to Σ.

In the following, M is a Riemannian G-manifold that admits sections. Let x be a regular point of M, and Σ the section of M through x. Recall that a small enough neighborhood U of x in Σ is a slice at x and so intersects each orbit near Gx in a unique point. Also recall that G_x acts trivially on Σ.

In general given a closed subset S of M we let $N(S)$ denote the closed subgroup $\{g \in G | gS = S\}$ of G, the largest subgroup of G which induces an action on S, and we let $Z(S)$ denote the kernel of this induced action, i.e., $Z(S) = \{g \in G | gs = s, \forall s \in S\}$ is the intersection of the isotropy subgroups G_s, $s \in S$. Thus $N(S)/Z(S)$ is a Lie group acting effectively on S. In particular when S is a section Σ then we denote $N(\Sigma)/Z(\Sigma)$ by $W = W(\Sigma)$ and call it the *generalized Weyl group* of Σ.

5.6.13. Remark. If M is the compact Lie group G with the Adjoint action, then for a subgroup H of G, $N(H)$ and $Z(H)$ are respectively the normalizer and centralizer of H. If for H we take a maximal torus T of G (which is in fact a section of the Adjoint action) then $Z(T) = T$ and $W(T) = N(T)/T$ is the usual Weyl group of G.

5.6.14. Remark. Let x be a regular point, S a normal slice at x, and Σ a section at x. As remarked above $G_x \subseteq Z(S) \subseteq Z(\Sigma)$, and conversely from the definition of $Z(\Sigma)$ it follows that $Z(\Sigma) \subseteq G_x$, so $Z(\Sigma) = G_x$. Moreover if $g\Sigma = \Sigma$ then $g\Sigma$ is the section at the regular point gx. So $G_x = Z(\Sigma) = Z(g\Sigma) = G_{gx}$. Then it follows from $G_{gx} = gG_xg^{-1}$ that we have $N(\Sigma) \subseteq N(G_x)$ and $W(\Sigma) \subseteq N(G_x)/G_x$.

5.6.15. Proposition.   *The generalized Weyl group $W(\Sigma)$ of a section Σ is a discrete group. Moreover if Σ' is a second section for M then $W(\Sigma')$ is isomorphic to $W(\Sigma)$ by an isomorphism which is well determined up to inner automorphism.*

PROOF.   Let Σ be the section and S the normal slice at the regular point x. Then S is an open subset of Σ. If $g \in N(\Sigma)$ is near the identity then $gx \in S$. Since S meets every orbit near x at a unique point, $gx = x$, i.e., $g \in G_x = Z(\Sigma)$, so $Z(\Sigma)$ is open in $N(\Sigma)$ and hence $W(\Sigma)$ is discrete. If Σ' is a section then $\Sigma' = g_0 \Sigma$ and so $g \mapsto g_0 g g_0^{-1}$ clearly induces an isomorphism of $W(\Sigma)$ onto $W(\Sigma')$. ∎

5.6.16. Example.   The isotropy representation of the symmetric space $M = G/K$ at eK admits sections. In fact, let $\mathcal{G} = \mathcal{K} + \mathcal{P}$ be the orthogonal decomposition of the Lie algebra $\mathcal{G}$ of G as in Example 5.4.5 and $\mathfrak{A}$ a maximal abelian subalgebra in $\mathcal{P}$. Then $\mathfrak{A}$ is a section and the generalized Weyl group W is the standard Weyl group associated to the symmetric space G/K. These representations have the following remarkable properties:
  (i) Given $p \in \mathcal{P}$, the slice representations of $\mathcal{P}$ again admits sections.
  (ii) $\mathfrak{A}/W \simeq \mathcal{P}/K$.
  (iii) Chevalley Restriction Theorem ([He],[Wa]): Let $R[\mathcal{P}]^G$ be the algebra of G-invariant polynomials on $\mathcal{P}$, and $R[\mathfrak{A}]^W$ the algebra of W-invariant polynomials on $\mathfrak{A}$. Then the restriction map $R[\mathcal{P}]^G \to R[\mathfrak{A}]^W$ defined by $f \mapsto f|\mathfrak{A}$ is an algebra isomorphism.

Following J. Dadok we shall say that an orthogonal representation space is *polar* if it admits sections. The following theorem of Dadok [Da] says that the isotropy representations of symmetric spaces are "essentially" the only polar representations.

5.6.17. Theorem. ·   *Let $\rho : H \to O(n)$ be a polar representation of a compact connected Lie group. Then there exists an n-dimensional symmetric space $M = G/K$ and a linear isometry $A : R^n \to TM_{eK}$ mapping H-orbits onto K-orbits.*

5.6.18. Corollary.   *If ρ is a finite dimensional polar representation, then the corresponding generalized Weyl group is a classical Weyl group.*

5.6.19. Definition.   A G-manifold M is called *polar* if the G-action is proper, Fredholm, isometric, and admits sections.

5.6.20. Remark.   The generalized Weyl group of a polar G-manifold is not a Weyl group in general. In fact we will now construct examples with an arbitrary finite group as the generalized Weyl group. Given any compact group G, a closed subgroup H of G, a finite subgroup W of $N(H)/H$, and a smooth manifold Σ such that W acts faithfully on Σ, we let $\pi : N(H) \to N(H)/H$ be the natural projection map, and $K = \pi^{-1}(W)$, so K acts naturally on Σ. Let

$$M = G \times_K \Sigma = \{(g,\sigma)| \; g \in G, \; \sigma \in \Sigma\}/\sim,$$

where the equivalence relation $\sim$ is defined by $(g,\sigma) \sim (gk^{-1}, k\sigma)$, and define the G-action on M by $\gamma(g,\sigma) = (\gamma g, \sigma)$. Now suppose ds^2 is a metric on M such that $ds^2|\Sigma$ and $ds^2|\nu(\Sigma)$ are K-invariant. Then G acts on M isometrically with $e \times \Sigma$ as a section, (H) as the principal orbit type, and W as the generalized Weyl group.

Note that any finite group W can be embedded as a subgroup of some $SO(n)$. Thus taking $G = SO(n)$, $H = e$, and $\Sigma = S^{n-1}$ in the the above construction gives a G-manifold admitting sections and having W as its generalized Weyl group. This makes it seem unlikely that there can be a good structure theory for polar actions in complete generality. Nevertheless, Dadok's theorem 5.6.17 gives a classification for the polar actions on S^n, and it would be interesting to classify the polar actions for other special classes of Riemannian manifolds, say for arbitrary symmetric spaces.

Although a general structure theory for polar actions is unlikely, we will now see that the special properties in Example 5.6.16, for the isotropy representations of symmetric spaces, continue to hold for all polar actions.

5.6.21. Theorem. *If M is a polar G-manifold and $p \in M$, then the slice representation at p is also polar. In fact, if Σ is a section for M through p then $T\Sigma_p$ is a section and $W(\Sigma)_p = \{\varphi \in W(\Sigma)| \; \varphi(p) = p\}$ is the generalized Weyl group for the slice representation at p.*

PROOF. Let $V = \nu(Gp)_p$ be the space of the slice representation, and $V_0 = T\Sigma_p$. Then, by definition of a section, V_0 is a linear subspace of V. Suppose B is a small ball centered at the origin in V, $S = \exp_p(B)$ is a normal slice at p, and $x = \exp_p(v) \in S$. By Corollary 5.1.13, $G_x \subseteq G_p$ for all $x \in S$. So the isotropy subgroup of the linear G_p-action on V at x is G_x. From this follows the well-known fact that the G_p-orbit of x in V has the same codimension as the G-orbit of x in M. By Proposition 5.6.12 it suffices to show that for each $u \in V_0$ and ξ in the Lie algebra of G_p, $\langle \xi(u), v \rangle = 0$ for all $v \in V_0$. Let g^s be the one parameter subgroup on G_p generated by ξ, and $u(t) = \exp_p(tu)$. Choose $v(t) \in T\Sigma_{u(t)}$ such that as $t \to 0$ $v(t) \to v$ in $T\Sigma$. Since Σ is a section,

$$\langle \xi(u(t)), v(t) \rangle_{u(t)} = 0, \tag{5.6.1}$$

where $\langle\ ,\ \rangle_{u(t)}$ is the inner product on $TM_{u(t)}$. Note that the vector field ξ for the G_p-action on V is given by

$$\xi(v) = \lim_{s \to 0} dg_p^s(u)$$

$$= \lim_{s \to 0} \lim_{t \to 0} g^s(u(t))$$

$$= \lim_{t \to 0} \lim_{s \to 0} g^s(u(t)) = \lim_{t \to 0} \xi(u(t)).$$

Letting $t \to 0$ in (5.6.1), we obtain $\langle \xi(u), v \rangle = 0$.

It remains to prove that $W(V_0) = W(\Sigma)_p$. To see this note that $N(V_0) = N(\Sigma) \cap G_p$ and $Z(V_0) = Z(\Sigma) \cap G_p = Z(\Sigma)$, so $W(V_0) \subseteq W(\Sigma)_p$. Conversely if $gZ(\Sigma) \in W(\Sigma)_p$, then $gp = p$, which implies that $W(\Sigma)_p \subseteq W(V_0)$. ∎

5.6.22. Corollary.   *Let M be a polar G-manifold. If M has a fixed point then the generalized Weyl group of M is a Weyl group.*

5.6.23. Corollary.   *If M is a polar G-manifold then for any $p \in M$, G_p acts transitively on the set of sections of M that contains p.*

PROOF.   Let Σ_1 and Σ_2 be sections through p and let x be a regular point of Σ_1 near p. We may regard Σ_2 as a section for the slice representation at p, so it meets $G_p x$, i.e., there exists $g \in G_p$ such that $gx \in \Sigma_2$. Since $g\Sigma_1$ and Σ_2 are both sections of M containing the regular point gx they are equal by Theorem 5.6.7 (5). ∎

5.6.24. Corollary.   *Let M be a polar G-manifold, Σ a section of M, and $W = W(\Sigma)$ its generalized Weyl group. Then for $x \in \Sigma$ we have $Gx \cap \Sigma = Wx$.*

PROOF.   It is obvious that $Wx \subseteq Gx \cap \Sigma$. Conversely suppose $x' = gx \in \Sigma$. Then $g\Sigma$ is a section at x', so by Corollary 5.6.23 there is $\gamma \in G_{x'}$ such that $\gamma g\Sigma = \Sigma$. Thus $\gamma g \in N(\Sigma)$ so $x' = \gamma x' = \gamma gx$ is in $N(\Sigma)x = Wx$. ∎

For a K-manifold N, we let $C^0(N)^K$ and $C^\infty(N)^K$ denote the space of all continuous and smooth K-invariant functions on N. As a consequence of Corollary 5.6.24 we see that if M is a polar G-manifold with Σ as a section and W as its generalized Weyl group, then the restriction map r from $C^0(M)^G$ to $C^0(\Sigma)^W$ defined by $r(f) = f|\Sigma$ is an isomorphism. Moreover, it follows from Theorem 5.6.21, Corollary 5.6.18, and a theorem of G. Schwarz

[Sh] (if G is a subgroup of $O(n)$ then every smooth G-invariant function on R^n can be written as a smooth functions of invariant polynomials) that the Chevalley restriction theorem can be generalized to smooth invariant functions of a polar action, i.e.,

5.6.25. Theorem [PT2].   *Suppose M is a polar G-manifold, Σ is a section, and $W = W(\Sigma)$ is its generalized Weyl group. Then the restriction map $C^\infty(M)^G \to C^\infty(\Sigma)^W$ defined by $f \mapsto f|\Sigma$ is an isomorphism.*

5.7 Submanifold geometry of orbits.

One important problem in the study of submanifolds of $N^p(c)$ is to determine submanifolds which have simple local invariants. The submanifolds with the simplest invariants are the totally umbillic submanifolds, and these have been completely classified (see section 2.2). Another interesting class consists of the compact submanifolds with parallel second fundamental forms. It is not surprising that the first examples of the latter arise from group theory. Ferus ([Fe]) noted that if M is an orbit of the isotropy representation of a symmetric space G/K and if M is itself a symmetric space with respect to the metric induced on it as a submanifold of the Euclidean space $T(G/K)_{eK}$, then the second fundamental form of M is parallel with respect to the induced normal connection (defined in section 2.1). Conversely, Ferus ([Fe]) showed that these are the only submanifolds of Euclidean spaces (or spheres) whose second fundamental forms are parallel. These results might lead one to think that orbits of isometric action on S^n may not be too difficult to characterize in terms of their local geometric invariants as submanifolds. But in fact, this turns out to be a rather complicated problem.

Let N be a Riemannian G-manifold, and $M = Gx_0$ a principal orbit in N. If v is a G-equivariant normal field on M, then by Proposition 5.4.12 (3), $M_v = \{\exp(v(x)) \mid x \in M\}$ is the G-orbit through $x = \exp_{x_0}(v(x_0))$. The map $M \to M_v$ defined by $gx_0 \to \exp_{gx_0}(v(gx_0))$ is a fibration. Moreover every orbit is of the form M_v for some equivariant normal field v. So in order to understand the submanifold geometry of orbits of N, it suffices to consider principal orbits.

It follows from Proposition 5.4.12, Corollary 5.4.11 and Theorem 5.6.7 that we have

5.7.1. Theorem.   *Suppose M is a principal orbit of an isometric polar G-action on N. Then*

(1) a G-equivariant normal field is parallel with respect to the induced normal connection,

(2) $\nu(M)$ is globally flat,

(3) if v is a parallel normal field on M then the shape operators $A_{v(x)}$ and $A_{v(y)}$ are conjugate for all $x, y \in M$, i.e., the principal curvatures of M along parallel normal field v are constant,

(4) there exists $r > 0$ such that

$$\mathcal{U} = \{\exp_x(v) \mid x \in M, \, v \in \nu(M)_x, \, \|v\| < r\}$$

is a tubular neighborhood of M,

(5) if S_0 is the normal slice at x_0, $\{\exp_{x_0}(v) \mid v \in \nu(M)_{x_0}, \, \|v\| < r\}$, with the induced metric from N, then the map $\pi : \mathcal{U} \to S_0$, defined by $\pi(\exp_x(v(x))) = \exp_{x_0}(v(x_0))$ for $x \in M$ and v a parallel normal field, is a Riemannian submersion,

(6) $\{M_v \mid v$ is a parallel normal vector field of $M\}$ is a singular foliation of N.

Note that the local invariants (normal and principal curvatures) of principal orbits of polar actions of N are quite simple. So, both from the point of view of submanifold geometry and that of group actions, it is natural to make the following definition:

5.7.2. Definition. A submanifold M of N is called *isoparametric* if $\nu(M)$ is flat and the principal curvatures along any parallel normal field of M are constant.

5.7.3. Example. If N is a polar G-manifold, then the principal G-orbits are isoparametric in N. In particular the principal orbits of the isotropy representation of a symmetric space U/K are isoparametric in the Euclidean space $T(U/K)_{eK}$. But unlike the case of totally umbilic submanifolds and submanifolds with parallel second fundamental forms of $N^p(c)$, there are many isoparametric submanifolds of $N^p(c)$ which are *not* orbits. These submanifolds are far from being classified, but there is a rich theory for such manifolds (for example properties (4-6) of Theorem 5.7.1 hold for these submanifolds), and this will be developed in the later chapters.

Next we will discuss the submanifold geometry of a general Riemannian G-manifold. It again follows from previous discussions that we have

5.7.4. Theorem. *Suppose M is a principal orbit of an isometric G-manifold N. Then*

(1) there exist a tubular neighborhood U of M, a Riemannian manifold B and a Riemannian submersion $\pi : U \to B$ having M as a fiber,

(2) if v is a π-parallel normal field on M then the shape operators $A_{v(x)}$ and $A_{v(y)}$ are conjugate for all $x, y \in M$, i.e., the principal curvatures of M along a π-parallel normal field v are constant,

(3) if v is a π-parallel normal field on M then M_v is an embedded submanifold of N and the map $M \to M_v$ defined by $x \to \exp_x(v(x))$ is a fibration,

(4) $\{M_v \mid v$ is a $\pi-$parallel normal vector field of $M\}$ is the orbit foliation on N given by G.

This leads us to make the following definition:

5.7.5. Definition.   An embedded submanifold M of N is *orbit-like* if

(i) there exist a tubular neighborhood U of M in N, a Riemannian manifold B and a Riemannian submersion $\pi : U \to B$ having M as a fiber,

(ii) if v is a π-parallel normal field on the fiber $M_b = \pi^{-1}(b)$ then the shape operators $A_{v(x)}$ and $A_{v(y)}$ of M_b are conjugate for all $x, y \in M_b$, i.e., the principal curvatures of M_b along parallel normal field v are constant.

Then the following are some natural questions and problems:

(1) Let M be an orbit-like submanifold of $N^p(c)$, and suppose its Riemannian submersion π is defined on $U = N^p(c)$. Is there a subgroup G of $\mathrm{Iso}(N^p(c))$ such that all G-orbits are principal and π is the orbit map?

(2) Do conditions (3) and (4) of Theorem 5.7.4 hold for orbit-like submanifolds? If $\|v\|$ is small then it follows from Definition 5.7.5 that (3) and (4) are true. But it is unknown for large v.

(3) Suppose M^n is a submanifold of $N^{n+k}(c)$ with a global normal frame field $\{e_\alpha\}$ such that the principal curvatures of M along e_α are constant. Are there a "good" necessary and sufficient condition on M that guarantee M is orbit-like.

(4) Develop a theory of isoparametric submanifolds of symmetric spaces.

5.8   Infinite dimensional examples.

First we review and set some terminology for manifolds of maps. Let M be a compact Riemannian n-manifold. Then for all k

$$(u, v)_k = \int_M ((I + \Delta)^{\frac{k}{2}} u, \, v) dx$$

defines an inner product on the space $C^\infty(M, \mathbf{R}^m)$ of smooth maps from M to $\mathbf{R}^m$, where dx is the volume element of M and $(,)$ is the standard inner product on $\mathbf{R}^m$. Let $H^k(M, \mathbf{R}^m)$ denote the completion of $C^\infty(M, \mathbf{R}^m)$ with respect to the inner product $(\, , \,)_k$. It follows from the Sobolev embedding theorem [GT] that if $k > \frac{n}{2}$ then $H^k(M, \mathbf{R}^m)$ is contained in $C^0(M^n, \mathbf{R}^m)$ and the inclusion map is compact. Let N be a complete Riemannian manifold isometrically embedded in the Euclidean space $\mathbf{R}^m$. If $k > \frac{n}{2}$ then

$$H^k(M, N) = \{u \in H^k(M, \mathbf{R}^m) \mid u(M) \subseteq N\}$$

is a Hilbert manifold (for details see [Pa6]). In particular, $H^k(S^1, N)$ is a Hilbert manifold if $k > \frac{1}{2}$.

Let G be a simple compact connected Lie group, T a maximal torus of G, $\mathcal{G}$, $\mathcal{T}$ the corresponding Lie algebras, and b the Killing form on $\mathcal{G}$. Then $(u, v) = -b(u, v)$ defines an inner product on $\mathcal{G}$. Let ξ denote the trivial principal G-bundle on S^1. Then the Hilbert group $\hat{G} = H^1(S^1, G)$ is the gauge group, and the Hilbert space $V = H^0(S^1, \mathcal{G})$ is the space of H^0-connections of ξ. The group $\hat{G}$ acts on V by the gauge transformations:

$$g \cdot u = gug^{-1} + g'g^{-1}.$$

5.8.1. Theorem. *Let G be a compact Lie group, T a maximal torus of G, and $\mathcal{G}$, $\mathcal{T}$ the corresponding Lie algebras. Let $\hat{G} = H^1(S^1, G)$ act on $V = H^0(S^1, \mathcal{G})$ by*

$$g \cdot u = gug^{-1} + g'g^{-1}.$$

Then this $\hat{G}$-action is isometric, proper, Fredholm, and admits section. In fact, $\hat{T} = $ the set of constant maps in V with value in T, is a section, and the associate generalized Weyl group $W(\hat{T})$ is the affine Weyl group $W \times_s \Lambda$, where $\Lambda = \{t \in \mathcal{T} \mid \exp(t) = e\}$ and

$$(w_1, \lambda_1) \cdot (w_2, \lambda_2) = (w_1 w_2, \lambda_2 + w_2(\lambda_1)).$$

PROOF.   Since the Killing form on $\mathcal{G}$ is $Ad(G)$ invariant, the $\hat{G}$-action is isometric (by affine isometries). To see that it is proper, suppose $g_n \cdot u_n \to v$ and $u_n \to u$. Since G is compact, $g_n \cdot u \to v$, i.e., $g_n u g_n^{-1} + g_n' g_n^{-1} \to v$, which implies that $\|u_n + g_n^{-1} g_n'\|_0$ is bounded. So $\|g_n^{-1} g_n'\|$ is bounded. Since G is compact, $\|g_n\|_0$ is bounded. Hence $\|g_n\|_1$ is bounded. It follows from the Sobolev embedding theorem and Rellich's lemma that the inclusion map $H^1(S^1, G) \hookrightarrow C^0(S^1, G)$ is a compact operator, so there exists a subsequence (still denoted by g_n) converging to g_0 in $H^0(S^1, G)$. But

$$\|g_n u g_n^{-1} + g_n' g_n^{-1} - v\|_0 = \|g_n u + g_n' - v g_n\|_0 \to 0,$$

so $g_n \to g_0$ in $H^1(S^1, G)$.

The differential P of the orbit map $g \mapsto gx$ at e is

$$P : H^1(S^1, \mathcal{G}) \to H^0(S^1, \mathcal{G}), \quad u \mapsto u' + [u, x],$$

which is elliptic. So it follows form the standard elliptic theory [GT] that P is Fredholm. This proves that the $\hat{G}$-action is Fredholm.

Next we show that $\hat{T}$ meets every $\hat{G}$-orbit. Let $\Phi : H^0(S^1, \mathcal{G}) \to G$ be the holonomy map, i.e., given $u \in H^0(S^1, \mathcal{G})$, let $f : R \to G$ be the unique solution for $f' f^{-1} = u$ and $f(0) = e$, then $\Phi(u) = f(2\pi)$. Given $u \in H^0(S^1, \mathcal{G})$, by the maximal torus theorem there exist $s \in G$ and $a \in T$ such that $s\Phi(u)s^{-1} = \exp(2\pi a)$. Let $\hat{a}$ denote the constant map $\hat{a}(t) = a$. Then $\hat{a} \in \hat{G} \cdot u$. To see this, let $h(t) = \exp(ta)sf^{-1}(t)$, then $h(0) = h(2\pi) = s$, i.e., $h \in H^1(S^1, G)$ and $h \cdot u = \hat{a}$.

It remains to prove that $\hat{T}$ is orthogonal to every $\hat{G}$-orbit. Given $t \in T$, we let $\hat{t} \in H^0(S^1, \mathcal{G})$ denote the constant map with value t. Let $\hat{t}_0 \in \hat{T}$. Then

$$T(\hat{G} \cdot \hat{t}_0)_{\hat{t}_0} = \{v' + [v, \hat{t}_0] | \ v \in H^1(S^1, \mathcal{G})\}.$$

Given any $\hat{t} \in \hat{T}$, we have

$$(\hat{t}, v' + [v, \hat{t}_0])_0 = \int_{S^1} (t, v'(\theta) + [v(\theta), t_0]) \, d\theta$$

$$= \int_{S^1} (t, v'(\theta)) \, d\theta + \int_{S^1} (t, [v(\theta), t_0]) \, d\theta$$

$$= 0 + \int_{S^1} ([t_0, t], v(\theta)) \, d\theta = 0$$

So $\hat{T}$ is a section. ∎

There is little known about the classification of polar actions on Hilbert spaces.

Chapter 6.

Isoparametric Submanifolds.

In section 5.7, we defined a submanifold of a space form to be isoparametric if its normal bundle is flat and if the principal curvatures along any parallel normal vector field are constant (Definition 5.7.2). These submanifolds arise naturally in representation theory for, as we saw, an orbit of an orthogonal representation is isoparametric if and only if it is a principal orbit of a polar representations, so in particular principal coadjoint orbits are isoparametric. And because their local invariants are so simple, isoparametric manifolds are also natural models to use in the classification theory of submanifolds. Although the principal orbits of a polar action are isoparametric, not all isoparametric submanifolds in R^m and S^m are orbits. Nevertheless, as we will see in this chapter, every isoparametric submanifold of R^m or S^m has associated to it a singular, orbit-like foliation, and this foliation has many of the same remarkable properties of the orbit foliations of polar actions. Thus isoparametric submanifolds can be viewed as a geometric generalization of principal orbits of polar actions.

There is an interesting history of this subject, which explains the origin of the name "isoparametric". A hypersurface is always given locally as the level set of some smooth function f, and then $\|\nabla f\|^2$, $\triangle f$ are called the first and second differential parameters of the hypersurface. So it is natural to make the following definition: a smooth function $f : R^{n+1} \to R$ is called *isoparametric* if $\|\nabla f\|^2$ and $\triangle f$ are functions of f. The family of the level hypersurfaces of f is then called an isoparametric family, since clearly the first and second differential parameters are constant on each hypersurface of the family. It is not difficult to show that an isoparametric family in R^n must be either parallel hyperplanes, concentric spheres, or concentric spherical cylinders. This was proved by Levi-Civita [Lc] for $n = 2$, and by B. Segré [Se] for arbitrary n. Shortly after this work of Levi-Civita and Segré, É. Cartan ([Ca3]-[Ca5]) considered isoparametric functions f on space forms, and discovered many interesting examples for S^{n+1}. Among other things Cartan showed that the level hypersurfaces of f have constant principal curvatures. And conversely, he showed that if M is a hypersurface of $N^{n+1}(c)$ with constant principal curvatures, then there is at least a local isoparametric function having M as a level set. Cartan called such hypersurfaces isoparametric. In the past dozen years, many people carried forward this research. Around mid 1970's, Münzner [Mü1,2] completed a beautiful structure theory of isoparametric hypersurfaces in spheres, reducing their classification to a difficult, but purely algebraic problem. Although many people have subsequently made significant contributions to this classification problem, including

Abresch [Ab], Ferus, Karcher, Münzner [FKM], Ozeki and Takeuchi [OT1,2], it is still far from being completely solved. There have also been applications of isoparametric hypersurface theory to harmonic maps [Ee] and minimal hypersurfaces ([No],[FK]). Recently, with the purpose in mind of constructing harmonic maps, Eells [Ee] gave a definition of isoparametric map that generalizes the concept of isoparametric function. Carter and West [CW2] also gave a definition of isoparametric maps $S^{n+k} \to R^k$; their purpose being to generalize Cartan's work to higher codimension. Using their definition, they showed that the regular level of an isoparametric map is an isoparametric submanifold. They also showed that there is a Coxeter group associated to each codimension two isoparametric submanifold of a sphere, but they did not obtain a similar result for higher codimension. This work led Terng [Te2] to the definition used in this section.

6.1 Isoparametric maps.

6.1.1. Definition.   A smooth map $f = (f_{n+1}, \ldots, f_{n+k}) : N^{n+k}(c) \to R^k$ is called *isoparametric* if

(1) f has a regular value,

(2) $\langle \nabla f_\alpha, \nabla f_\beta \rangle$ and $\triangle f_\alpha$ are functions of f for all α, β,

(3) $[\nabla f_\alpha, \nabla f_\beta]$ is a linear combination of $\nabla f_{n+1}, \ldots, \nabla f_{n+k}$, with coefficients being functions of f, for all α and β.

This definition agrees with Cartan's when $k = 1$. In the following we will proceed to prove that regular level submanifolds of an isoparametric map are isoparametric.

Hereafter we will use the notation introduced in Chapter 2. Suppose $f : N^{n+k}(c) \to R^k$ is isoparametric. Applying the Gram-Schmidt process to $\{\nabla f_\alpha\}$ we may assume that at any regular point of f, there is a local orthonormal frame field $e_1, \ldots, e_{n+k}$ with dual coframe $\omega_1, \ldots, \omega_{n+k}$ such that

$$df_\alpha = \sum_\beta c_{\alpha\beta}\, \omega_\beta, \qquad (6.1.1)$$

with $\text{rank}(c_{\alpha\beta}) = k$, and where the $c_{\alpha\beta}$ are functions of f. So

$$dc_{\alpha\beta} \equiv 0 \mod (\omega_{n+1}, \ldots, \omega_{n+k}). \qquad (6.1.2)$$

It is obvious that $\omega_\alpha = 0$ defines the level submanifolds of f. Condition (3) implies that the *normal distribution* defined by $\omega_i = 0$ on the set of regular points of f is completely integrable.

6.1.2. Proposition. *Let $f : N^{n+k}(c) \to R^k$ be isoparametric, $b = f(q)$ a regular value, $M = f^{-1}(b)$, and F the leaf of the normal distribution through q. Then*

(i) F is totally geodesic,

(ii) $\nu(M)$ is flat and has trivial holonomy group.

PROOF. Take the exterior differential of (6.1.1), and using the structure equations, we obtain

$$\sum_\beta dc_{\alpha\beta} \wedge \omega_\beta + \sum_{\beta i} c_{\alpha\beta}\omega_{\beta i} \wedge \omega_i + \sum_{\beta\gamma} c_{\alpha\beta}\omega_{\beta\gamma} \wedge \omega_\gamma = 0. \qquad (6.1.3)$$

From (6.1.2), since the coefficient of $\omega_i \wedge \omega_\gamma$ in (6.1.3) is zero, we obtain

$$\sum_\beta c_{\alpha\beta}(-\omega_{\beta i}(e_\gamma) + \omega_{\beta\gamma}(e_i)) = 0. \qquad (6.1.4)$$

But rank$(c_{\alpha\beta}) = k$, hence:

$$\omega_{\beta i}(e_\gamma) = \omega_{\beta\gamma}(e_i).$$

From condition (3) of Definition 6.1.1, we have

$$[e_\alpha, e_\beta] = \sum_\gamma u_{\alpha\beta\gamma}e_\gamma = \nabla_{e_\alpha} e_\beta - \nabla_{e_\beta} e_\alpha$$

$$= \sum_i (\omega_{\beta i}(e_\alpha) - \omega_{\alpha i}(e_\beta))e_i + \sum_\gamma (\omega_{\beta\gamma}(e_\alpha) - \omega_{\alpha\gamma}(e_\beta))e_\gamma.$$

Hence

$$\omega_{\beta i}(e_\alpha) = \omega_{\alpha i}(e_\beta),$$

$$\omega_{\beta\gamma}(e_\alpha) - \omega_{\alpha\gamma}(e_\beta) = u_{\alpha\beta\gamma},$$

where $u_{\alpha\beta\gamma}$ is a function of f. In particular, we have

$$\omega_{\beta\alpha}(e_\alpha) = u_{\alpha\beta\alpha},$$

is a function of f. Using (6.1.4), we have

$$\omega_{\beta i}(e_\alpha) = \omega_{\beta\alpha}(e_i)$$

$$= \omega_{\alpha i}(e_\beta) = \omega_{\alpha\beta}(e_i) = -\omega_{\beta\alpha}(e_i).$$

So $\omega_{\alpha\beta}(e_i) = 0$ and $\omega_{\alpha i}(e_\beta) = 0$, i.e., $\omega_{\alpha\beta} = 0$ on M, and $\omega_{\alpha i} = 0$ on F. This implies that the e_α are parallel normal fields on M and that F is totally geodesic.

Note that e_α on M can be obtained by applying the Gram-Schmidt process to $\nabla f_{n+1}, \ldots, \nabla f_{n+k}$, so e_α is a global parallel normal frame on M, hence the holonomy of $\nu(M)$ is trivial. ∎

6.1.3. Corollary. *With the same assumption as in Proposition 6.1.2,*
(i) $\nabla f_\alpha | M$ is a parallel normal field on M for all $n + 1 \leq \alpha \leq n + k$,
(ii) if v is a parallel normal field on M, then there exists $t_0 > 0$ such that $\{exp_x(tv(x)) | \ x \in M\}$ is a regular level submanifold of f for $|t| < t_0$.

In order to prove that a regular level submanifold of an isoparametric map is isoparametric we need the following simple and direct generalization of Theorem 3.4.2 on Bonnet transformations.

6.1.4. Proposition. *Suppose $X : M^n \to N^{n+k}(c)$ is an isometric immersion with flat normal bundle, and v is a unit parallel normal field. Then $X^* = aX + bv$ is an immersion if and only if $(aI - bA_v)$ is non-degenerate on M. Here A_v is the shape operator of M in the direction v, and $(a, b) = (1, t)$ for $c = 0$, is $(\cos t, \sin t)$ for $c = 1$, and is $(\cosh t, \sinh t)$ for $c = -1$. Moreover:*
(i) $\nu(M^*)$ is flat,
(ii) $TM_q = TM^*_{q^*}$, $\nu(M)_q = \nu(M^*)_{q^*}$, where $q^* = X^*(q)$,
(iii) $v^* = -cbX + av$ is a parallel normal field on M^*,
(iv) $A^*_{v^*} = (cbI + aA_v)(aI - bA_v)^{-1}$.

PROOF. We will prove only the case $c = 0$, the other cases being similar. Let e_A be an adapted frame on M. Taking the differential of X^* we obtain

$$dX^* = I - tA_v + t(\nabla^\nu v).$$

Since v is parallel, we have $dX^* = I - tA_v$. Hence X^* is an immersion if and only if $(I - tA_v)$ is invertible. So e_A is also an adapted frame for M^*, and the dual coframe is $\omega_i^* = \sum_j (\delta_{ij} - t(A_v)_{ij})\omega_j$. Moreover, $\omega_{i\alpha}^* = \omega_{i\alpha}$, so we have $A^*_{v^*} = A_v(I - tA_v)^{-1}$. ∎

6.1.5. Proposition. *With the same assumptions as in Proposition 6.1.2:*
(i) the mean curvature vector of M is parallel,
(ii) the principal curvatures of M along a parallel normal field are constant.

PROOF. We choose a local orthonormal frame e_A as in the proof of Proposition 6.1.2. Let

$$d(f_\alpha) = \sum_A (f_\alpha)_A \omega_A,$$

$$\nabla^2 f_\alpha = \sum_{AB} (f_\alpha)_{AB} \, \omega_A \otimes \omega_B.$$

Using (6.1.1) we have

$$(f_\alpha)_i = 0, \quad (f_\alpha)_\beta = c_{\alpha\beta}.$$

Now (1.3.6) gives

$$(f_\alpha)_{ii} = -\sum_\beta c_{\alpha\beta} h_{i\beta i},$$

$$(f_\alpha)_{\beta\beta} = dc_{\alpha\beta}(e_\beta) + \sum_\gamma c_{\alpha\gamma}\omega_{\gamma\beta}(e_\beta),$$

so we have

$$\triangle f_\alpha = \sum_\beta dc_{\alpha\beta}(e_\beta) - \sum_\beta c_{\alpha\beta} H_\beta + \sum_{\beta\gamma} c_{\alpha\gamma}\omega_{\gamma\beta}(e_\beta),$$

where $H_\beta = \sum_i h_{i\beta i}$ is the mean curvature of level submanifolds of f in the direction of e_β. Since $\triangle f_\alpha$, $c_{\alpha\beta}$ and $\omega_{\gamma\beta}(e_\beta)$ are all functions of f, $\sum_\beta c_{\alpha\beta} H_\beta$ is a function of f. However $\mathrm{rank}(c_{\alpha\beta}) = k$, and hence the H_α are functions of f, i.e., each H_α's is constant on M. But the e_α are parallel normal fields on M, so (i) is proved.

To prove (ii) we use the method used by Nomizu [No] in codimension one. Let X be the position function of M in R^{n+k}. By Corollary 6.1.3, there exists $t_0 > 0$ such that $X^* = X + te_\alpha$ is an immersion if $|t| < t_0$, and $X^*(M)$ is a regular level of f. Then, by (i), the mean curvature H_α^* of X^* in the direction of $e_\alpha^* = e_\alpha$ is constant. Using Proposition 6.1.4 (iv) and the identity:

$$A(I - tA)^{-1} = A \sum_{m=0}^{\infty} t^m A^m = \sum_{m=0}^{\infty} A^{m+1} t^m,$$

we have

$$H_\alpha^* = \sum_{m=0}^{\infty} (\mathrm{tr}(A_{e_\alpha}^{m+1})) t^m. \qquad (6.1.5)$$

Note that H_α^* is independent of $x \in M$, so the right hand side of (6.1.5) is a function of t alone. Hence $\mathrm{tr}(A_{e_\alpha}^{m+1})$ is a function of t for all m and this implies that the eigenvalues of A_{e_α} are constant on M. ∎

As a consequence of Propositions 6.1.2 and 6.1.5, we have

6.1.6. Theorem.   Let $f : N^{n+k}(c) \to R^k$ be isoparametric, b a regular value, and $M = f^{-1}(b)$. Then M is isoparametric.

6.2 Curvature distributions.

In this section we assume that M^n is an immersed isoparametric submanifold of R^{n+k}. Since $\nu(M)$ is flat, by Proposition 2.1.2, $\{A_v|\ v \in \nu(M)_q\}$ is a family of commuting self-adjoint operators on TM_q, so there exists a common eigendecomposition $TM_q = \bigoplus_{i=1}^p E_i(q)$. Let $\{e_\alpha\}$ be a local orthonormal parallel normal frame. By definition of isoparametric, $A_{e_\alpha(x)}$ and $A_{e_\alpha(q)}$ have same eigenvalues. So E_i's are smooth distributions and $TM = \bigoplus E_i$. The E_i's are characterized by the equation

$$A_{e_\alpha}|E_i = n_{i\alpha} \mathrm{id}_{E_i},$$

together with the conditions that if $i \neq j$ then there exists α_0 with $n_{i\alpha_0} \neq n_{j\alpha_0}$. Note that the $E_i(q)$ are the common eigenspaces of all the shape operators at q, so they are independent of the choice of the e_α, and are uniquely determined up to a permutations of their indices. These distributions E_i are called the *curvature distributions* of M.

We will make the following standing assumptions:

(1) M has p curvature distributions $E_1, \ldots, E_p$, and $m_i = \mathrm{rank}(E_i)$.

(2) Let $\{e_i\}$ be a local orthonormal tangent frame for M such that E_i is spanned by $\{e_j|\mu_{i-1} < j \leq \mu_i\}$, where $\mu_i = \sum_{s=1}^i m_s$. So we have

$$\omega_{\alpha\beta} = 0, \tag{6.2.1}$$

$$\omega_{i\alpha} = \lambda_{i\alpha}\omega_i, \tag{6.2.2}$$

where $\lambda_{i\alpha}$ are constant. In fact, $\lambda_{i\alpha} = n_{j\alpha}$ if $\mu_{j-1} < i \leq \mu_j$.

(3) Let $v_i = \sum_\alpha n_{i\alpha} e_\alpha$. Then

$$A_v|E_i = \langle v, v_i \rangle \mathrm{id}_{E_i}, \tag{6.2.3}$$

for any normal field v. Clearly (6.2.3) characterizes the v_i, so in particular v_i is independent of the choice of e_α, i.e., each v_i is a well-defined normal field associated to E_i. In fact, if $\bar{e}_\alpha$ is another local parallel normal frame on M and $\bar{n}_{i\alpha}$ the eigenvalues of $A_{\bar{e}_\alpha}$ then

$$v_i = \sum_\alpha \bar{n}_{i\alpha} \bar{e}_\alpha = \sum_\alpha n_{i\alpha} e_{i\alpha}.$$

We call v_i the *curvature normal* of M associated to E_i.

(4) Let $n_i = (n_{in+1}, \ldots, n_{in+k})$.

If M is isoparametric in R^{n+k} then M is also isoparametric in R^{n+k+1}. To avoid this redundancy, we make the following definition:

6.2.1. Definition. A submanifold M of R^{n+k} is *full* if M is not included in any affine hyperplane of R^{n+k}.

6.2.2. Definition. An immersed, full, isoparametric submanifold M^n of R^{n+k} is called a rank k isoparametric submanifold in R^{n+k}.

6.2.3. Proposition. *An immersed isoparametric submanifold M^n of R^{n+k} is full if and only if the curvature normals $v_1, \ldots, v_p$ spans $\nu(M)$. In particular, if M^n is full and isoparametric in R^{n+k} then $k \leq n$.*

PROOF. Note that $v_1, \ldots, v_p$ span $\nu(M)$ if and only if the rank of the $k \times p$ matrix $N = (n_{i\alpha})$ is k. Suppose M is contained in a hyperplane normal to a constant unit vector $u_0 \in R^{n+k}$. Then we can choose $e_{n+1} = u_0$, so $n_{in+1} = 0$ for all i, and $\text{rank}(N) < k$. Conversely, if $\text{rank}(N) < k$ then there exists a unit vector $c = (c_\alpha) \in R^k$ such that $\langle c, n_i \rangle = 0$ for all $1 \leq i \leq p$. We claim that $v = \sum_\alpha c_\alpha e_\alpha$ is a constant vector b in R^{n+k}. To see this, we note that the eigenvalues of A_v are $\langle v, v_i \rangle = \langle c, n_i \rangle = 0$, i.e., $A_v = 0$. But $dv = -A_v$, so v is constant on M. Then it follows that

$$d(\langle X, b \rangle) = \langle dX, b \rangle = \sum_i \omega_i \langle e_i, b \rangle = 0.$$

Hence $\langle X, b \rangle = c_0$ a constant, i.e., M is contained in a hyperplane. ∎

Recall that the endpoint map $Y : \nu(M) \to R^{n+k}$ is defined by $Y(v) = x + v$ for $v \in \nu(M)_x$. Using the frame e_A, we can write

$$Y = Y(x, z) = x + \sum_\alpha z_\alpha e_\alpha(x).$$

The differential of Y is

$$dY = dX + \sum_\alpha z_\alpha de_\alpha + \sum_\alpha dz_\alpha e_\alpha$$

$$= \sum_{i=1}^p (1 - \langle z, n_i \rangle) id_{E_i} + \sum_\alpha dz_\alpha e_\alpha.$$

Now recall also that a point y of R^{n+k} is called a focal point of M if it is a singular value of Y, that is if it is of the form $y = Y(v)$ where dY_v has rank less than $n + k$. The set Γ of all focal points of M is called the focal set of M

6.2.4. Proposition. *Let M be an immersed isoparametric submanifold M^n of R^{n+k} and Γ its focal set. For each $q \in M$ let Γ_q denote the intersection* of Γ with the normal plane $q + \nu(M)_q$ to M at q. Then Γ, is the union of the Γ_q, and each Γ_q is the union of the p hyperplanes $\ell_i(q) = \{q + v \mid v \in \nu(M)_q,\ \langle v, v_i \rangle = 1\}$ in $q + \nu(M)_q$. These $\ell_i(q)$ are called the focal hyperplane associated to E_i at q.

6.2.5. Corollary.

(1) The curvature normal $v_i(q)$ is normal to the focal hyperplane $\ell_i(q)$ in $q + \nu(M)_q$.

(2) The distance $d(q, \ell_i(q))$ from q to $\ell_i(q)$ is $1/\|v_i\|$.

6.2.6. Proposition.

Let $X : M^n \to R^{n+k}$ be an immersed isoparametric submanifold, and v a parallel normal field. Then $X + v$ is an immersion if and only if $\langle v_i, v \rangle \neq 1$ for all $1 \leq i \leq p$. Moreover,

(i) the parallel set M_v defined by v, i.e., the image of $X + v$, is an immersed isoparametric submanifold,

(ii) let $q^* = q + v(q)$, then $TM_q = T(M_v)_{q^*}$, $\nu(M)_q = \nu(M_v)_{q^*}$, and $q + \nu(M)_q = q^* + \nu(M_v)_{q^*}$.

(iii) if $\{e_\alpha\}$ is a local parallel normal frame on M then $\{\bar{e}_\alpha\}$ is a local parallel normal frame on M_v, where $\bar{e}_\alpha(q^*) = e_\alpha(q)$,

(iv) $E_i^*(q^*) = E_i(q)$ are the curvature distributions of M_v, and the corresponding curvature normals are given by

$$v_i^*(q^*) = v_i(q)/(1 - \langle v, v_i \rangle),$$

(v) the focal hyperplane $\ell_i^*(q^*)$ of M_v associated to E_i^* is the same as the focal hyperplane $\ell_i(q)$ of M associated to E_i.

PROOF. Since v is parallel, there exist constants z_α such that $v = \sum_\alpha z_\alpha e_\alpha$. The differential of $X + v$ is

$$d(X + v) = dX + \sum_\alpha z_\alpha de_\alpha$$

$$= \sum_i \omega_i e_i - \sum_{i,\alpha} z_\alpha \omega_{i\alpha} e_i$$

$$= \sum_i (1 - \sum_\alpha z_\alpha \lambda_{i\alpha}) \omega_i e_i.$$

So we may choose the following local frame on M_v:

$$e_A^* = e_A, \quad \omega_i^* = (1 - \sum_\alpha z_\alpha \lambda_{i\alpha}) \omega_i.$$

Then $\omega^*_{AB} = \langle de^*_A, e^*_B \rangle = \omega_{AB}$. In particular, we have

$$\omega^*_{\alpha\beta} = 0,$$

$$\omega^*_{i\alpha} = \lambda_{i\alpha}\omega_i = \frac{\lambda_{i\alpha}}{1 - \sum_\beta z_\beta \lambda_{i\beta}}\omega^*_i,$$

which proves the proposition. ∎

Next we will prove that the curvature distributions are integrable. First we need some formulas for the Levi-Civita connection of M in terms of E_i. Using (6.2.1), (6.2.2) and the structure equations, we have

$$d\omega_{i\alpha} = d(\lambda_{i\alpha}\omega_i) = \lambda_{i\alpha}d\omega_i = \lambda_{i\alpha}\sum_j \omega_{ij} \wedge \omega_j$$

$$= \sum_j \omega_{ij} \wedge \omega_{j\alpha} = \sum_j \lambda_{j\alpha}\omega_{ij} \wedge \omega_j,$$

so

$$\sum_j (\lambda_{i\alpha} - \lambda_{j\alpha})\omega_{ij} \wedge \omega_j = 0.$$

Suppose $\omega_{ij} = \sum_m \gamma_{ijm}\omega_m$, then we have

$$\sum_{j,m} (\lambda_{i\alpha} - \lambda_{j\alpha})\gamma_{ijm}\omega_m \wedge \omega_j = 0.$$

This implies that

6.2.7. Proposition. *Let* $\omega_{ij} = \sum_m \gamma_{ijm}\omega_m$. *Then*

$$(\lambda_{i\alpha} - \lambda_{j\alpha})\gamma_{ijm} = (\lambda_{i\alpha} - \lambda_{m\alpha})\gamma_{imj}, \quad \text{if } j \neq m.$$

In particular, if e_i, $e_m \in E_{i_1}$, $e_j \in E_{i_2}$, *and* $i_1 \neq i_2$, *then* $\gamma_{ijm} = 0$.

6.2.8. Theorem. *Let* M^n *be an immersed isoparametric submanifold of* R^{n+k}. *Then each curvature distribution* E_i *is integrable.*

PROOF. For simplicity, we assume $i = 1$ and $m = m_1$. E_1 is defined by the following 1-form equations on M:

$$\omega_i = 0, \quad m < i \leq n.$$

Using the structure equation, we have

$$d\omega_i = \sum_{j=1}^m \omega_{ij} \wedge \omega_j = \sum_{j,s=1}^m \gamma_{ijs}\omega_s \wedge \omega_j.$$

Since $\omega_{ij} = -\omega_{ji}$, $\gamma_{ijs} = -\gamma_{jis}$, which is zero by Proposition 6.2.7. So E_1 is integrable. ∎

6.2.9. Theorem. *Let M^n be a complete, immersed, isoparametric submanifold of R^{n+k}, E_i the curvature distributions, v_i the corresponding curvature normals, and $\ell_i(q)$ the focal hyperplane associated to E_i at $q \in M$. Let $S_i(q)$ denote the leaf of E_i through q.*

(1) If $v_i \neq 0$ then

(i) $E_i(x) \oplus Rv_i(x)$ is a fixed $(m_i + 1)$-plane ξ_i in R^{n+k} for all $x \in S_i(q)$,

(ii) $x + (v_i(x)/\|v_i(x)\|^2)$ is a constant $c_0 \in \xi_i$ for all $x \in S_i(q)$,

(iii) $S_i(q)$ is the standard sphere of $c_0 + \xi_i$ with radius $1/\|v_i\|$ and center at c_0,

(iv) $E_i(x) \oplus \nu(M)_x$ is a fixed $(m_i + k)$-plane η_i in R^{n+k} for all $x \in S_i(q)$,

(v) $\ell_i(x) = \ell_i(q)$ for all $x \in S_i(q)$, which is the $(k-1)$-plane perpendicular to $c_0 + \xi_i$ in $c_0 + \eta_i$ at c_0,

(vi) given $y \in \ell_i(q)$ we have $\|x - y\| = \|q - y\|$ for all $x \in S_i(q)$.

(2) If $v_i = 0$ then $E_i(x) = E_i(q)$ is a fixed m_i-plane for all $x \in S_i(q)$ and $S_i(q)$ is the plane parallel to $E_i(q)$ passes through q.

PROOF. It suffices to prove this theorem for E_1. Let $m = m_1$. To obtain (1), we compute the differential of the map $f = e_1 \wedge \ldots \wedge e_m \wedge v_1$ from $S_1(q)$ to the Grassman manifold $Gr(m+1, n+k)$. Since

$$e_1 \wedge \ldots \wedge e_m \wedge dv_1 = 0$$

on $S_1(q)$, we have

$$d(e_1 \wedge \ldots \wedge e_m \wedge v_1) =$$

$$\sum_{i \leq m} e_1 \wedge \ldots \wedge \left(\sum_{j > m} \omega_{ij} e_j + \sum_\alpha \omega_{i\alpha} e_\alpha \right) \wedge e_{i+1} \wedge \ldots \wedge e_m \wedge v_1.$$

Using Proposition 6.2.7, we have $\omega_{ij} = \sum_{s \leq m} \gamma_{ijs} \omega_s = 0$ if $i \leq m$ and $j > m$. So we have

$$df = \sum_{i \leq m, \alpha} e_1 \wedge \ldots \wedge \omega_{i\alpha} e_\alpha \wedge e_{i+1} \wedge \ldots \wedge e_m \wedge v_1$$

$$= \sum_{i \leq m, \alpha, \beta} e_1 \wedge \ldots \wedge n_{1\alpha} \omega_i e_\alpha \wedge e_{i+1} \wedge \ldots \wedge e_m \wedge n_{1\beta} e_\beta$$

$$= \sum_{i \leq m, \alpha, \beta} n_{1\alpha} n_{1\beta} \omega_i e_1 \wedge \ldots e_\alpha \wedge e_{i+1} \ldots e_m \wedge e_\beta = 0,$$

which proves (1)(i). Similarly one can prove (1)(iv) by showing that

$$d(e_1 \wedge \ldots \wedge e_m \wedge e_{n+1} \ldots \wedge e_{n+k}) = 0$$

on $S_1(q)$. Next we calculate the differential of $X + (v_1/\|v_1\|^2)$ on $S_1(q)$:

$$d\left(X + \frac{v_1}{\|v_1\|^2}\right) = Id_{E_1} - \frac{1}{\|v_1\|^2}A_{v_1}|E_1.$$

Since $A_v|E_i = \langle v, v_i \rangle id_{E_i}$, (1)(ii) follows, and (1)(iii) is a direct consequence. Note that $v_1(x)$ is normal to $\ell_1(x)$ in $c_0 + \xi_1$, so it follows from (1)(i) and (1)(iv) that $\ell_1(x)$ is perpendicular to $c_0 + \xi_1$ in $c_0 + \eta_1$ at c_0 for all $x \in S_1(q)$. Hence (1)(v) and (vi) follow.

If $v_1 = 0$ then $\omega_{i\alpha} = 0$ for $i \leq m$. By Proposition 6.2.7, $\omega_{ij} = 0$ on $S_1(q)$ if $i \leq m$ and $j > m$. So $d(e_1 \wedge \ldots \wedge e_m) = 0$ on $S_1(q)$, which proves (2). ∎

Because an m_0-plane is not compact, we have

6.2.10. Corollary.   *If M^n is a compact, immersed, full isoparametric submanifold of R^{n+k}, then all the curvature normals of M are non-zero.*

6.2.11. Proposition.   *Let $\omega_{ij} = \sum_m \gamma_{ijm}\omega_m$. Then*
(i) $(\lambda_{i\alpha} - \lambda_{j\alpha})\gamma_{ijm} = h_{i\alpha jm}$,
(ii) if $e_i \in E_{i_1}$, $e_j \in E_{i_2}$ and $i_1 \neq i_2$, then $\gamma_{ijj} = 0$.

PROOF. Using (2.1.19), we obtain

$$0 = \sum_m h_{j\alpha jm}\omega_m,$$

$$(\lambda_{i\alpha} - \lambda_{j\alpha})\omega_{ij} = \sum_m h_{i\alpha jm}\omega_m. \quad \blacksquare$$

6.3   Coxeter groups associated to isoparametric submanifolds.

In this section we assume that $X : M^n \to R^{n+k}$ is an immersed full isoparametric submanifold. Let $E_0, E_1, \ldots, E_p$ be the curvature distributions, v_i the corresponding curvature normals, and $\ell_i(q)$ the focal hyperplane in $q + \nu(M)_q$ associated to E_i. We may assume $v_0 = 0$, so $v_i \neq 0$ for all $i > 0$. We will use the following standing notations:
(1) $\nu_q = q + \nu(M)_q$.
(2) R_i^q denotes the reflection of ν_q across the hyperplane $\ell_i(q)$.

(3) T_i^q denotes the linear reflection of $\nu(M)_q$ along $v_i(q)$, i.e.,

$$T_i^q(v) = v - 2\frac{\langle v, v_i(q)\rangle}{\|v_i\|^2}v_i(q).$$

(3) Let φ_i be the diffeomorphism of M defined by $\varphi_i(q) =$ the antipodal point of q in the leaf sphere $S_i(q)$ of E_i for $i > 0$. Note that φ_i^2 is clearly the identity map of M; we call it the involution associated to E_i.

(4) S_p will denote the group of permutations of $\{1, \ldots, p\}$.

It follows from (1) of Theorem 6.2.9 that

$$\varphi_i = X + 2\frac{v_i}{\|v_i\|^2},$$

$$\varphi_i(q) = R_i^q(q).$$

Since φ_i is a diffeomorphism it follows from Proposition 6.2.6 that:

6.3.1. Proposition.   *If $v_i \neq 0$ then $1 - 2(\langle v_i, v_j\rangle/\|v_i\|^2)$ never vanishes for $0 \leq j \leq p$.*

6.3.2. Theorem.   *There exist permutations $\sigma_1, \ldots, \sigma_p$ in S_p such that*
(1) $E_j(\varphi_i(q)) = E_{\sigma_i(j)}(q)$, i.e., $\varphi_i^*(E_j) = E_{\sigma_i(j)}$, in particular we have $m_j = m_{\sigma_i(j)}$,
(2) $v_{\sigma_i(j)}(q) = \left(1 - 2\frac{\langle v_i, v_{\sigma_i(j)}\rangle}{\|v_i\|^2}\right) v_j(\varphi_i(q))$,
(3) $T_i^q(v_j(q)) = \left(1 - 2\frac{\langle v_i, v_{\sigma_i(j)}\rangle}{\|v_i\|^2}\right)^{-1} v_{\sigma_i(j)}(q)$.

PROOF.   It suffices to prove the theorem for E_1. Note that $\varphi_1 = X + v$ and $M_v = \varphi_1(M) = M$, where $v = 2v_1/\|v_1\|^2$ is parallel. So by Proposition 6.2.6 (iv), there exists $\sigma \in S_p$ such that (1) is true.

By Proposition 6.2.6 (iii), $\bar{e}_\alpha(x) = e_\alpha(\varphi_1(x))$ gives a parallel normal frame on M. So the two parallel normal frames $\bar{e}_\alpha$ and e_α differ by a constant matrix C in $O(k)$. To determine C, we parallel translate $e_\alpha(q)$ with respect to the induced normal connection of M in R^{n+k} to $q^* = \varphi_1(q)$. Let ξ_1, η_1, c_0 be as in Theorem 6.2.9. Then the leaf $S_1(q)$ of E_1 at q is the standard sphere in the $(m_1 + 1)$-plane $c_0 + \xi_1$, which is contained in the $(m_1 + k)$-plane $c_0 + \eta_1$, and $e_\alpha|S_1(q)$ is a parallel normal frame of $S_1(q)$ in $c_0 + \eta_1$. In particular, the normal parallel translation of $e_\alpha(q)$ to q^* on $S_1(q)$ in $c_0 + \eta_1$ is the same as the normal parallel translation on M in R^{n+k}. Note that the normal planes of $S_1(q)$ at q and q^* in $c_0 + \eta_1$ are the same. Let π denote the parallel translation in the normal bundle of $S_1(q)$ in $c_0 + \eta$ from q to q^*.

Then it is easy to see that $\pi(v_1(q)) = -v_1(q)$ and $\pi(u) = u$ if u is a normal vector at q perpendicular to $v_1(q)$, i.e., π is the linear reflection R_1^q of $\nu(M)_q$ along $v_1(q)$. So

$$e_\alpha(q^*) = T_1^q(e_\alpha(q)) = T_1^q(\bar{e}_\alpha(q^*)).$$

Since $(T_1^q)^{-1} = T_1^q$,

$$\bar{e}_\alpha(q^*) = T_1^q(e_\alpha(q^*)) = e_\alpha(q^*) - 2\frac{\langle v_1(q), e_\alpha(q^*)\rangle}{\|v_1\|^2} v_1(q).$$

But $v_1(q^*) = -v_1(q)$, so we have

$$\begin{aligned}
\bar{e}_\alpha &= e_\alpha - 2\frac{\langle e_\alpha, v_1\rangle}{\|v_1\|^2} v_1 \\
&= \sum_{\alpha,\beta}(\delta_{\alpha\beta} - 2\frac{n_{1\alpha}n_{1\beta}}{\|n_1\|^2})e_\beta.
\end{aligned} \tag{6.3.1}$$

Let $\lambda_{i\alpha}$ and $\bar{\lambda}_{i\alpha}$ be the eigenvalues of A_{e_α} and $A_{\bar{e}_\alpha}$ on E_i respectively. Then (6.3.1) implies that

$$\bar{\lambda}_{i\alpha} = \sum_\beta(\delta_{\alpha\beta} - 2\frac{n_{1\alpha}n_{1\beta}}{\|n_1\|^2})\lambda_{i\beta}.$$

We have proved that $E_i(q^*) = E_{\sigma(i)}(q)$, so using Proposition 6.2.6 (iv) we have

$$\lambda_{i\alpha} = \frac{\lambda_{\sigma(i)\alpha}}{1 - 2\frac{\langle v_1, v_{\sigma(i)}\rangle}{\|v_1\|^2}}. \tag{6.3.2}$$

Note that

$$\begin{aligned}
v_i(\varphi_1(q)) &= T_1^q(v_i(q)), \quad \text{since } v_i \text{ is parallel} \\
&= (v_i - 2\frac{\langle v_1, v_i\rangle}{\|v_1\|^2}v_1)(q) \\
&= \sum_\alpha \bar{\lambda}_{i\alpha}\bar{e}_\alpha(\varphi_1(q)) = \sum_\alpha \bar{\lambda}_{i\alpha}e_\alpha(q), \quad \text{by (6.3.2)} \\
&= \sum_\alpha \frac{\lambda_{\sigma(i)\alpha}}{1 - 2\frac{\langle v_1, v_{\sigma(i)}\rangle}{\|v_1\|^2}}e_\alpha(q), \\
&= \left(1 - 2\frac{\langle v_1, v_{\sigma(i)}\rangle}{\|v_1\|^2}\right)^{-1} v_{\sigma(i)}(q). \blacksquare
\end{aligned}$$

As a consequence of Theorem 6.3.2 (3) and Corollary 5.3.7, we have

6.3.3. Corollary. *The subgroup W^q of $O(\nu(M)_q)$ generated by the linear reflections $T_1^q, \ldots, T_p^q$ is a finite Coxeter group.*

From the fact that the curvature normals are parallel we have:

6.3.4. Proposition. *Let $\pi_{q,q'} : \nu(M)_q \to \nu(M)_{q'}$ denote the parallel translation map. Then $\pi_{q,q'}$ conjugates the group W^q to $W^{q'}$. In particular, we have associated to M a well-defined Coxeter group W.*

6.3.5. Theorem. $R_i^q(\ell_j(q)) = \ell_{\sigma_i(j)}(q)$.

PROOF. It suffices to prove the theorem for $i = 1, j = 2$. We may assume that $\sigma_1(2) = 3$. In our proof q is a fixed point of M, so we will drop the reference to q whenever there is no possibility of confusion. Let $\ell = R_1(\ell_2)$. Since v_i is normal to ℓ_i, it follows from Theorem 6.3.2(3) that ℓ is parallel to ℓ_3. Choose $q' \in \ell_3$ and $Q \in \ell$ such that $\|q - q'\| = d(q, \ell_3) = c$ and $\|q - Q\| = d(q, \ell)$ respectively. Let $1/a = \|v_1\|$ and $1/b = \|v_2\|$. By Theorem 6.2.9, $\overrightarrow{qq'} = v_3/\|v_3\|^2$. Note that 6.3.2(3) gives

$$T_1(v_2) = \left(1 - 2\frac{\langle v_1, v_3 \rangle}{\|v_1\|^2}\right)^{-1} v_3. \tag{6.3.3}$$

We claim that $\overrightarrow{qq'} = \overrightarrow{qQ}$, which will prove that $\ell = \ell_3$. It is easily seen that $\overrightarrow{qq'}$ and $\overrightarrow{qQ}$ are parallel. We divide the proof of the claim into four cases:

(Case i) $\ell_1 \parallel \ell_2$ and v_1, v_2 are in the opposition directions.

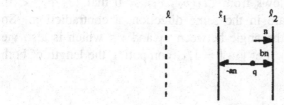

Let n be the unit direction of v_2. Then $\overrightarrow{qQ} = -(2a + b)n$. Note that v_3 is equal to $(\epsilon/c)n$ for $\epsilon = 1$ or -1. Using (6.3.3), we have

$$T_1(v_2) = T_1\left(\frac{1}{b}n\right) = -\frac{1}{b}n$$

$$= \left(1 - 2\frac{\langle v_1, v_3 \rangle}{\|v_1\|^2}\right)^{-1} v_3 = \left(1 + 2\frac{\epsilon a}{c}\right)^{-1} \frac{\epsilon}{c}n.$$

So

$$-1/b = \frac{\epsilon}{c}(1 + 2a\epsilon/c)^{-1}. \qquad (6.3.4)$$

If $\epsilon = 1$ then the right hand side of (6.3.4) is positive, a contradiction. So $\epsilon = -1$. Then (6.3.4) implies $c = 2a + b$, which proves the claim.

(Case ii) $\ell_1 \cap \ell_2 \neq \emptyset$, and $\langle v_1, v_2 \rangle < 0$.

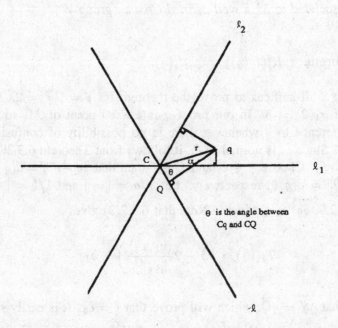

θ is the angle between Cq and CQ

Note that $\langle T_1(v_2), v_1 \rangle = \langle v_2, T_1(v_1) \rangle = \langle v_2, -v_1 \rangle > 0$ and $T_1(v_2)$ and $\overrightarrow{qQ}$ are in the same direction. We claim that v_3 and $T_1(v_2)$ are in the same direction. If not then it follows from $\langle T_1(v_2), v_1 \rangle > 0$ that $\langle v_3, v_1 \rangle < 0$. By (6.3.3), $T_1(v_2)$ and v_3 are in the same direction, a contradiction. So $\langle v_1, v_3 \rangle > 0$. Let θ denote the angle between v_1 and v_3, which is also the angle between v_1 and $T_1(v_2)$. Let $\|v_3\| = 1/c$; computing the length of both sides of (6.3.3) gives

$$1/b = \frac{1}{c}(1 - 2a\cos\theta/c)^{-1},$$

i.e., $c = b + 2a\cos\theta$. Let α and γ be the angles shown in the diagram. Then

$$\begin{aligned}
\overrightarrow{qQ} &= r\sin(\theta + \alpha) \\
&= r(\sin(\theta - \alpha) + 2\cos\theta\sin\alpha) \\
&= b + 2a\ \cos\theta.
\end{aligned}$$

This proves $\overrightarrow{qQ} = \overrightarrow{qq'}$.

The proofs of the claim for the following two cases are similar to those for (i) and (ii) respectively and are left to the reader.

(iii) $\ell_1 \parallel \ell_2$ and v_1, v_2 are in the same direction.

(iv) $\ell_1 \cap \ell_2 \neq \emptyset$, and $\langle v_1, v_2 \rangle \geq 0$. ∎

As a consequence of Corollary 5.3.7, we have:

6.3.6. Corollary. *If M^n is a rank k isoparametric submanifold of R^{n+k}, then:*

(i) the subgroup of isometries of $\nu_q = q + \nu(M)_q$ generated by the reflections R_i^q in the focal hyperplanes $\ell_i(q)$ is a finite rank k Coxeter group, which is isomorphic to the Coxeter group W associated to M,

(ii) $\bigcap\{\ell_i(q)|1 \leq i \leq p\}$ consists of one point.

Let $\triangle_q$ be the connected component of $\nu_q - \bigcap\{\ell_i|1 \leq i \leq p\}$ containing q. Then the closure $\bar{\triangle}_q$ is a simplicial cone and a fundamental domain of W and $\{v_i| \ell_i(q)$ contains a $(k-1) - $ simplex of $\bar{\triangle}_q\}$ is a simple root system for W. If $\varphi \in W$ and $\varphi(\ell_i) = \ell_j$ then by Theorem 6.3.2 we have $m_i = m_j$. So we have

6.3.7. Corollary. *We associate to each rank k isoparametric submanifold M^n of R^{n+k} a well-defined marked Dynkin diagram with k vertices, namely the Dynkin diagram of the associated Coxeter group with multiplicities m_i.*

6.3.8. Examples. Let G be a compact, rank k simple Lie group, and $\mathcal{G}$ its Lie algebra with inner product $\langle \ , \ \rangle$, where $-\langle \ , \ \rangle$ is the Killing form of G. Let $\mathcal{T}$ be a maximal abelian subalgebra of $\mathcal{G}$, $a \in \mathcal{T}$ a regular point, and $M = Ga$ the principal orbit through a. Since this orthogonal action is polar ($\mathcal{T}$ is a section), M is isoparametric in $\mathcal{G}$ of codimension k. Note that

$$TM_x = \{[\xi, x]| \ \xi \in \mathcal{G}\},$$

$$\nu(M)_{gag^{-1}} = g\mathcal{T}g^{-1}.$$

Given $b \in \mathcal{T}$, $\hat{b}(gag^{-1}) = gbg^{-1}$ is a well-defined normal field on M. Since $d\hat{b}_a([\xi, a]) = [\xi, b]$ and

$$\langle [\xi, b], t \rangle = \langle -[b, \xi], t \rangle = \langle \xi, [b, t] \rangle = \langle \xi, 0 \rangle = 0$$

for all $t \in \nu(M)_a$, $\hat{b}$ is parallel and the shape operator is

$$A_{\hat{b}}([\xi, a]) = -[\xi, b]. \qquad (6.3.5)$$

To obtain the common eigendecompositions of $\{A_{\hat{b}}\}$, we recall that if $\triangle^+$ is a set of positive roots of $\mathcal{G}$ then there exist x_α, y_α in $\mathcal{G}$ for each $\alpha \in \triangle^+$ such that

$$\mathcal{G} = \mathcal{T} \oplus \{Rx_\alpha \oplus Ry_\alpha \mid \alpha \in \triangle^+\},$$

$$[t, x_\alpha] = \alpha(t)y_\alpha, \quad [t, y_\alpha] = -\alpha(t)x_\alpha, \tag{6.3.6}$$

where $\alpha(t) = \langle \alpha, t \rangle$, and $t \in \mathcal{T}$. Using (6.3.5) and (6.3.6), we have

$$A_b(x_\alpha) = -\frac{\alpha(b)}{\alpha(a)}x_\alpha, \quad A_b(y_\alpha) = -\frac{\alpha(b)}{\alpha(a)}y_\alpha.$$

This implies that the curvature distributions of M are given by $E_\alpha = Rx_\alpha \oplus Ry_\alpha$ for $\alpha \in \triangle^+$ and the curvature normals are given by $v_\alpha = -\alpha/\langle \alpha, a \rangle$. So the Coxeter group associated to M as an isoparametric submanifold is the Weyl group of G, and all the multiplicities are equal to 2.

If $v \in \nu(M)_q$, then $q + v \in \ell_i(q)$ if and only if $\langle v, v_i \rangle = 1$, so as a consequence of Corollary 6.3.6 (ii), we have:

6.3.9. Corollary. If M^n is a rank k isoparametric submanifold of R^{n+k}, then there exists $a \in R^k$ such that $\langle a, n_i \rangle = 1$ for all $1 \leq i \leq p$.

6.3.10. Corollary. Suppose $X : M^n \to R^{n+k}$ is a rank k immersed isoparametric submanifold and all the curvature normals are non-zero. Then there exist vectors $a \in R^k$ and $c_0 \in R^{n+k}$ such that M is contained in the sphere of radius $\|a\|$ centered at c_0 in R^{n+k} so that

$$X + \sum_\alpha a_\alpha e_\alpha = c_0.$$

In particular, we have

$$\bigcap \{\ell_i(q) \mid q \in M, 1 \leq i \leq p\} = \{c_0\}.$$

PROOF. By Corollary 6.3.9, there exists $a \in R^k$ such that $\langle a, n_i \rangle = 1$. We claim that the map $X + \sum_\alpha a_\alpha e_\alpha$ is a constant vector $c_0 \in R^{n+k}$ on M, because

$$d\left(X + \sum_\alpha a_\alpha e_\alpha\right) = \sum_{i=1}^p (1 - \langle a, n_i \rangle) \, id_{E_i} = 0.$$

So we have

$$\|X - c_0\|^2 = \|\sum_\alpha a_\alpha e_\alpha\|^2 = \|a\|^2. \quad \blacksquare$$

6.3.11. Corollary.   *The following statements are equivalent for an immersed isoparametric submanifold M^n of R^{n+k}:*
(i) M is compact,
(ii) all the curvature normals of M are non-zero.
(iii) M is contained in a standard sphere in R^{n+k}.

6.3.12. Corollary.   *If M^n is a rank k isoparametric submanifold of R^{n+k} and zero is one of the curvature normals for M corresponding to the curvature distribution E_0, then there exists a compact rank k isoparametric submanifold M_1 of R^{n+k-m_0} such that $M = E_0 \times M_1$.*

PROOF.   By Corollary 6.3.9, there exists $a \in R^k$ such that $\langle a, n_i \rangle = 1$ for all $1 \le i \le p$. Consider the map $X^* = X + \sum_\alpha a_\alpha e_a : M \to R^{n+k}$. Then

$$dX^* = \sum_{i=0}^p (1 - \langle a, n_i \rangle) \, id_{E_i} = id_{E_0},$$

and $M^* = X^*(M)$ is a flat m_0-plane of R^{n+k}. So $X^* : M \to M^*$ is a submersion, and in particular the fiber is a smooth submanifold of M. But the tangent plane of the fiber is $\bigoplus_{i=1}^p E_i$, so it is integrable. On the other hand $\bigoplus_{i=1}^p E_i$ is defined by

$$\omega_i = 0, \quad i \le m_0,$$

so we have

$$0 = d\omega_i = \sum_{j > m_0} \omega_{ij} \wedge \omega_j = \sum_{j,m > m_0} \gamma_{ijm} \, \omega_m \wedge \omega_j.$$

Hence

$$\gamma_{ijm} = \gamma_{imj}, \quad \text{for } i \le m_0, \ j \ne m > m_0. \tag{6.3.7}$$

By Proposition 6.2.7, we have

$$\lambda_{j\alpha}\gamma_{ijm} = \lambda_{m\alpha}\gamma_{imj}, \quad \text{for } i \le m_0. \tag{6.3.8}$$

If $e_j, e_m \in E_s$ for some $s > 0$, then Proposition 6.2.7 imply that $\gamma_{ijm} = 0$. If e_j and e_m belong to different curvature distributions, then there exists α_0 such that $\lambda_{j\alpha_0} \ne \lambda_{m\alpha_0}$. So (6.3.7) and (6.3.8) implies that $\gamma_{ijm} = 0$. Therefore we have proved that

$$\omega_{ij} = 0, \quad \omega_{i\alpha} = 0, \quad i \le m_0, \ j > m_0,$$

on M. Let $M_1 = (X^*)^{-1}(q^*)$. Then both M and $M_1 \times E_0$ have flat normal bundles and the same first, second fundamental forms. So the fundamental theorem of submanifolds (Corollary 2.3.2) implies that $M = M_1 \times E_0$. ∎

Next we discuss the irreduciblity of the associated Coxeter group of an isoparametric submanifold, which leads to a decomposition theorem for isoparametric submanifolds.

If $M_i^{n_i}$ is isoparametric in $R^{n_i+k_i}$ with Coxeter group W_i on R^{k_i} for $i = 1, 2$, then $M_1 \times M_2$ is isoparametric in $R^{n_1+n_2+k_1+k_2}$ with Coxeter group $W_1 \times W_2$ on $R^{k_1} \times R^{k_2}$. The converse is also true.

6.3.13. Theorem. *Let M^n be a compact rank k isoparametric submanifold of R^{n+k}, and W its associated Coxeter group. Suppose $R^k = R^{k_1} \times R^{k_2}$ and $W = W_1 \times W_2$, where W_i is a Coxeter group on R^{k_i}. Then there exist two isoparametric submanifolds M_1, M_2 with Coxeter groups W_1, W_2 respectively such that $M = M_1 \times M_2$.*

PROOF. We may assume that $n_i \in R^{k_1} \times 0$, $R_i \in W_1$ for $i \leq p_1$, and $n_j \in 0 \times R^{k_2}$, $R_j \in W_2$ for $j > p_1$. Since W_1 is a finite Coxeter group, there exists a constant vector $a \in R^{k_1} \times 0$ such that $\langle a, n_i \rangle = 1$ for all $i \leq p_1$. Consider $X^* = X + \sum_\alpha a_\alpha e_\alpha$. Since $\langle a, n_j \rangle = 0$ for all $j > p_1$, we have

$$dX^* = \sum_{j>p_1}^{p} id_{E_j}.$$

So an argument similar to that in Corollary 6.3.12 implies that $V = \bigoplus_{i \leq p_1} E_i$ and $H = \bigoplus_{j>p_1}^{p} E_j$ are integrable, and that M is the product of a leaf of V and a leaf of H. ∎

6.3.14. Definition. An isoparametric submanifold M^n of R^{n+k} is called *irreducible*, if M is not the product of two lower dimensional isoparametric submanifolds.

As a consequence of Theorem 6.3.13, we have:

6.3.15. Proposition. *An isoparametric submanifold of Euclidean space is irreducible if and only if its associated Coxeter group is irreducible.*

Since every Coxeter group can be written uniquely as the product of irreducible Coxeter groups uniquely up to permutation, we have:

6.3.16. Theorem. *Every isoparametric submanifold of Euclidean space can be written as the product of irreducible ones, and such decomposition is unique up to permutation.*

As a consequence of Corollary 6.3.11 and the following proposition we see that the set of compact, isoparametric submanifolds of Euclidean space coincides with the set of compact isoparametric submanifolds of standard spheres.

6.3.17. Proposition. *If M^n is an isoparametric submanifold of S^{n+k} then M is an isoparametric submanifold of R^{n+k+1}.*

PROOF. Let $X : M \to S^{n+k}$ be the immersion, and $\{e_A\}$ the adapted frame for X, ω_i the dual coframe, and ω_{AB} the Levi-Civita connection 1-form. We may assume that e_α's are parallel, i.e., $\omega_{\alpha\beta} = 0$ for $n < \alpha,\ \beta \le n+k$. Set $e_{n+k+1} = X$, then $\{e_1, \ldots, e_{n+k+1}\}$ is an adapted frame for M as an immersed submanifold of R^{n+k+1}. Since

$$de_{n+k+1} = dX = \sum \omega_i e_i,$$

we have $\omega_{n+k+1,\alpha} = 0$ and $A_{e_{n+k+1}} = -id$. This implies that M is isoparametric in R^{n+k+1}. ∎

6.4 Existence of isoparametric polynomial maps.

In this section, given an isoparametric submanifold M^n of R^{n+k}, we will construct a polynomial isoparametric map on R^{n+k} which has M as a level submanifold. This construction is a generalization of the Chevalley Restriction Theorem in Example 5.6.16.

By Corollary 6.3.11 and 6.3.12, we may assume that M^n is a compact, rank k isoparametric submanifold of R^{n+k}, and $M \subseteq S^{n+k-1}$. Let W be the Coxeter group associated to M, and p the number of reflection hyperplanes of W, i.e., M has p curvature normals. In the following we use the same notation as in section 6.2.

Given $q \in M$, there is a simply connected neighborhood U of q in M such that U is embedded in R^{n+k}. Let e_α be a parallel normal frame, v_i the curvature normals $\sum_\alpha n_{i\alpha} e_\alpha$, and $n_i = (n_{in+1}, \ldots, n_{in+k})$. Let $Y : U \times R^k \to R^{n+k}$ be the endpoint map, i.e., $Y(x, z) = x + \sum_\alpha z_\alpha e_\alpha(x)$. Then there is a small ball B centered at the origin in R^k such that $Y|U \times B$ is a local coordinate system for R^{n+k}. In particular, $z \cdot n_i < 1$ for all $z \in B$ and $1 \le i \le p$. We denote $Y(U \times B)$ by $\mathcal{O}$. In fact, $\mathcal{O}$ is a tubular neighborhood of M in R^{n+k}. Since $M \subseteq S^{n+k-1}$, by Corollary 6.3.10 there exists a vector $a \in R^k$ such that

$$X = \sum_\alpha a_\alpha e_\alpha.$$

Then

$$Y = X + \sum_\alpha z_\alpha e_\alpha = \sum_\alpha (z_\alpha - a_\alpha) e_\alpha.$$

Let $y = z - a$ (note that $y = 0$ corresponds to the origin of R^{n+k} and the W-action on $q + \nu(M)_q$ induces an action on R^k which is linear in y). Then y_α is a smooth function defined on the tubular neighborhood $\mathcal{O}$ of M in R^{n+k}. It is easily seen that any W-invariant smooth function u on R^k can be extended uniquely to a smooth function f on $\mathcal{O}$ that is constant on all the parallel submanifolds of the form M_v, where v is a parallel normal field on M with $v(q) \in B$. That is, we extend f by the formula $f(Y(x,z)) = u(z-a) = u(y)$. We will call this f simply the *extension* of u.

In order to construct a global isoparametric map for M, we need the following two lemmas.

6.4.1. Lemma.   *If $u : R^k \to R$ is a W-invariant homogeneous polynomial of degree k, then the function*

$$\varphi(y) = \sum_{i=1}^{p} m_i \frac{\nabla u(y) \cdot n_i}{y \cdot n_i}$$

is a W-invariant homogeneous polynomial of degree $k - 2$.

PROOF.   Let R_i denote the reflection of R^k along the vector n_i. Since $u(R_i y) = u(y)$, $\nabla u(R_i(y)) = R_i(\nabla u(y))$. We claim that $\nabla u(y) \cdot n_i = 0$ if $y \cdot n_i = 0$. For if $y \cdot n_i = 0$ then $R_i(y) = y$, so $\nabla u(y) = R_i(\nabla u(y))$, i.e., $\nabla u(y) \cdot n_i = 0$. Therefore $\varphi(y)$ is a homogeneous polynomial of degree $k - 2$. To check that φ is W-invariant, we note that

$$\varphi(R_i(y)) = \sum_j m_j \frac{\nabla u(R_i(y)) \cdot n_j}{R_i(y) \cdot n_j}$$

$$= \sum_j m_j \frac{R_i(\nabla u(y)) \cdot n_j}{R_i(y) \cdot n_j}$$

$$= \sum_j m_j \frac{\nabla u(y) \cdot R_i(n_j)}{y \cdot R_i(n_j)}.$$

Then the lemma follows from Theorem 6.3.2 (3).   ∎

6.4.2. Lemma.   *Let $u : R^k \to R$ be a W-invariant homogeneous polynomial of degree k, and $f : \mathcal{O} \to R$ its extension. Then*
  (i) $\triangle f$ is the extension of a W-invariant homogeneous polynomial of degree $(k - 2)$ on R^k,

(ii) $\|\nabla f\|^2$ is the extension of a W-invariant homogeneous polynomial of degree $2(k-1)$ on R^k.

PROOF. Since

$$dY = \sum_i (1 - z \cdot n_i)\, id_{E_i} + \sum_\alpha dz_\alpha e_\alpha$$

$$= \sum_i y \cdot z_i\, id_{E_i} + \sum_\alpha dy_\alpha e_\alpha,$$

we may choose a local frame field $e_A^* = e_A$ on $\mathcal{O} \subset R^{n+k}$, and the dual coframe is

$$\omega_j^* = (y \cdot n_i)\omega_j, \quad \text{if } \sum_{r=1}^{i-1} m_r < j \le \sum_{r=1}^{i} m_r,$$

$$\omega_a^* = dy_\alpha.$$

The Levi-Civita connection 1-form on $\mathcal{O}$ is $\omega_{AB}^* = \omega_{AB}$. Then by (1.3.6) we have

$$\triangle y_\alpha = -\sum_{i=1}^p \frac{m_i n_{i\alpha}}{y \cdot n_i}.$$

Since $f(x,y) = u(y)$, we have

$$df = \sum_\alpha u_\alpha \omega_\alpha^*, \quad \|\nabla f\|^2 = \|\bar{\nabla} u\|^2,$$

$$\triangle f = \bar{\triangle} u + \sum_i m_i \frac{\bar{\nabla} u \cdot n_i}{y \cdot n_i},$$

where $\bar{\triangle}, \bar{\nabla}$ are the standard Laplacian and gradient on R^k. Then (i) follows from Lemma 6.4.1. To prove (ii), we note that $\bar{\nabla} u(R_i(y)) = R_i(\bar{\nabla} u(y))$, so $\|\bar{\nabla} u\|^2$ is a W-invariant polynomial of degree $2(k-1)$ on R^k. ∎

6.4.3. Theorem. *Let M^n be a rank k isoparametric submanifold in R^{n+k}, W the associated Coxeter group, q a point on M, and $\nu_q = q + \nu(M)$ the affine normal plane at q. If $u : \nu_q \to R$ is a W-invariant homogeneous polynomial of degree m, then u can be extended uniquely to a homogeneous degree m polynomial f on R^{n+k} such that f is constant on M.*

PROOF. We may assume $\nu_q = R^k$. We prove this theorem on $\mathcal{O}$ by using induction on the degree k of u. The theorem is obvious for $m = 0$. Suppose it is true for all $\ell < m$. Given a degree m W-invariant homogeneous polynomial u on R^k, by Lemma 6.4.2, $\|df\|^2$ is again the extension of a W-invariant homogeneous polynomial of degree $2k-2$ on R^k. Applying Lemma 6.4.2 repeatedly, we have $\triangle^{m-1}(\|df\|^2)$ is the extension of a degree zero W-invariant polynomial, hence it is a constant. Therefore

$$0 = \triangle^m(\|df\|^2)$$
$$= \sum_{r=0}^{m} \sum_{\substack{s+s'=m-r \\ i,i_1,\dots,i_r}} c_{r,s}(\triangle^s f)_{i,i_1,\dots,i_r} (\triangle^{s'} f)_{i,i_1,\dots,i_r},$$

where $c_{r,s}$ are constants depending on r and s. We claim that

$$(\triangle^s f)_{i,i_1,\dots,i_r}(\triangle^{s'} f)_{i,i_1,\dots,i_r}, \quad s' = m-r-s,$$

is zero if $r < m$. For we may assume that $s \geq m-r-s$, i.e., $s \geq s'$, so $s \geq 1$. By Lemma 6.4.2, $\triangle^s f$ is the extension of a degree $m-2s$ W-invariant polynomial on R^k. By the induction hypothesis, $\triangle^s f$ is a homogeneous polynomial on $\mathcal{O} \subset R^{n+k}$ of degree $m-2s$, hence all the partial derivatives of order bigger than $m-2s$ will be zero. We have $r+1 > r \geq m-2s$ by assumption, so we obtain

$$0 = \sum_{i,i_1,\dots,i_m} f^2_{i,i_1,\dots,i_m},$$

i.e., $D^\alpha f = 0$ in $\mathcal{O}$ for $|\alpha| = k+1$. This proves that f is a homogeneous polynomial of degree k in $\mathcal{O}$. There is a unique polynomial extension on R^{n+k}, which we still denote by f. ∎

By Theorem 5.3.18 there exist k homogeneous W-invariant polynomials $u_1,\dots,u_k$ on R^k such that the ring of W-invariant polynomials on R^k is the polynomial ring $R[u_1,\dots,u_k]$.

6.4.4. Theorem. *Let M, W, q, ν_q be as in Theorem 6.4.3, and let $u_1,\dots,u_k$ be a set of generators of the W-invariant polynomials on ν_q. Then $u = (u_1,\dots,u_k)$ extends uniquely to an isoparametric polynomial map $f : R^{n+k} \to R^k$ having M as a regular level set. Moreover,*

(1) each regular set is connected,
(2) the focal set of M is the set of critical points of f,
(3) $\nu_q \cap M = W \cdot q$,
(4) $f(R^{n+k}) = u(\nu_q)$,
(5) for $x \in \nu_q$, $f(x)$ is a regular value if and only if x is W-regular,
(6) $\nu(M)$ is globally flat.

PROOF. Let $f_1,\dots,f_k$ be the extended polynomials on R^{n+k}. Because $u_1,\dots,u_k$ are generators, $f = (f_1,\dots,f_k)$ will automatically satisfies condition (1) and (2) of Definition 6.1. Since y_α are part of local coordinates, $[y_\alpha, y_\beta] = 0$. But f is a function of y, so condition (3) of Definition 6.1.1 is satisfied. Then (1)-(5) follow from the fact that $u_1, \ldots, u_k$ separate the orbits of W and that regular points of the map $u = (u_1, \ldots, u_k)$ are just the W-regular points. Finally, since $\{\nabla f_1, \ldots, \nabla f_k\}$ is a global, parallel, normal frame for M, $\nu(M)$ is globally flat. ∎

6.4.5. Corollary. *Let M^n be an immersed isoparametric submanifold of R^{n+k}. Then*

(i) M is embedded,

(ii) $\nu(M)$ is globally flat.

The above proof also gives a constructive method for finding all compact irreducible isoparametric submanifolds of Euclidean space. To be more specific, given an irreducible Coxeter group W on R^k with multiplicity m_i for each reflection hyperplane ℓ_i of W such that $m_i = m_j$ if $g(\ell_i) = \ell_j$ for some $g \in W$, i.e., given a marked Dynkin diagram. Suppose W has p reflection hyperplanes $\ell_1, \ldots, \ell_p$. Let a_i be a unit normal vector to ℓ_i. Set $n = \sum_{i=1}^p m_i$. Let $u_1, \ldots, u_k$ be a fixed set of generators for the ring of W-invariant polynomials on R^k, which can be chosen to be homogeneous of degree k_i. Then there are polynomials V_i, Φ_i, U_{ij}, and Ψ_{ijm} on R^k such that

$$\Delta u_i = V_i(u), \quad \nabla u_i \cdot \nabla u_j = U_{ij}(u),$$

$$\sum_j m_j \frac{\nabla u_i \cdot a_j}{y \cdot a_j} = \Phi_i(u), \quad [\nabla u_i, \nabla u_j] = \sum_m \Psi_{ijm}(u) \nabla u_m.$$

Then any polynomial solution $f = (f_1, \ldots, f_k) : R^{n+k} \to R^k$, with f_i being homogeneous of degree k_i, of the following system is an isoparametric map:

$$\Delta f_i = V_i(f) + \Phi_i(f),$$

$$\nabla f_i \cdot \nabla f_j = U_{ij}(f), \tag{6.4.1}$$

$$[\nabla f_i, \nabla f_j] = \sum_m \Psi_{ijm}(f) \nabla f_m.$$

Moreover, if M is any regular level submanifold of such an f, then the associated Coxeter group and multiplicities of M are W and m_i respectively.

Since $u_1 : R^k \to R$ can be chosen to be $\sum_{i=1}^k x_i^2$, the extension f_1 on R^{n+k} is $\sum_{i=1}^{n+k} x_i^2$. So (6.4.1) is a system of equations for $(k-1)$ functions. Because both the coefficients and the admissible solutions for (6.4.1) are homogeneous polynomials, the problem of classifying isoparametric submanifolds becomes a purely algebraic one.

6.4.6. Remark. Theorem 6.4.4 was first proved by Münzner in [Mü1,2] for the case of isoparametric hypersurfaces of spheres, i.e., for rank 2 isoparametric submanifolds of Euclidean space. Suppose W is the dihedral group of $2p$ elements on R^2. Then W has p reflection lines in R^2, and we may choose $a_j = (\cos(j\pi/p), \sin(j\pi/p))$ for $0 \leq j < p$. By Theorem 6.3.2, all m_i's are equal to some integer m if p is odd, and $m_1 = m_3 = \cdots$, $m_2 = m_4 = \cdots$ if p is even. So we have $n = pm$ if p is odd, and $n = p(m_1 + m_2)/2$ if p is even. It is easily seen that we can choose

$$u_1(x,y) = x^2 + y^2, \quad u_2(x,y) = Re((x + iy)^p).$$

Let $f_i : R^{n+2} \to R$ be the extensions. Then $f_1(x) = \|x\|^2$. Let $F = f_2$. Then it follows from a direct computation that (6.4.1) becomes the equations given by Münzner in [Mü1,2]:

$$\Delta F(x) = c\|x\|^{p-2}$$
$$\|\nabla F(x)\|^2 = p^2\|x\|^{2p-2},$$

where $c = 0$ if p is odd and $c = (m_2 - m_1)p^2/2$ if p is even.

6.5   Parallel foliations and The Slice Theorem.

In this section we will prove that the family of parallel sets of an isoparametric submanifold of R^m gives an orbit-like singular foliation on R^m. Moreover we for this foliation have an analogue of the Slice Theorem for polar actions (5.6.21), and this provides us with an important inductive method for the study of such submanifolds.

Let $X : M^n \to S^{n+k-1} \subseteq R^{n+k}$ be a rank k isoparametric submanifold, $q \in M$, and $\nu_q = q + \nu(M)_q$. Note that ν_q contains the origin. Then $\nu_q - \bigcap_{i=1}^p \ell_i(q)$ has $|W|$ (the order of W) connected components. The closure Δ of each component is a simplicial cone, and is a fundamental domain of W, called a chamber of W on ν_q.

Let σ be a simplex of Δ. We define the following:

$I(q, \sigma) = \{j \mid \sigma \subseteq \ell_j(q)\}$,
$V(q, \sigma) = \bigcap\{\ell_j(q) \mid j \in I(q, \sigma)\}$,
$\xi(q, \sigma) = $ the orthogonal complement of $V(q, \sigma)$ in ν_q through q,
$\eta(q, \sigma) = \xi(q, \sigma) \oplus \bigoplus\{E_j(q) \mid j \in I(x, \sigma)\}$,
$m_{q,\sigma} = \sum\{m_j \mid j \in I(q, \sigma)\}$,
$W_{q,\sigma} = $ the subgroup of W generated by the reflections $\{R_j^q \mid j \in I(q, \sigma)\}$.

6.5.1. Proposition.   *Let $X : M^n \to S^{n+k-1} \subseteq R^{n+k}$ be isoparametric, $q \in M$, and Δ the chamber on ν_q containing q. Let σ be a simplex of Δ, and*

v a parallel normal field on M such that $q + v(q) \in \sigma$. Let $f : R^{n+k} \to R^k$ be the isoparametric polynomial map constructed in Theorem 6.4.4. Then

(i) The map $\pi_v = X + v : M \to R^{n+k}$ has constant rank $n - m_\sigma$, so the parallel set $M_v = (X + v)(M)$ is an immersed submanifold of dimension $n - m_\sigma$ and $\pi_v : M \to M_v$ is a fibration.

(ii) $M_v = f^{-1}(f(q + v(q))$, i.e., M_v is a level set of f, so M_v is an embedded submanifold of R^{n+k}.

PROOF.   Note that

$$d\pi_v = d(X + v) = \sum_{i=1}^{p} (1 - \langle v, v_i \rangle) \, id_{E_i}. \qquad (6.5.1)$$

If $\dim(\sigma) = k$, i.e., $q + v(q) \in \sigma = (\triangle)^0$ the interior of $\triangle$, then π_v is an immersion and all the results follows from Proposition 6.2.6. If $\dim(\sigma) < k$, then $i \in I(q, \sigma)$ if and only if $1 = \langle v, v_i \rangle$. So $\text{rank}(\pi_v) = n - m_\sigma$, which proves (i). (ii) follows from the way f is constructed.   ∎

6.5.2. Corollary.   With the same notation as in Proposition 6.5.1:

(i) If v, w are two parallel normal fields on M such that $q + v(q)$ and $q + w(q)$ are distinct points in $\triangle$, then $M_v \cap M_w = \emptyset$.

(ii) $\{M_v | \ q + v(q) \in \triangle\}$ gives an orbit-like singular foliation on R^{n+k}, that we call the parallel foliation of M on R^{n+k}.

PROOF.   Let $u_1, \cdots, u_k$ be a set of generators for the ring of W-invariant polynomials on ν_q. Then $u_1, \cdots, u_k$ separates W-orbits. Since $q + v(q)$ and $q + w(q)$ are two distinct points in $\triangle$, there exists i such that $u_i(q + v(q)) \neq u_i(q + w(q))$. But the isoparametric polynomial f is the extension of $u = (u_1, \ldots, u_k)$ to R^{n+k}, $f(q + v(q)) \neq f(q + w(q))$. But $M_v = f^{-1}(f(q + v(q))$. So (i) follows.

Given $y \in R^{n+k}$, let $f_y : M \to R$ be the Euclidean distance function defined by $f_y(x) = \|x - y\|^2$. Since M is compact, there exists $x_0 \in M$ such that $f_y(x_0)$ is the absolute minimum of f_y. So the index of f_y at x_0 is zero, and it follows from Theorem 4.2.6 (iv) that y is in the chamber $\triangle_{x_0}$ on ν_{x_0} containing x_0. Let v be the unique parallel normal field on M such that $x_0 + v(x_0) = y$. Then $q + v(q) \in \triangle$ and $y \in M_v$, which prove (ii).   ∎

6.5.3. Corollary.   With the same notation as in Proposition 6.5.1, let B denote $\triangle \cap S^{k-1}$, where S^{k-1} is the unit sphere of ν_q centered at the origin (note that $0 \in \nu_q$). Then $\{M_v | \ q + v(q) \in B\}$ gives an orbit-like singular foliation on S^{n+k-1}, which will be called the parallel foliation of M on S^{n+k-1}.

Given $x \in R^{n+k}$, we let M_x denote the unique leaf of the parallel foliation of M that contains x. Then the parallel foliation of M in S^{n+k-1} and R^{n+k} can be rewritten as

$$\{M_x| \; x \in \Delta\},$$

$$\{M_x| \; x \in \Delta \cap S^{k-1}\},$$

respectively.

6.5.4. Proposition.   If $M^n \subset S^{n+k-1} \subset R^{n+k}$ is isoparametric, then for all $r \neq 0$ we have
(i) rX is isoparametric for all $r \neq 0$,
(ii) rM_v is again a parallel submanifold of M if M_v is.

6.5.5. Corollary.   Let $M^n \subset S^{n+k-1} \subset R^{n+k}$ be isoparametric, and $\mathcal{F}$ the parallel foliation of M in S^{n+k-1}. Then the parallel foliation of M in R^{n+k} is $\{rF| \; r \geq 0, \; F \in \mathcal{F}\}$.

6.5.6. Examples.   Let G/K be a compact, rank k symmetric space. Since the isotropy representation of G/K at eK is polar (Example 5.6.16), the principal K-orbits are codimension k isoparametric submanifolds of $\mathcal{P}$. Let $\mathcal{G} = \mathcal{K} + \mathcal{P}$ be the orthogonal decomposition with respect to the Killing form on $\mathcal{G}$, and M a principal K-orbit in $\mathcal{P}$. Then M is of rank k as an isoparametric submanifold, $\nu(M)_x$ is just the maximal abelian subalgebra through x in $\mathcal{P}$, and the associated Coxeter group W and chambers on ν_x are the standard ones for the symmetric space. If v is a parallel normal field on M then the parallel submanifold M_v is the K-orbit through $x + v(x)$, i.e., the parallel foliation of M is the orbit foliation of the K-action on $\mathcal{P}$. If $y_i = x + v(x)$ lies on one and only one reflection hyperplane $\ell_i(x)$ then the orbit through y_i is subprincipal (i.e., if $gK_zg^{-1} \subset K_{y_i}$ and $gK_zg^{-1} \neq K_{y_i}$ then z is a regular point), and the differences of dimensions between Kx and Ky_i is m_i. Therefore the marked Dynkin diagram associated to M can be computed explicitly, and this will be done later.

In the following we calculate the mean curvature vector for each M_v.

6.5.7.   Theorem.   Let $M^n \subseteq S^{n+k-1} \subseteq R^{n+k}$ be isoparametric, and let $\{v_i| \; i \in I\}$ be its set of curvature normal fields. Let $q \in M$, Δ a chamber on ν_q, σ a simplex of Δ, and v a parallel normal field such that $q^* = q + v(q) \in \sigma$. Let H^* denote the mean curvature vector field of M_v in R^{n+k}. If $x^* = x + v(x) \in M_v$, then

(i)

$$H^*(x^*) = \sum_{i \in I \setminus I(q,\sigma)} \frac{m_i v_i}{1 - v \cdot v_i}(x),$$

(ii) $H^*(x^*) \in V(x, \sigma)$.

In particular we have the following identities:

$$\sum_{i \in I \setminus I(q,\sigma)} \frac{m_i \, v_i \cdot v_j}{1 - v \cdot v_i} = 0, \quad j \in I(q,\sigma).$$

PROOF. Let $I(\sigma)$ denote $I(q, \sigma)$. It follows from (6.5.1) that

$$\nu(M_v)_{x^*} = \nu(M)_x \oplus \bigoplus \{E_i(x) \mid i \in I(\sigma)\},$$

$$T(M_v)_{x^*} = \bigoplus \{E_i(x) \mid i \in I \setminus I(\sigma)\}.$$

We may assume $e_1(x), \ldots, e_r(x)$ span $T(M_v)_{x^*}$ (where $r = n - m(q,\sigma)$), so $\{e_\alpha(x) \mid n+1 \le \alpha \le n+k\} \cup \{e_j(x) \mid j > r\}$ spans $\nu(M_v)_{x^*}$. Let ω_A^* and ω_{AB}^* be the dual coframe and connection 1-forms on M_v. Then by (6.5.1),

$$\omega_i^* = (1 - v \cdot v_s)\omega_i, \quad \text{if } e_i \in E_s,$$

$$\omega_{AB}^* = \omega_{AB}.$$

So the projection of $H^*(x^*)$ onto $\nu(M)_x$ is

$$\sum_{i \in I \setminus I(\sigma)} \frac{m_i v_i(x)}{1 - v \cdot v_i}.$$

Let $\omega_{ij}^* = \sum_m \gamma_{ijm}\omega_m$. Then by Proposition 6.2.7, $\gamma_{iji} = 0$ if $i \le r$ and $j > r$. This proves (i).

For (ii), we need to show that $H^*(x^*) \cdot v_i(x) = 0$ for all $i \in I(\sigma)$. It suffices to show that $H^*(x^*) \cdot v$ is a constant vector for all unit vector $v \in E_i(x) \oplus Rv_i(x)$ (because $H^*(x^*) \in \nu(M)_x$, so $H^*(x^*) \cdot e = 0$ if $e \in E_i(x)$). To prove this, we note that from $v_i/\|v_i\|$ defines a diffeomorphism from the leaf $S_i(x)$ of E_i to the unit sphere of $E_i(x) \oplus Rv_i(x)$, and the principal curvature of M_v in these directions can be calculated as follows:

$$de_j^*(x^*) \cdot v_i(x)/\|v_i\| = de_j(x) \cdot v_i(x)/\|v_i\| = \frac{v_s \cdot v_i}{\|v_i\|}\omega_j$$

$$= \frac{v_s \cdot v_i}{1 - v \cdot v_i}\omega_j,$$

is a constant if $e_j \in E_s$ and $j \leq r$. ∎

6.5.8. Corollary. *With the same notation as in Theorem 6.5.7, let τ denote the intersection of σ and the unit sphere of ν_q. Then*
(i) $L_\tau = \bigcup \{M_x \mid x \in \tau\}$ is a smooth submanifold of S^{n+k-1},
(ii) let $x_0 \in \tau$, then L_τ is diffeomorphic to $M_{x_0} \times \tau$,
(iii) the mean curvature vector of M_x in L_τ is equal to the mean curvature vector of M_x in S^{n+k-1}.

PROOF. (i) and (ii) are obvious, and (iii) is a consequence of Theorem 6.5.7 (ii). ∎

6.5.9. The Slice Theorem. *Let $X : M^n \to R^{n+k}$ be a rank k isoparametric submanifold, W its Coxeter group, and m_i its multiplicities. Let $q \in M$, and let σ be a simplex of a chamber $\triangle$ of W on ν_q. Let ξ_σ, η_σ, m_σ, and W_σ denote $\xi(q, \sigma)$, $\eta(q, \sigma)$, $m_{q,\sigma}$, and $W_{q,\sigma}$ respectively. Let v be a parallel normal field on M such that $q + v(q) \in \sigma$, and let $\pi_v : M \to R^{n+k}$ denote the fibration $X + v$ as in Proposition 6.5.1. Then:*
(i) The connected component $S_{q,v}$ of the fiber of π_v through q is a m_σ-dimensional isoparametric submanifold of rank $k - \dim(\sigma)$ in the Euclidean space η_σ.
(ii) The normal plane at q to $S_{q,v}$ in η_σ is ξ_σ, the associated Coxeter group of $S_{q,v}$ is the W_σ, and the reflection hyperplanes of $S_{q,v}$ at q are $\{\ell_i(q) \cap \xi_\sigma \mid i \in I(q, \sigma)\}$.
(iii) If v^* is a parallel normal field on M such that $q + v^*(q) \in \sigma$, then $S_{q,v} = S_{q,v^*}$, and will be denoted by $S_{q,\sigma}$.
(iv) If u is a parallel normal field on M such that $q + u(q) \in \xi_\sigma$, then $u|S_{q,\sigma}$ is a parallel normal field of $S_{q,\sigma}$ in η_σ.
(v) Given $z \in V(q, \sigma)$ we have $\|x - z\| = \|q - z\|$ for all $x \in S_{q,\sigma}$.
(vi) If $x \in S_{q,\sigma}$ then σ is a simplex in ν_x and $V(x, \sigma) = V(q, \sigma)$.

PROOF. (iii) follows from the fact that $\mathrm{Ker}(d\pi_v) = \mathrm{Ker}(d\pi_{v^*})$, which is

$$\bigoplus \{E_i \mid i \in I(q, \sigma)\}.$$

Let $q^* = \pi_v(q)$. It is easily seen that we have

$$T(M_v)_{q^*} = \bigoplus \{E_j \mid j \text{ is not in } I(q, \sigma)\},$$

$$\nu(M_v)_{q^*} = \nu(M)_x \oplus \bigoplus \{E_i(x) \mid i \in I(q, \sigma)\}, \tag{6.5.2}$$

$$T(S_{q,\sigma})_x = \bigoplus \{E_i(x) \mid i \in I(q, \sigma)\},$$

for all $x \in S_{q,\sigma}$. Note that the left hand side of (6.5.2) is a fixed plane independent of $x \in S_{q,\sigma}$ and the right hand side of (6.5.2) always contains the tangent plane of $S_{q,\sigma}$. Hence $S_{q,\sigma} \subset q + \nu(M_v)_{q^*}$. Moreover, if u is a parallel normal field on M such that $u(q)$ is tangent to $V(q,\sigma)$ then $\langle u, v_i \rangle = 0$ for all $i \in I(q,\sigma)$. This implies that $d(u|S_{q,\sigma}) = 0$, i.e., $V(q,\sigma)$ is perpendicular to $S_{q,\sigma}$. So it remains to prove that $S_{q,\sigma}$ is isoparametric. But this follows from the fact that M is isoparametric. ∎

6.5.10. Example. Suppose M^n is an isoparametric submanifold of R^{n+3}, M is contained in S^{n+2}, B_3 is the associated Coxeter group, and its marked Dynkin diagram is:

B_3 ○———◎═══○
$\quad\quad m_1 \quad m_2 \quad m_3$

Then the fundamental domain of W on S^2 of ν_q is the following geodesic triangle on S^2:

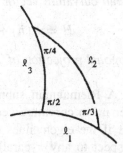

ℓ_3 $\quad$ $\pi/4$ $\quad$ ℓ_2

$\pi/2$

$\pi/3$

ℓ_1

6.6 Applications to minimal submanifolds.

In this section we will give a generalization of the Hsiang-Lawson [HL] cohomogeneity method for minimal submanifolds. For details see [PT1].

Let $\pi : E \to B$ be a Riemannian submersion. A submanifold N of E will be called *projectable* if $N = \pi^{-1}(M)$ for some submanifold M of B. A deformation F_t of N is called projectable if each $F_t(N)$ is projectable, and

F_t is called *horizontal* if each curve $F_t(x)$ is horizontal, or equivalently if the deformation vector field of F_t is horizontal. F_t is a π-invariant deformation of N if it is both projectable and horizontal. Clearly if $f_t : M \to B$ is a deformation of M then there is a unique π-invariant lifting $F_t : N \to E$ of f_t; namely for each x in N, $F_t(x)$ is the horizontal lift of the curve $f_t(\pi(x))$ through x, and the deformation field of F_t is the horizontal lift of the deformation field of f_t. Thus there is a bijective correspondence between the π-invariant deformations of $N = \pi^{-1}(M)$ in E and deformations of M in B.

6.6.1. Definition. The *fiber mean curvature vector field* h of a Riemannian submersion $\pi : E \to B$ is defined as follows: $h(x)$ is the mean curvature vector at x of the fiber $\pi^{-1}(\pi(x))$ in E.

The following proposition follows by a straightforward calculation.

6.6.2. Proposition.   Let $\pi : E \to B$ be a Riemannian submersion, M a submanifold of B, and $N = \pi^{-1}(M)$. Let H denote the mean curvature of M in B, $\hat{H}$ the mean curvature of N in E, H^* the horizontal lifting of H to N, and h the fiber mean curvature vector field in E. Then

$$\hat{H} = P(h) + H^*,$$

where P_x is the orthogonal projection of TE_x onto $\nu(N)_x$.

6.6.3. Definition.   A Riemannian submersion $\pi : E \to B$ is called *h-projectable* if the fiber mean curvature vector field h is projectable. We call π *quasi-homogeneous* if the eigenvalues of the shape operator of any fiber $F = \pi^{-1}(b)$ with respect to any π-parallel field ξ are constant (depending only on $d\pi(\xi)$, not on x in F).

6.6.4. Remark.   It is immediate from the above formula defining h that a quasi-homogeneous Riemannian submersion is h-projectable.

6.6.5. Example.   Let E be a Riemannian G-manifold. If E has a single orbit type, then the orbit space $B = E/G$ is a smooth manifold and there is a unique metric on B such that the orbit map $\pi : E \to B$ is a Riemannian submersion. Then π is quasi-homogeneous.

6.6.6. Theorem.   Let $\pi : E \to B$ be a h-projectable Riemannian submersion, and M a submanifold of B. Then a submanifold $N = \pi^{-1}(M)$ of E is a minimal submanifold of E if and only if N is a stationary point of the area functional A with respect to all the π-invariant deformations of N in E.

PROOF.   We have $\hat{H} = P(h) + H^*$ by Proposition 6.6.2. Since h is projectable and $d\pi_x(\nu(N)_x) = \nu(M)_{\pi(x)}$, $P(h)$ and therefore $\hat{H}$ is projectable.

Let ξ denote the normal field $d(\hat{H})$ of M in B. Then $f_t(x) = exp_x(t\xi(x))$ defines a deformation of M in B with ξ as deformation field. Let f_t^* be the induced π-invariant deformation of N in E. Then the deformation field of f_t^* is $\hat{H}$. Let $A(t) =$ the area of $f_t^*(N)$, then

$$A'(0) = \int_M \|\hat{H}\|^2 dv.$$

If N is a critical point of A with respect to all π-invariant deformations, then $A'(0) = 0$, hence $\hat{H} = 0$. ∎

Let $\pi : E^{n+k} \to B^k$ be an h-projectable Riemannian submersion. Then the above theorem implies that the minimal equation for finding $(n + r)$ dimensional π-invariant minimal submanifolds in E is reduced to an equation in r independent variables. To be more specific, if the fiber of π is compact we define $v : B \to \mathbf{R}$ by $v(b) =$ the volume of $\pi^{-1}(b)$. Then the volume of $\pi^{-1}(M)$ is the integral of the positive function v with respect to the induced metric on M. Hence we have:

6.6.7.   Theorem.   *Suppose* $\pi : (E, g_0) \to (B, g)$ *is an h-projectable Riemannian submersion. Then* $\pi^{-1}(M)$ *is minimal in* E *if and only if* M *is minimal in* (B, g^*), *where* $g^* = v^{2/r}g$, $v(b) =$*the volume of* $\pi^{-1}(b)$, *and* $r = \dim(M)$.

6.6.8. Remark.   If π is h-projectable, then the vector equation $\hat{H} = P(h) + H^*$ is equivalent to the equation $d\pi(\hat{H}) = d\pi(P(h)) + H$. Hence one can reduce the problem of finding π-invariant minimal submanifolds $N = \pi^{-1}(M)$ of E to the problem of finding a submanifold M of B with the prescribed mean curvature vector $H = -d\pi(P(h))$. We can also reduce the problem of finding constant mean curvature hypersurfaces N in E to the problem of finding a hypersurface M of B with the prescribed mean curvature $H = -\|d\pi(P(h))\| + c$, for some constant c.

6.6.9. Definition.   Suppose E is a complete Riemannian manifold, and $B = \bigcup_\alpha B_\alpha$ is a stratified set such that each B_α is a Riemannian manifold. A continuous map $\pi : E \to B$ is called a stratified submersion if $E_\alpha = \pi^{-1}(B_\alpha)$ is a stratification of E, and $\pi_\alpha = \pi|E_\alpha : E_\alpha \to B_\alpha$ is a submersion for each α. Then π is called a stratified Riemannian submersion if each π_α is a Riemannian submersion, and π is called h-projectable (resp. quasi-homogeneous) if the mean curvature vector of $\pi_\alpha^{-1}(b)$ in E_α is the mean curvature vector of $\pi_\alpha^{-1}(b)$ in E for all α and b in B_α, and each π_α is h-projectable (resp. quasi-homogeneous).

6.6.10. Definition. M is a stratified subset of a stratified set B if $M \cap B_\alpha$ is a submanifold of B_α, for each stratum B_α. A deformation $f_t : M \to B$ is strata preserving if $f_t(M \cap B_\alpha)$ is contained in B_α for each α. A submanifold N of E is π-invariant if N is of the form $\pi^{-1}(M)$ for some stratified subset M of B. Given a strata preserving deformation f_t of M into B, then there is a unique horizontal strata preserving lifting F_t of f_t. We call such a deformation of N a π-invariant strata preserving deformation. Then the following is a straightforward generalization of Theorem 6.6.6 above.

6.6.11. Theorem.   *Let $\pi : E \to B$ be a stratified h-projectable Riemannian submersion. Then a π-invariant submanifold N of E is minimal in E if and only if N is a critical point of the area functional with respect to all the π-invariant strata preserving deformations of N in E.*

6.6.12. Example.   Let G be a compact Lie group acting isometrically on a complete Riemannian manifold E. The mean curvature vector field H of an orbit Gx in E is clearly a G-equivariant normal field, and hence $H(x)$ lies in the fixed point set of the isotropy representation at x. But this fixed point set is the tangent space of the union of the orbits of type (G_x). Then the orbit space E/G is naturally stratified by the orbit types, and each stratum has a natural metric such that the projection map $\pi : E \to E/G$ is a quasi-homogeneous Riemannian submersion. Theorems 6.6.6 and 6.6.11 for this case were proved in [HL].

6.6.13. Example.   Let $M^n \subseteq S^{n+k-1} \subseteq R^{n+k}$ be isoparametric, W the Coxeter group associated to M, and $\triangle_q$ the Weyl chamber of W on $\nu_q = q + \nu(M)_q = \nu(M)_q$ containing q. Since $\triangle_q$ is a simplicial cone, $B =$ the intersection of the unit sphere S^{k-1} of ν_q and $\triangle_q$, has a natural stratification. In fact, each stratum of B is given by the intersection of some simplex of $\triangle_q$ with S^{k-1}. Let M_x denote the unique submanifold through $x \in B$ and parallel to M. By Corollary 6.5.3 we have:

$$\bigcup \{M_x | x \in B\} = S^{n+k-1}.$$

If σ is a stratum of B and $x_0 \in B$, then

$$E_\sigma = \bigcup \{M_x | x \in \sigma\}$$

is diffeomorphic to $M_{x_0} \times \sigma$. So the stratification on B induces one on S^{n+k-1}. Let $\pi : S^{n+k-1} \to B$ be defined by $\pi(y) = x$ if $y \in M_x$. Then by Corollary 6.5.8, π is a stratified quasi-homogeneous Riemannian submersion. Let σ be a stratum of B, $x \in B$, and $n_\sigma = \dim(M_x)$. Then the function $A_\sigma : \bar\sigma \to R$ defined by $A_\sigma(x) = n_\sigma$-dimensional volume of M_x, is continuous. If x lies on the boundary $\partial\sigma$ then $\dim(M_x) < n_\sigma$. So A_σ restricts to $\partial\sigma$ is zero. Since $\bar\sigma$ is compact, there exists $x_\sigma \in \sigma$ which is the maximum of A_σ.

So by Theorem 6.6.6, M_{x_σ} is a minimal submanifold of S^{n+k-1}. For $k=3$, the minimal equation of π-invariant N^n in S^{n+1} is an ordinary differential equation on S^2/W, which depends only on the multiplicities m_i. Hence the construction of cohomogeneity 1 minimal hyperspheres in S^{n+1} given in [Hs2,3] (with W being of rank 2), produces more minimal hyperspheres in S^{n+1}, which are not of cohomogeneity one [FK].

Chapter 7.

Proper Fredholm Submanifolds in Hilbert Spaces.

In this chapter we generalize the submanifold theory of Euclidean space to Hilbert space. In order to use results the infinite dimensional differential topological we restrict ourself to the class of proper Fredholm immersions (defined below).

7.1 Proper Fredholm immersions.

Let M be an immersed submanifold of the Hilbert space V (i.e., TM_x is a closed linear subspace of V), and let $\nu(M)_x = (TM_x)^\perp$ denote the normal plane of M at x in V. Using the same argument as in chapter 2, we conclude that, given a smooth normal field v on M and $u \in TM_{x_0}$, the orthogonal projection of $dv_{x_0}(u)$ onto TM_{x_0} depends only on $v(x_0)$ and not on the derivatives of v at x_0; it be denoted by $-A_{v(x_0)}(u)$ (the shape operator of M with respect to the normal vector $v(x_0)$). The first and second fundamentals forms I, II and the normal connection ∇^ν on M can be defined in the same (invariant) manner as in the finite dimensional case, i.e.,

$$I(x) = \langle \; , \; \rangle | TM_x,$$

$$\langle II(x)(u_1, u_2), v \rangle = \langle A_v(u_1), u_2 \rangle,$$

$$\nabla^\nu v = \text{the orthogonal projection of } dv \text{ onto } \nu(M).$$

Since all these local invariants for M are well-defined, the method of moving frame is valid here (because when we expand well-defined tensor fields in terms of local orthonormal frame field, then the infinite series are convergent). Arguing the same way as in the finite dimensional case, we can prove that I, II and the induced normal connection ∇^ν satisfy the Gauss, Codazzi and Ricci equations. Moreover, the Fundamental Theorem 2.3.1 is valid for immersed submanifolds of Hilbert space. As a consequence of the Ricci equation, we also have the analogue of Proposition 2.1.2:

7.1.1. Proposition. *Suppose M is an immersed submanifold of the Hilbert space V and the normal bundle $\nu(M)$ is flat. Then the family $\{A_v | \; v \in \nu(M)_x\}$ of shape operators is a commuting family of operators on TM_x.*

Although these elementary parts of the theory of submanifold geometry work just as in the finite dimensional case, many of the deeper results are not true in general. For example, the infinite dimensional differential topology developed by Smale and infinite dimensional Morse theory developed by Palais and Smale will not work for general submanifolds of Hilbert space without further restrictions. Recall also that the spectral theory of the shape operators and the Morse theory of the Euclidean distance functions of submanifolds of R^n are closely related and play essential roles in the study of the geometry and topology of submanifolds of R^n. Here again, without some restrictions important aspects of these theories will not carry over to the infinite dimensional setting. One of the main goals of this section is to describe a class of submanifolds of Hilbert space for which the techniques of infinite dimensional geometry and topology can be applied to extend some of the deeper parts of the theory of submanifold geometry.

The end point map $Y : \nu(M) \to V$ for an immersed submanifold M of a Hilbert space V is defined just as in Definition 4.1.7; i.e., $Y : \nu(M) \to V$ is given by $Y(v) = x + v$ for $v \in \nu(M)_x$.

7.1.2. Definition. An immersed finite codimension submanifold M of V is *proper Fredholm* (PF), if

(i) the end point map Y is Fredholm,

(ii) the restriction of Y to each normal disk bundle of finite radius r is proper.

Since the basic theorems of differential calculus and local submanifold geometry work for PF submanifolds just as for submanifolds of R^n, Proposition 4.1.8 is valid for PF submanifolds of Hilbert spaces. In particular, we have

$$dY_v = (I - A_v, id), \qquad (7.1.1)$$

which implies that

7.1.3. Proposition.   *The end point map Y of an immersed submanifold M of a Hilbert space V is Fredholm if and only if $I - A_v$ is Fredholm for all normal vector v of M.*

7.1.4. Remarks.

(i) An immersed submanifold M of R^n is PF if and only if the immersion is proper.

(ii) If M is a PF submanifold of V, and M is contained in the sphere of radius r with center x_0 in V, then $v(x) = x_0 - x$ is a normal field on M with length r, and $Y(x, v(x)) = x_0$. Since Y is proper on the r-disk normal bundle, M is compact. This implies that M must be finite dimensionional. It follows that PF submanifolds of an infinite dimensional Hilbert space V cannot lie on a hypersphere of V. In particular, the unit sphere of V is not PF.

7.1.5. Examples.

(1) A finite codimension linear subspace of V is PF.

(2) Let $\varphi : V \to V$ be a self-adjoint, injective, compact operator. Then the hypersurface

$$M = \{x \in V | \langle \varphi(x), x \rangle = 1\}$$

is PF. To see this we note that $v(x) = \varphi(x)/\|\varphi(x)\|$ is a unit normal field to M, and $A_{v(x)}(u) = -\varphi(u)^{TM_x}/\|\varphi(x)\|$ is a compact operator on TM_x, where $\varphi(u)^{TM_x}$ denote the orthogonal projection of $\varphi(u)$ onto TM_x. So it follows from Proposition 7.1.3 that the end point map of M is Fredholm. Next assume that $x_n \in M$, $\{\lambda_n \varphi(x_n)\}$ is bounded, and $Y(x_n, \lambda_n \varphi(x_n)) = x_n + \lambda_n \varphi(x_n) \to y$. Then x_n is bounded, and $\langle x_n + \lambda_n \varphi(x_n), x_n \rangle = \|x_n\|^2 + \lambda_n$ is bounded, which implies that λ_n is bounded. Since φ is compact and $\{\lambda_n x_n\}$ is bounded, $\varphi(\lambda_n x_n)$ has a convergent subsequence, and so $\{x_n\}$ has a convergent subsequence.

7.1.6. Theorem.
Suppose G is an infinite dimensional Hilbert Lie group, G acts on the Hilbert space V isometrically, and the action is proper and Fredholm. Then every orbit Gx is an immersed PF submanifold of V.

PROOF.   First we prove that the end point map Y of $M = Gx$ is Fredholm. Because every isometry of V is an affine transformation, we have

$$(I - A_v)(\xi(x)) = (\xi(x + v))^{T_x},$$

where $\xi \in \mathcal{G}$, $v \in \nu(M)_x$, and u^{T_x} denotes the tangential component of u with respect to the decomposition $V = TM_x \oplus \nu(M)_x$. It follows from the definition of Fredholm action that the differential of the orbit map at e is Fredholm. So the two maps $\xi \mapsto \xi(x)$ and $\xi \mapsto \xi(x + v)$ are Fredholm maps from $\mathcal{G}$ to V. In particular, $T(Gx)_x$ and $T(G(x + v))_{x+v}$ are of finite codimension. So the map $P : T(G(x+v))_{x+v} \to T(Gx)_x$ defined by $P(u) = u^{T_x}$ is Fredholm. Hence $I - A_v$ is Fredholm, i.e., Y is Fredholm. Next we assume that $x_n \in M$, $v_n \in \nu(M)_{x_n}$, $\|v_n\| \le r$, and $Y(x_n, v_n) \to y$. Then there exist linear isometry φ_n of V and $c_n \in V$ such that $g_n = \varphi_n + c_n \in G$ and $x_n = g_n(x)$. Note that $dg_n = \varphi_n$, $u_n = \varphi_n^{-1}(v_n) \in \nu(M)_x$, and

$$Y(g_n x, v_n) = \varphi_n(x) + c_n + \varphi_n(u_n) = \varphi_n(x + u_n) + c_n = g_n(x + u_n) \to y.$$

Since $\{u_n\}$ is a bounded sequence in the finite dimensional Euclidean space $\nu(M)_x$, there exists a convergent subsequence $u_{n_i} \to u$. So we have $g_{n_i}(x + u_{n_i}) \to y$ and $x + u_{n_i} \to x + u$. It then follows from the definition of proper action that g_{n_i} has a convergent subsequence in G, which implies that x_{n_i} has a convergent subsequence in M.   ∎

7.1.7. Proposition.   *Let M be an immersed PF submanifold of V, $x \in M$, $v \in \nu(M)_x$, and A_v the shape operator of M with respect to v. Then:*

(1) A_v has no residual spectrum,

(2) the continuous spectrum of A_v is either $\{0\}$ or empty,

(3) the eigenspace corresponding to a non-zero eigenvalue of A_v is of finite dimension,

(4) A_v is compact.

PROOF.   Since A_v is self-adjoint, it has no residual spectrum. Note that the eigenspace of A_v with respect to a non-zero eigenvalue λ is

$$\mathrm{Ker}(\lambda I - A_v) = \mathrm{Ker}(I - \frac{1}{\lambda} A_v) = \mathrm{Ker}(I - A_{\frac{v}{\lambda}}).$$

So (3) follows from Proposition 7.1.3. Now suppose $\lambda \neq 0$, $\mathrm{Ker}(A_v - \lambda I) = 0$, and $\mathrm{Im}(A_v - \lambda I)$ is dense in TM_x. Since $A_v - \lambda I$ is Fredholm, $\mathrm{Im}(A_v - \lambda I)$ is closed and equal to TM_x, i.e., $A_v - \lambda I$ is invertible, which proves (2). To prove (4) it suffices to show that if λ_i is a sequence of distinct real numbers in the discrete spectrum of A_v and $\lambda_i \to \lambda$ then $\lambda = 0$. But if $\lambda \neq 0$, then the self-adjoint Fredholm operator $P = I - A_{v/\lambda}$ induces an isomorphism $\tilde{P}$ on $V/\mathrm{Ker}(P)$, so $\tilde{P}$ is bounded. Let δ denote $\|\tilde{P}\|$. Then $|(1 - \lambda_i/\lambda)^{-1}| \leq \delta$, and hence $|\lambda - \lambda_i|/|\lambda| \geq 1/\delta > 0$, contradicting $\lambda_i \to \lambda$. ∎

It follows from (7.1.1) that $e \in \nu(M)_x$ is a regular point of Y if and only if $I - A_e$ is an isomorphism. Moreover, the dimension of $\mathrm{Ker}(I - A_e)$ and $\mathrm{Ker}(dY_e)$ are equal, which is finite by Proposition 7.1.3. Hence the Definition 4.2.1 of focal points and multiplicities makes sense for PF submanifolds.

7.1.8. Definition.   Let $e \in \nu(M)_x$. The point $a = Y(e)$ in V is called a non-focal point for a PF submanifold M of V with respect to x if dY_e is an isomorphism. If $m = \dim(\mathrm{Ker}(dY_e)) > 0$ then a is called a focal point of multiplicity m for M with respect to x.

The set Γ of all the focal points of V is called the *focal set* of M in V, i.e., Γ is the set of all critical values of the normal bundle map Y. So applying the Sard-Smale Transversality theorem [Sm2] for Fredholm maps to the end point map Y of M, we have:

7.1.9. Proposition.   *The set of non-focal points of a PF submanifold M of V is open and dense in V.*

By the same proof as in Proposition 4.1.5, we have:

7.1.10. Proposition.   *Let M be an immersed PF submanifold of V, and $a \in V$. Let $f_a : M \to R$ denote the map defined by $f_a(x) = \|x - a\|^2$. Then:*

(i) $\nabla f_a(x) = 2(x-a)^{T_x}$, the projection of $(x-a)$ onto TM_x, so in particular x_0 is a critical point of f_a if and only if $(x_0 - a) \in \nu(M)_{x_0}$,

(ii) $\frac{1}{2}\nabla^2 f_a(x_0) = I - A_{(a-x_0)}$ at the critical point x_0 of f_a,

(iii) f_a is non-degenerate if and only if a is a non-focal point of M in V,

It follows from Propositions 7.1.9 and 7.1.10 that:

7.1.11. Corollary.   *If M is an immersed PF submanifold of V, then f_a is non-degenerate for all a in an open dense subset of V.*

As a consequence of Proposition 7.1.7 and 7.1.10:

7.1.12. Proposition.   *Let M be an immersed PF submanifold of V. Suppose x_0 is a critical point of f_a and V_λ is the eigenspace of $A_{(a-x_0)}$ with respect to the eigenvalue $\lambda \neq 0$. Then:*

(i) $\dim(V_\lambda)$ is finite,

(ii) $\mathrm{Index}(f_a, x_0) = \sum\{\dim(V_\lambda) | \lambda > 1\}$, which is finite.

Morse theory relates the homology of a smooth manifold to the critical point structure of certain smooth functions. This theory was extended to infinite dimensional Hilbert manifolds in the 1960's by Palais and Smale ([Pa2],[Sm1]) for the class of smooth functions satisfying Condition C (see Part II, chapter 1).

7.1.13. Theorem.   *Let M be an immersed PF submanifold of a Hilbert space V, and $a \in V$. Then the map $f_a : M \to R$ defined by $f_a(x) = \|x-a\|^2$ satisfies condition C.*

PROOF.   We will write f for f_a. Suppose

$$|f(x_n)| \leq c, \quad \|\nabla f(x_n)\| \to 0.$$

Let u_n be the orthogonal projection of $(x_n - a)$ onto TM_{x_n}, and v_n the projection of $(x_n - a)$ onto $\nu(M)_{x_n}$. Since $\|x_n - a\|^2 \leq c$ and $u_n \to 0$, $\{v_n\}$ is bounded (say by r). So $(x_n, -v_n)$ is a sequence in the r-disk normal bundle of M, and

$$Y(x_n, -v_n) = x_n - v_n = (x_n - a) - v_n + a = u_n + a \to a.$$

Since M is a PF submanifold, $(x_n, -v_n)$ has a convergent subsequence in $\nu(M)$, which implies that x_n has a convergent subsequence in M.   ∎

7.1.14. Remark. Let M be an immersed submanifold of V (not necessarily PF). Then the condition that all f_a satisfy condition C is equivalent to the condition that the restriction of the end point map to the unit disk normal bundle is proper.

7.2   Isoparametric submanifolds in Hilbert spaces.

In this section we will study the geometry of isoparametric submanifolds of Hilbert spaces. They are defined just as in R^n.

7.2.1. Definition.   An immersed PF submanifold M of a Hilbert space $(V, \langle \ , \ \rangle)$ is called isoparametric if

(i) $\nu(M)$ is globally flat,

(ii) if v is a parallel normal field on M then the shape operators $A_{v(x)}$ and $A_{v(y)}$ are orthogonally equivalent for all $x, y \in M$.

7.2.2. Remark.   Although Definition 5.7.2 seems weaker than Definition 7.2.1 (where we only assume that $\nu(M)$ is flat), if $V = R^n$, we have proved in Theorem 6.4.4 that $\nu(M)$ *is* globally flat. So these two definitions agree when V is a finite dimensional Hilbert space.

7.2.3. Definition.   An immersed submanifold $f : M \to V$ is *full*, if $f(M)$ is not included in any affine hyperplane of V. M is a rank k immersed isoparametric submanifold of V if M is a full, codimension k, isoparametric submanifold of V.

7.2.4. Remarks.

(i) Since PF submanifolds of V have finite codimension, an isoparametric submanifold of V is of finite codimension.

(ii) It follows from Remark 7.1.4 that if M is a full isoparametric submanifold of V and M is contained in the sphere of radius r centered at c_0, then both M and V must be of finite dimension.

Since compact operators have eigen-decompositons and the normal bundle of an isoparametric submanifold of V is flat, it follows from Proposition 7.1.1 and 7.1.7 that:

7.2.5. Proposition.   *If M is an isoparametric PF submanifold of a Hilbert space V, then there exist E_0 and a family of finite rank smooth distributions $\{E_i \mid i \in I\}$ such that $TM = E_0 \bigoplus \{E_i \mid i \in I\}$ is the common eigen-decomposition for all the shape operators A_v of M and $A_v|E_0 = 0$.*

Since A_v is linear for $v \in V$, there exist smooth sections λ_i of $\nu(M)^*$ such that

$$A_v|E_i = \lambda_i(v)id_{E_i},$$

for all $i \in I$. Identifying $\nu(M)^*$ with $\nu(M)$ by the induced inner product from V, we obtain smooth normal fields v_i on M such that

$$A_v|E_i = \langle v, v_i \rangle id_{E_i}, \qquad (7.2.1)$$

for all $i \in I$. These E_i's, λ_i's and v_i's are called the curvature distributions, principal curvatures, and curvature normals for M respectively. If v is a parallel normal field on an isoparametric submanifold M then A_v has constant eigenvalues. So it follows from (7.2.1) that each curvature normal field v_i is parallel.

7.2.6. Proposition.   *If M is a rank k isoparametric PF submanifold of Hilbert space, and $\{v_i|\ i \in I\}$ are its curvature normals, then there is a positive constant c such that $\|v_i\| \leq c$ for all $i \in I$.*

PROOF.   Let F denote the continuous function defined on the unit sphere S^{k-1} of the normal plane $\nu(M)_q$ by $F(v) = \|A_v\|$. Since S^{k-1} is compact, there is a constant $c > 0$ such that $F(v) \leq c$. Since the eigenvalues of A_v are $\langle v, v_i \rangle$, we have $|\langle v, v_i \rangle| \leq c$ for all $i \in I$ and all unit vector $v \in \nu(M)_q$. ∎

7.2.7. Proposition.   *Let M be a rank k immersed isoparametric subman-ifold of Hilbert space, $\nu_q = q + \nu(M)_q$ the affine normal plane at q, and $\Gamma_q = \Gamma \cap \nu_q$ the set of focal points for M with respect to q. Then:*
(i) $\Gamma_q = \bigcup \{\ell_i(q)|\ i \in I\}$, where $\ell_i(q)$ is the hyperplane in ν_q defined by

$$\ell_i(q) = \{q + v|\ v \in \nu(M)_q, \ \langle v, v_i(q) \rangle = 1\}.$$

(ii) $\mathcal{H} = \{\ell_i(q)|\ i \in I\}$ is locally finite, i.e., given any point $p \in \nu_q$ there is an open neighborhood U of p in ν_q such that $\{i \in I|\ell_i(q) \cap U \neq \emptyset\}$ is finite.

PROOF.   Let Y be the end point map of M. By (7.2.1), $x = q + e \in \Gamma_q$ if and only if 1 is an eigenvalue of A_e. So there exists $i_0 \in I$ such that $1 = \langle e, v_{i_0} \rangle$, i.e., $x \in \ell_{i_0}(q)$. This proves (i).

Let $J(x) = \{i \in I|\ x \in \ell_i(q)\}$ for $x = q + e \in \nu_q$. Then the eigenspace V_1 of A_e corresponding to eigenvalue 1 is $\bigoplus \{E_j|j \in J(x)\}$. Since A_e is compact and $\{\langle e, v_i \rangle|\ i \in I\}$ are the eigenvalues of A_e, the set $J(x)$ is finite and there exist $\delta > 0$ such that $|1 - \langle e, v_i \rangle| > \delta$ for all i not in $J(x)$. By analytic geometry, if i is not in $J(x)$ then

$$d(x, \ell_i(q)) = \frac{|1 - \langle e, v_i \rangle|}{\|v_i\|^2} > \frac{\delta}{c},$$

where c is the upper bound for $\|v_i\|$ as in Proposition 7.2.6. So we conclude that the ball $B(x, \delta/c)$ of radius δ/c and center x meets only finitely many $\ell_i(q)$ (in fact it intersects $\ell_i(q)$ only for $i \in J(x)$). ∎

We next note the following:
  (i) the Frobenius integrability theorem is valid for finite rank distributions on Banach manifolds,
  (ii) the proof of the existence of a Coxeter group in Chapter 6 depended only on the facts that all the curvature distributions and $\nu(M)$ are of finite rank and the family of focal hyperplanes $\{\ell_i \mid i \in I\}$ is locally finite.
So it is not difficult to see that most of the results in sections 6.2 and 6.3 for isoparametric submanifolds of R^n can be generalized to the infinite dimensional case. In particular the statements from 6.2.3 to 6.2.9, from 6.3.1 to 6.3.5, and the Slice Theorem 6.5.9 are all valid if we replace M by a rank k isoparametric submanifold of a Hilbert space and the index set $1 \leq i \leq p$ of curvature normals by $\{i \mid i \in I\}$. In particular the analogues of Theorem 6.3.2 and 6.3.5 for infinite dimensional isoparametric submanifolds give:

7.2.8. Theorem.   Let φ_i be the involution associated to the curvature distribution E_i.
  (i) There exists a bijection $\sigma_i : I \to I$ such that $\sigma_i(i) = i$, $\varphi_i^*(E_j) = E_{\sigma_i(j)}$ and $m_j = m_{\sigma_i(j)}$.
  (ii) Let R_i^q denote the reflection of ν_q in $\ell_i(q)$. Then

$$R_i^q(\ell_j(q)) = \ell_{\sigma_i(j)}(q),$$

i.e., R_i^q permutes $\mathcal{H} = \{\ell_i(q) \mid i \in I\}$.

Note that R_i^q permutes hyperplanes in $\mathcal{H}$ and $\mathcal{H}$ is locally finite, so by Theorem 5.3.6 the subgroup of isometries of $\nu_q = q + \nu(M)_q$ generated by $\{R_i^q \mid i \in I\}$ is a Coxeter group.

7.2.9. Theorem.   Let M be an immersed isoparametric submanifold in the Hilbert space V, E_i the curvature normals, and $\{v_i \mid i \in I\}$ the set of curvature normals. Let W^q be the subgroup of the group of isometries of the affine normal plane $\nu_q = q + \nu(M)_q$ generated by reflections φ_i in $\ell_i(q)$. Then W^q is a Coxeter group. Moreover, let $\pi_{q,q'} : \nu(M)_q \to \nu(M)_{q'}$ denote the parallel translation with respect to the induced normal connection, then the map $P_{q,q'} : \nu_q \to \nu_{q'}$, defined by $P_{q,q'}(q + u) = q' + \pi_{q,q'}(u)$, conjugates W^q to $W^{q'}$ for any q and q' in M.

7.2.10. Corollary.   *Let M be a rank k immersed isoparametric subman-ifold of the infinite dimensional Hilbert space V, $\{E_i | i \in I\}$ the curvature distributions, and $\{\ell_i(q) | i \in I\}$ the curvature normal vectors at $q \in M$. Then associated to M there is a Coxeter group W with $\{\ell_i(q) | i \in I\}$ as its root system.*

7.2.11. Corollary.   *Let M be an isoparametric submanifold of the infinite dimensional Hilbert space V, $\{E_i | i \in I\}$ the curvature distributions, and $\{v_i | i \in I\}$ the curvature normals. Suppose $0 \in I$ and $v_0 = 0$.*
  (1) If I is a finite set, then
    (i) there exists a constant vector $c_0 \in V$ such that $\bigcap \{\ell_i(q) | i \in I\} = \{c_0\}$ for all $q \in M$,
    (ii) the Coxeter group associated to M is a finite group,
    (iii) the rank of E_0 is infinite,
    (iv) $\tilde{E} = \bigoplus \{E_i | i \neq 0, \ i \in I\}$ is integrable,
    (v) $M \simeq S \times E_0$, where S is an integral submanifold of $\tilde{E}$.
  (2) If I is an infinite set, then the Coxeter group associated to M is an infinite group.

Let Δ_q be the connected component of $\nu_q \setminus \bigcap \{\ell_i \mid i \in I\}$ containing q. If I is an infinite set, then W is an affine Weyl group, the closure $\bar{\Delta}_q$ is a fundamental domain of W and its boundary has $k + 1$ faces. If $\varphi \in W$ and $\varphi(\ell_i) = \ell_j$ then $m_i = m_j$. It follows that:

7.2.12. Corollary.   *Let M be a rank k isoparametric submanifold of an infinite dimensional Hilbert space having infinitely many curvature distribu-tions. Then there is associated to M a well-defined marked Dynkin diagram with $k + 1$ vertices, namely the Dynkin diagram of the associated affine Weyl group with multiplicities m_i.*

7.2.13. Example.   Let $\hat{G}$ be the H^1-loops on the compact simple Lie group G, V the Hilbert space of H^0-loops on the Lie algebra $\mathcal{G}$ of G, and let $\hat{G}$ act on V by gauge transformations as in Example 5.8.1. This action is polar, so the principal $\hat{G}$-orbits in V are isoparametric. In the following we calculate the basic local invariants of these orbits as submanifolds of V. Let Δ^+ denote the set of positive roots of G. Then there exist x_α and y_α in $\mathcal{G}$ for all $\alpha \in \Delta^+$ such that

$$\mathcal{G} = T \bigoplus \{Rx_a \oplus Ry_\alpha | \alpha \in \Delta^+\},$$

$$[h, x_\alpha] = \alpha(h)y_\alpha, \ [h, y_\alpha] = -\alpha(h)x_\alpha,$$

for all $h \in \mathcal{T}$. If rank$(G) = k$ and $\{t_1, \ldots, t_k\}$ is a bases of $\mathcal{T}$, then the union of the following sets

$$\{x_\alpha \cos n\theta, \ y_\alpha \cos n\theta | \ \alpha \in \Delta^+, \ n \geq 0 \text{ an integer}\},$$

$$\{x_\alpha \sin m\theta, \ y_\alpha \sin m\theta | \ \alpha \in \Delta^+, m > 0 \text{ an integer}\},$$

$$\{t_i \cos n\theta, \ t_i \sin m\theta | \ 1 \leq i \leq k, \ n \geq 0, \ m > 0, \text{ are integers}\}$$

is a separable basis for V. An orbit $M = \hat{G}\hat{t}_0$ is principal if and only if $\alpha(t_0) + n \neq 0$ for all $\alpha \in \Delta^+$ and $n \in \mathbf{Z}$. Let $\hat{t}_1 \in \hat{\mathcal{T}}$ be a regular point. Then the shape operator of M along the direction $\hat{t}_1$ is

$$A_{\hat{t}_1}(v' + [v, \hat{t}_0]) = [v, \hat{t}_1].$$

Using the above separable basis for V, it is easily seen that $A_{\hat{t}_1}$ is a compact operator, the eigenvalues are

$$\{\alpha(t_1)/\alpha(t_0) + n \ | \ \alpha \in \Delta^+, \ n \in \mathbf{Z}\},$$

and each has multiplicity 2. So the associated Coxeter group of M as an isoparametric submanifold is the affine Weyl group $W(\mathcal{T}^0)$ of the section $\mathcal{T}^0$, and all the multiplicities $m_i = 2$.

Chapter 8.

Topology of Isoparametric Submanifolds.

In this chapter we use the Morse theory developed in part II to prove that any non-degenerate distance function on an isoparametric submanifold of Hilbert space is of linking type, and so it is perfect. We also give some restriction for the possible marked Dynkin diagrams of these submanifolds. As a byproduct we are able to generalize the notion of tautness to proper Fredholm immersions of Hilbert manifolds into Hilbert space.

8.1 Tight and taut immersions in R^n.

Let $\varphi : M^n \to R^m$ be an immersed compact submanifold, and $\nu^1(M)$ the the bundle of unit normal vectors of M. The restriction of the normal map N of M to $\nu^1(M)$ will still be denoted by N, i.e., $N : \nu^1(M) \to S^{m-1}$ is defined be $N(v) = v$. There is a natural volume element $d\sigma$ on $\nu^1(M)$. In fact, if dV is a $(m - n - 1)$-form on $\nu^1(M)$ such that dV restricts to each fiber of $\nu^1(M)_x$ is the volume form of the sphere of $\nu(M)_x$, then $d\sigma = dv \wedge dV$, where dv is the volume element of M. Let da be the standard volume form on S^{n+k-1} normalized so that $\int_{S^{n+k-1}} da = 1$. Then the Gauss-Kronecker curvature of an immersed surface in R^3 can be generalized as follows:

8.1.1. Definition. The Lipschitz-Killing curvature at the unit normal direction v of an immersed submanifold M^n in R^m is defined to be the determinant of the shape operator A_v.

8.1.2. Definition. The total absolute curvature of an immersion $\varphi : M^n \to R^m$ is

$$\tau(M, \varphi) = \int_{\nu^1(M)} |\det(A_v)| \, d\sigma,$$

where $d\sigma$ is the volume element of $\nu^1(M)$, and A_v is the shape operator of M in the unit normal direction v.

8.1.3. Definition. An immersion $\varphi_0 : M^n \to R^m$ is called *tight* if

$$\tau(M, \varphi) \geq \tau(M, \varphi_0),$$

for any immersions $\varphi : M \to R^s$.

Chern and Lashof began the study of tight immersions in the 1950's [CL1,2]. They proved the following theorem:

8.1.4. Theorem. *If* $\varphi : M^n \to R^m$ *is an immersion then, for any field F,*

$$\tau(M, \varphi) \geq \sum_i b_i(M, F),$$

where $b_i(M, F)$ is the ith Betti number of M with respect to F.

It is a difficult and as yet unsolved problem to determine which manifolds admit tight immersions. An important step towards the solution is Kuiper's [Ku2] reformulation of the problem in terms of the Morse theory of height functions. Given a Morse function $f : M \to R$, let

$$\mu_k(f) = \text{the number of critical points of } f \text{ with index } k,$$
$$\mu(f) = \sum_i \mu_k(f).$$

The Morse number $\gamma(M)$ of M is defined by

$$\gamma(M) = \inf\{\mu(f)| \ f : M \to R \text{ is a Morse function}\}.$$

Let $\varphi : M^n \to R^m$ be an immersion. By Proposition 4.1.8, $dN_v = (-A_v, id)$, so we have

$$N^*(da) = (-1)^n \det(A_v)d\sigma,$$

and the total absolute curvature $\tau(M, \varphi)$ is the total volume of the image $N(\nu^1(M))$, counted with multiplicities but ignoring orientation. Let h_p denote the height function as in section 4.1. Then it follows from Propositions 4.1.1 and 4.1.8 that $p \in S^{m-1}$ is a regular value of N if and only if the height function h_p is a Morse function. In this case $N^{-1}(p)$ is a finite set with $\mu(h_p)$ elements. But by the Morse inequalities we have $\mu(h_p) \geq \sum_i b_i(M, F)$, and in particular:

$$\tau(M, \varphi) \geq \sum_i b_i(M),$$

$$\tau(M, \varphi) \geq \gamma(M).$$

This proves the following stronger result of Kuiper:

8.1.5. Theorem.
 (i) $\gamma(M) = \inf\{\tau(M, \varphi)| \ \varphi : M \to R^m \text{ is an immersion }\}.$
 (ii) *An immersion $\varphi_0 : M \to R^m$ is tight if and only if every non-degenerate height functions h_p has $\gamma(M)$ critical points.*

Banchoff [Ba] studied the problem of finding all tight surfaces that lie in a sphere, and later this led to the study of taut immersions by Carter and West [CW1]. Note that if $\varphi : M^n \to R^m$ is a tight immersion and $\varphi(M)$ is contained in the unit sphere S^{m-1}, then the Euclidean distance function f_p and the height function h_p have the same critical point theory because $f_p = 1 + \|p\|^2 - 2h_p$. Taut immersions are "essentially" the spherical tight immersions.

A non-degenerate smooth function $f : M \to R$ is called a perfect Morse function if $\mu(f) = \sum b_i(M, F)$ for some field F. If we restrict ourself to the class of manifolds that satisfy the condition that $\gamma(M) = \sum b_i(M, F)$ for some field F, then an immersion $\varphi : M \to R^m$ is tight if and only if every non-degenerate height function h_a is perfect, and it is taut if and only if every non-degenerate Euclidean distance function f_a is perfect. There is a detailed and beautiful theory of tight and taut immersions for which we refer the reader to [CR2].

8.2   Taut immersions in Hilbert space.

In Theorem 7.1.13 we showed that the distance functions f_a of PF submanifolds in Hilbert space satisfy Condition C, so the concept of tautness can be generalized easily to PF immersions.

8.2.1. Definition.   A smooth function $f : M \to R$ on a Riemannian Hilbert manifold M is called a *Morse function* if f is non-degenerate, bounded from below, and satisfies Condition C.

For a Morse function f on M let

$$M_r(f) = \{x \in M|\ f(x) \le r\}.$$

Then it follows from Condition C that there are only finitely many critical points of f in $M_r(f)$. Let

$$\mu_k(f, r) = \text{ the number of critical points of index } k \text{ on } M_r(f),$$
$$\beta_k(f, r, F) = \dim(H_k(M_r(f), F)),$$

for a field F. Then the weak Morse inequalities gives

$$\mu_k(f, r) \ge \beta_k(f, r, F)$$

for all r and F.

8.2.2. Definition. A Morse function $f : M \to R$ is *perfect*, if there exists a field F such that $\mu_k(f, r) = \beta_k(f, r, F)$ for all r and k.

It follows from the standard Morse theory in part II that:

8.2.3. Theorem. *Let f be a Morse function. Then f is perfect if and only if there exists a field F such that the induced map on the homology*

$$i_* : H_*(M_r(f), F) \to H_*(M, F)$$

of the inclusion of $M_r(f)$ in M is injective for all r.

8.2.4. Definition. An immersed submanifold M of a Hilbert space is *taut* if M is proper Fredholm and every non-degenerate Euclidean distance function f_a on M is a perfect Morse function.

8.2.5. Remark. If M is properly immersed in R^n then the above definition is the same as section 8.1.

8.2.6. Remark. It is easy to see that the unit sphere S^{n-1} is a taut submanifold in R^n. But the unit hypersphere S of an infinite dimensional Hilbert space is not taut. First, S is contractible, but the non-degenerate distance function f_a has two critical points. Moreover S is not PF.

8.2.7. Example. We will see later that, given a simple compact connected group G, the orbits of the gauge group $H^1(S^1, G)$ acting on the space of connections $H^0(S^1, \mathcal{G})$ by gauge transformations as in section 5.8 are taut.

Let $R(f)$ denote the set of all regular values of f, and $C(f)$ denote the set of all critical points of f. The fact that the restriction of the end point map Y of M to the unit disk normal bundle is proper gives a uniform condition C for the Euclidean distance functions as we see in the following two propositions:

8.2.8. Proposition. *Let M be an immersed PF submanifold of V, and $a \in V$. Suppose $r < s$ and $[r, s] \subseteq R(f_a)$. Then there exists $\delta > 0$ such that if $\|b - a\| < \delta$ then $[r, s] \subseteq R(f_b)$.*

PROOF. If not, then there exist sequences b_n in V and x_n in M such that x_n is a critical point of f_{b_n} and

$$b_n \to a, \quad r \leq \|x_n - b_n\| \leq s.$$

It follows from Proposition 7.1.10 that $(x_n - b_n) \in \nu(M)_{x_n}$. Since the endpoint map Y of M restricted to the disk normal bundle of radius s is proper and $Y(x_n, b_n - x_n) = b_n \to a$, there is a subsequence of x_n converging to a point x_0 in M. Then it is easily seen that $r \leq \|x_0 - a\| \leq s$ and x_0 is a critical point of f_a, a contradiction. ∎

8.2.9. Proposition.   *Let M be an immersed PF submanifold of V, and $a \in V$. Suppose $r < s$ and $[r, s] \subseteq R(f_a)$. Then there exist $\delta_1 > 0$, $\delta_2 > 0$ such that if $\|b - a\| < \delta_1$ and $x \in M_s(f_b) \setminus M_r(f_b)$ then $\|\nabla f_b(x)\| \geq \delta_2$.*

PROOF.   By Proposition 8.2.8 there exists $\delta > 0$ such that $[r, s] \subseteq R(f_b)$ if $\|b - a\| < \delta$. Suppose no such δ_1 and δ_2 exist. Then there exist sequences b_n in V and x_n in M such that $b_n \to a$, $x_n \in M_s(f_{b_n}) \setminus M_r(f_{b_n})$, and $\|\nabla(f_{b_n})(x_n)\| \to 0$. Then

$$Y(x_n, -(x_n - b_n)^\nu) = x_n - (x_n - b_n)^\nu$$
$$= b_n + (x_n - b_n)^{TM_{x_n}} \to a,$$

and $\|x_n - b_n\| \leq s$. Since M is PF, x_n has a subsequence converging to a critical point x_0 of f_a in $M_s(f_a) - M_r(f_a)$, a contradiction. ∎

8.2.10. Proposition.   *Let M be an immersed, taut submanifold of a Hilbert space V, $a \in V$, and $r \in R$. Then the induced map on homology*

$$i_* : H_*(M_r(f_a), F) \to H_*(M, F)$$

of the inclusion of $M_r(f_a)$ in M is injective.

PROOF.   If a is a non-focal point (so f_a is non-degenerate), then it follows from the definition of tautness and Theorem 8.2.3 that i_* is an injection. Now suppose a is a focal point. If there is no critical value of f_a in $(r, r']$, then $M_r(f_a)$ is a deformation retract of $M_{r'}(f_a)$. So we may assume that r is a regular value of f_a, i.e., $r \in R(f_a)$. Then there exists $s > r$ such that $[r, s] \subseteq R(f_a)$. Choose $\delta_1 > 0$ and $\delta_2 > 0$ as in Proposition 8.2.9, and $\epsilon > 0$ such that $\epsilon < \min\{\delta_1, \delta_2, (s - r)/5\}$. Since the set of non-focal points of M in V is open and dense, there exists a non-focal point b such that $\|b - a\| < \epsilon$. Since f_b is non-degenerate, it follows from the definition of tautness that $i_* : H_*(M_t(f_b), F) \to H_*(M, F)$ is injective for all t. So it suffices to prove that $M_r(f_a)$ is a deformation retract of $M_r(f_b)$. Since $\epsilon < (s - r)/5$, there exist r_1, r_2, s_1 and s_2 such that $r_1 < s_1$, $r_2 < s_2$ and

$$r < r_1 - \epsilon < s_1 + \epsilon < s, \quad r_1 < r_2 - \epsilon < s_2 < s_2 + \epsilon < s_1.$$

From triangle inequality we have

$$M_{s_2}(f_b) - M_{r_2}(f_b) \subset M_{s_1}(f_a) - M_{r_1}(f_a) \subseteq M_s(f_b) - M_r(f_b).$$

Note that $\|\nabla f_a(x)\| \geq \delta_2$ if $x \in M_s(f_a) - M_r(f_a)$ and $\|\nabla f_b(x)\| > \delta_2$ if $x \in M_s(f_b) - M_r(f_b)$. Recall that $\nabla f_a(x) = (x-a)^T$ and $\nabla f_b(x) = (x-b)^T$. Since $\epsilon < \delta_2$, $(a-b)^T$ is the shortest side of the triangle with three sides $(x-a)^T$, $(x-b)^T$ and $(a-b)^T$ for all x in $M_{s_1}(f_a) - M_{r_1}(f_a)$. Using the cosine formula for the triangle we have

$$\langle \nabla f_a(x), \nabla f_b(x) \rangle > \frac{2\delta_2^2 - \epsilon^2}{2} > \frac{\epsilon^2}{2},$$

for x in $M_{s_1}(f_a) - M_{r_1}(f_a)$. Hence the gradient flow of f_a gives a deformation retract of $M_{s_1}(f_a)$ to $M_{s_2}(f_b)$. If $[r,s] \subseteq R(f)$, then $M_r(f)$ is a deformation retract of $M_t(f)$ for all $t \in [r,s]$, which proves our claim. ∎

8.2.11. Corollary.   *If M is connected and $\varphi : M \to V$ is a taut immersion then φ is an embedding.*

PROOF.   Since M is PF, $\varphi = Y|M \times 0$ is proper. So it suffices to prove that φ is one to one. Suppose $\varphi(p) = \varphi(q) = a$. If $p \neq q$ then there exists $\epsilon > 0$ such that $(0,\epsilon) \subseteq R(f_a)$ and p,q are in two different connected components of $M_\epsilon(f_a)$. This contradicts to the fact that $i_0 : H_0(M_\epsilon(f_a), F) \to H_0(M, F)$ is injective. ∎

8.2.12. Corollary.   *Suppose M is a connected taut submanifold of V, $a \in V$, and let $D_r(a)$ denote the closed ball of radius r and center a in V.*
  (i) For any $r \in R$ the set $M_r(f_a)$ is connected, or equivalently, $M \cap D_r(a)$ is connected.
  (ii) If x_o is an index 0 critical point of f_a then $f_a(x_o)$ is the absolute minimum of f_a; in particular a local minimum of f_a is the absolute minimum.
  (iii) If x_o is an isolate critical point of f_a with index 0 and $r_o = f_a(x_o)$, then $M_{r_o}(f_a) = \{x_o\}$, i.e., $\{x_o\} = M \cap D_{r_o}(a)$.

PROOF.   By Proposition 8.2.10, the map

$$i_0 : H_0(M_r(f_a), F) \to H_0(M, F)$$

is injective. Since $H_0(M, F) = F$, $M_r(f_a)$ is connected, which proves (i).

Next we prove (iii) for non-degenerate index 0 critical point. Let x_o be a non-degenerate index 0 critical point of f_a and $r_o = f_a(x_o)$. Then there is an open neighborhood U of x_o such that $M_{r_o}(f_a) \cap U = \{x_o\}$. Since $M_{r_o}(f_a)$ is connected, $M_{r_o}(f_a) = \{x_o\}$. In particular, r_o is the absolute minimum of f_a, i.e.,

$$\|x - a\| \geq \|x_o - a\|.$$

If x_o is a degenerate critical point, then there is $v \in \nu(M)_{x_o}$ such that $a = x_o + v$ and

$$\text{Hess}(f_a, x_o) = I - A_v \geq 0.$$

Let $a_t = a + tv$ for $0 < t < 1$. Then

$$\text{Hess}(f_{a_t}, x_o) = I - tA_v > 0.$$

So x_o is a non-degenerate index 0 critical point of f_{a_t}. But we have just shown that

$$\|x - a_t\| \geq \|x_o - a_t\| \tag{8.2.1}$$

for all $x \in M$. Letting $t \to 1$ in (8.2.1), we obtain (ii). ∎

8.3 Homology of isoparametric submanifolds.

In this section we use Morse theory to calculate the homology of isoparametric submanifolds of Hilbert spaces and prove that they are taut.

Let f be a Morse function on a Hilbert manifold M, q a critical point of f of index m. In Chapter 10 of Part II we define a pair (N, φ) to be a *Bott-Samelson cycle* for f at q if N is a smooth m-dimensional manifold and $\varphi : N \to M$ is a smooth map such that $f \circ \varphi$ has a unique and non-degenerate maximum at y_0, where $\varphi(y_0) = q$. (N, φ) is $\mathcal{R}$-orientable for a ring $\mathcal{R}$, if $H_m(N, \mathcal{R}) = \mathcal{R}$. We say f is of Bott-Samelson type with respect to $\mathcal{R}$ if every critical point of f has an $\mathcal{R}$-orientable Bott-Samelson cycle. Moreover if $\{q_i | i \in I\}$ is the set of critical points of f and (N_i, φ_i) is an $\mathcal{R}$-orientable Bott-Samelson cycle for f at q_i for $i \in I$ then $H_*(N, \mathcal{R})$ is a free module over $\mathcal{R}$ generated by the descending cells $(\varphi_i)_*([N_i])$, which implies that f is of linking type, and f is perfect.

8.3.1. Theorem.   *Let M be an isoparametric submanifold of the Hilbert space V, and x_0 a critical point of the Euclidean distance function f_a. Then*
   (i) there exist a parallel normal field v on M and finitely many curvature normals v_i such that $a = x_0 + v(x_0)$ and $\langle v, v_i \rangle > 1$,
   (ii) if

$$\langle v, v_r \rangle \geq \cdots \geq \langle v, v_1 \rangle > 1 > \langle v, v_{r+1} \rangle \geq \langle v, v_{r+2} \rangle \geq \cdots,$$

then
   (1) $\bigoplus \{E_i(x_0) | i \leq r\}$ is the negative space of f at x_0,
   (2) (N_r, u_r) is an $\mathcal{R}$-orientiable Bott-Samelson cycle at x_0 for f, where

$$N_r = \{(y_1, \ldots, y_r) | y_1 \in S_1(x_0), y_2 \in S_2(y_1), \ldots, y_r \in S_r(y_{r-1})\},$$

$$u_r : N_r \to M, \quad u_r(y_1, \ldots, y_r) = y_r,$$

and $S_i(x)$ is the leaf of E_i through x. Here $\mathcal{R} = \mathbf{Z}$ if all $m_i > 1$, and $\mathcal{R} = \mathbf{Z}_2$ otherwise.

PROOF. Since x_0 is a critical point of f_a, by Proposition 7.1.10 $a - x_0 \in \nu(M)_{x_0}$. Let v be the parallel normal field on M such that $v(x_0) = a - x_0$. Then (i) follows form the fact that the shape operator A_v is compact, the eigenvalues of A_v are $\langle v, v_i \rangle$, and $\nabla^2 f_a(x_0) = I - A_{(a-x_0)}$.

For (ii) it suffices to prove the following three statements:

(a) $y_0 = (x_0, \ldots, x_0)$ is the unique maximum point of $f \circ u_r$.

(b) $d(u_r)_{y_0}$ maps $T(N_r)_{y_0}$ isomorphically onto the negative space of f at x_0.

(c) If all $m_i > 1$, then (N_r, u_r) is $\mathbf{Z}$-orientable.

To see (b), let $N = N_r$, we note that N is contained in the product of r copies of M, $TN_{y_0} = \bigoplus\{F_i | i \le r\}$, where $F_i = (0, , \ldots, E_i(x_0), \ldots, 0)$ is contained in $\bigoplus\{TM_{x_0} | i \le r\}$ and $d(u_r)_{y_0}$ maps F_i isomorphically onto $E_i(x_0)$.

It follows from the definition of N_r that it is an iterated sphere bundle. The homotopy exact sequence for the fibrations implies that if the fiber and the base of a fibration are simply connected then the total space is also simply connected. Hence by induction the iterated sphere bundle N_r is simply connected, which proves (c).

Statement (a) follows from the lemma below. ∎

8.3.2. Lemma.   *We use the same notation as in Theorem 8.3.1. Then for any $q = (y_1, , \ldots, y_r)$ in N_r there is a continuous piecewise smooth geodesic α_q in V joining a to y_r such that the length of α_q is $\|x_0 - a\|$, and α_q is smooth if and only if $q = (x_0, \ldots, x_0)$.*

PROOF. Let $[xy]$ denote the line segment joining x and y in V. Let $\{z_i\} = \ell_i(x_0) \cap [ax_0]$. Then

$$[ax_0] = [az_1] \cup [z_1 z_2] \cup \ldots [z_r x_0].$$

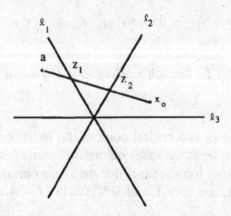

Let $a_i = (y_i + v(y_i))$, and $z_j(i) \in \ell_j(y_i) \cap [y_i a_i]$. Then $z_1(1) = z_1$ and $z_j(j-1) = z_j(j)$. Since $y_j \in S_j(y_{j-1})$ and $z_j(j-1) = z_j(j)$,

$$\alpha_q = [a z_1] \cup [z_1(1), z_2(1)] \cap [z_2(2), z_3(2)] \cup \ldots \cup [z_r(r), y_r]$$

satisfies the properties of the lemma.   ∎

8.3.3. Corollary.
Let M be an immersed isoparametric submanifold in a Hilbert space V with multiplicities m_i, and $a \in V$ a non-focal point of M. Then

(i) f_a is of Bott-Samelson type with respect to the ring $\mathcal{R} = \mathbf{Z}$, if all the multiplicities $m_i > 1$, and with respect to $\mathcal{R} = \mathbf{Z}_2$ otherwise,

(ii) M is taut.

It follows from Corollary 8.2.11 that:

8.3.4. Corollary.
An immersed isoparametric submanifold of a Hilbert space V is embedded.

To obtain more precise information concerning the homology groups of isoparametric submanifolds, we need to know the structure of the set of critical points of f_a. By The Morse Index Theorem (see Part II) we have

8.3.5. Proposition.
Let $M \subseteq V$ be isoparametric, and W its associated Coxeter group. Let $a \in V$, and let $C(f_a)$ denote the set of critical points of f_a.

(i) If $x_0 \in C(f_a)$ then $W \cdot x_0 \subseteq C(f_a)$, where $W \cdot x_0$ is the W-orbit through x_0 on $x_0 + \nu(M)_{x_0}$,

(ii) If $q \in C(f_a)$ then the index of f_a at q is the sum of the m_i's such that the open line segment (q, a) joining q to a meets $\ell_i(q)$.

Let $\nu_x = x + \nu(M)_x$. Then the closure of a connected component of $\nu_x \setminus \bigcup \{\ell_i(q) | i \in I\}$ is a Weyl chamber for the Coxeter group W-action on ν_x. In the following we let $\triangle_q$ denote the Weyl chamber of W on ν_q containing q. As a consequence of Proposition 8.3.5 and Corollary 8.2.12, we have

8.3.6. Proposition. *Suppose M is isoparametric in a Hilbert space and let $q \in M$. Let $\triangle_q$ be the Weyl chamber in $\nu_q = (q + \nu(M)q)$ containing q, and $a \in \triangle_q$.*
Then:

(i) q is a critical point of f_a with index 0,

(ii) $f_a(q)$ is the absolute minimum of f_a,

(iii) if a is non-focal with respect to q, then $f_a^{-1}(f_a(q)) = \{q\}$,

(ii) if a is a focal point with respect to q and a lies on the simplex σ of $\triangle_q$, then $f_a^{-1}(f_a(q)) = S_{q,\sigma}$ (as in the Slice Theorem 6.5.9).

8.3.7. Theorem. *Let M be an isoparametric submanifold of V, and $a \in \nu_q \cap \nu_{q'}$. Then a is non-focal with respect to q if and only if a is non-focal with respect to q', and $q' \in W \cdot q$.*

PROOF. There are $p \in W \cdot q$ and $p' \in W \cdot q'$ such that $a \in \triangle_p$ and $a \in \triangle_{p'}$. By Proposition 8.3.5 (ii), both p and p' are critical points of f_a with index 0. So by Proposition 8.3.6, $f_a(p) = f_a(p')$ is the absolute minimum of f_a. If a is non-focal with respect to q then by Proposition 8.3.6 (iii), we have $p = p'$ and a is non-focal with respect to p'. ∎

8.3.8. Corollary. *Let $M \subseteq V$ be isoparametric, W its associated Coxeter group. If $a \in V$ is non-focal with respect to q in M, then $C(f_a) = W \cdot q$.*

8.3.9. Corollary. *Let $M \subseteq V$ be isoparametric. Then $H_*(M, \mathcal{R}))$ can be computed explicitly in terms of the associated Coxeter group W and its multiplicities m_i. Here $\mathcal{R}$ is $\mathbf{Z}$ if all $m_i > 1$ and is $\mathbf{Z}_2$ otherwise.*

8.3.10. Corollary. *Let $M^n \subseteq R^{n+k}$ be isoparametric. Then*

$$\sum_i \text{rank}(H_i(M, \mathcal{R})) = |W|,$$

the order of W.

8.3.11. Corollary. *Let $M \subseteq V$ be isoparametric. A point $a \in V$ is non-focal with respect to $q \in M$ if and only if a is a regular point with respect to the W-action on ν_q.*

8.3.12. Corollary. *Let $M \subseteq V$ be isoparametric. If f_a has one non-degenerate critical point then f_a is non-degenerate, or equivalently if $a \in \nu_q$ is non-focal with respect to q then a is non-focal with respect to M.*

Let $a \in V$. Since f_a is bounded from below and satisfies condition C on M, f_a assumes its minimum, say at q. So $a \in \nu_q$, i.e.,

8.3.13. Proposition. *Let $M \subseteq V$ be isoparametric, and $Y : \nu(M) \to V$ the endpoint map. Then $Y(\nu(M)) = V$.*

8.4 Rank 2 isoparametric submanifolds in R^m.

In this section we will apply the results we have developed for isoparametric submanifolds of arbitrary codimension to a rank 2 isoparametric submanifold M^n of R^{n+2}. Because of Corollaries 6.3.12 and 6.3.11, we may assume that M is a hypersurface of S^{n+1}.

Let $X : M^n \to S^{n+1} \subseteq R^{n+2}$ be isoparametric, and e_{n+1} the unit normal field of M in S^{n+1}. Suppose M has p distinct principal curvatures $\lambda_1, \ldots, \lambda_p$ as a hypersurface of S^{n+1} with multiplicities m_i. Then

$$e_\alpha = e_{n+1}, \quad e_\beta = X$$

is a parallel normal frame on M, and the reflection hyperplanes $\ell_i(q)$ on $\nu_q = q + \nu(M)_q$ (we use q as the origin, $e_\alpha(q)$ and $e_\beta(q)$ are the two axes) are given by the equations:

$$\lambda_i z_\alpha - z_\beta = 1, \quad 1 \le i \le p.$$

The Coxeter group W associated to M is generated by reflections in ℓ_i. By the classification of rank 2 Coxeter groups, W is the Dihedral group of order $2p$. So we may assume that

$$\lambda_i = \cot\left(\theta_1 + \frac{(i-1)\pi}{p}\right), \quad 1 \le i \le p,$$

for some θ_1, where $-\pi/p < \theta_1 < 0$. This fact was proved by Cartan ([Ca3]). Let R_i denote the reflections of ν_q in $\ell_i(q)$. It is easily seen that

$$R_{i+1}(\ell_i) = \ell_{i+2},$$

if we let $\ell_{p+i} = \ell_i$ for $1 \le i < p$. By Theorem 6.3.2, we obtain the following result of Münzner:

$$m_1 = m_3 = \ldots,$$
$$m_2 = m_4 = \ldots.$$

In particular, if p is odd then all the multiplicities are equal. So the possible marked Dynkin diagrams for a rank 2 isoparametric submanifold of the Euclidean space are

$Z_2 \times Z_2$ ∘   ∘
                       m_1 m_2

A_2 ∘—∘
                      m_1 m_2

B_2 ∘═∘
                      m_1 m_2

G_2 ∘≡∘
                      m_1 m_2

Note that the intersection of $\ell_i(q)$ and the normal geodesic circle of M in S^{n+1} at q has exactly two points, which will be denoted by x_i and y_i, i.e.,

$$x_i = \cos\theta_i\, q + \sin\theta_i\, e_\alpha(q),$$
$$y_i = \cos(\pi + \theta_i)\, q + \sin(\pi + \theta_i)\, e_\alpha(q),$$

where $\theta_i = \theta_1 + \frac{(i-1)\pi}{p}$.

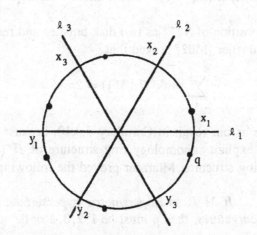

Let $\triangle_q$ denote the Weyl chamber of W on ν_q containing q. Then the intersection of $\triangle_q$ and the normal geodesic circle of M in S^{n+1} at q is the arc joining x_1 to y_p. Let M_t denote the parallel submanifold of M through $\cos t\, q + \sin t\, e_\alpha(q)$. Then

$$\bigcup \{M_t \mid -\pi/p + \theta_1 \leq t \leq \theta_1\} = S^{n+1}.$$

Note that M_t is diffeomorphic to M and is an embedded isoparametric hypersurface of S^{n+1} if $-\pi/p + \theta_1 \leq t \leq \theta_1$. And the focal set Γ of M in S^{n+1} has exactly two sheets, $M_1 = M_{\theta_1}$ and $M_2 = M_{(-\pi/p+\theta_1)}$, so they are also called the focal submanifolds of M. The dimension of M_i is $n - m_i$ for $i = 1, 2$. Let v_i be the parallel normal fields on M such that $x_1 = q + v_1(q)$, $y_p = q + v_2(q)$. Then $M_i = M_{v_i}$ the parallel submanifold. So by Proposition 6.5.1, $\pi_i : M \to M_i$ defined by $\pi_i(x) = x + v_i(x)$ is a fibration and M is a S^{m_i}-bundle over M_i.

Let B_i be the normal disk bundle of radius r_i of M_i in S^{n+1}, where $r_1 = \theta_1$ and $r_2 = \pi/p - \theta_1$. So

$$B_i = \{\cos t\, x + \sin t\, v(x) \mid |t| \leq r_i,\ v \text{ is normal to } M \text{ in } S^{n+1}\},$$

and $\partial B_i = M$. Next we claim that $B_1 \cup B_2 = S^{n+1}$. To see this, let $a \in S^{n+1}$. Since M is compact, f_a assumes minimum, say at x_0. So $a = \cos t\, x_0 + \sin t\, e_\alpha(x_0)$ for some t. Because x_0 is the minimum of f_a, a must lie in the Weyl chamber $\triangle_{x_0}$ of W on ν_{x_0}, i.e., $-r_2 < t < r_1$. So $a \in B_1$ if $0 \leq t \leq r_1$ and $a \in B_2$ if $-r_2 \leq t \leq 0$. This proves the following results of Münzner [Mü1]:

$$B_1 \cup B_2 = S^{n+1},$$
$$\partial B_1 = \partial B_2 = B_1 \cap B_2 = M,$$
$$B_i \text{ is a } (m_i + 1) - \text{disk bundle over } M_i.$$

Using this decomposition of S^{n+1} as two disk bundles and results from algebraic topology, Münzner [Mü2] proved that

$$\sum_i \text{rank}(H_i(M)) = 2p,$$

which is the same as our result in Corollary 8.3.10, because $|W| = 2p$. He also obtained the explicit cohomology ring structure of $H^*(M, \mathbf{Z}_2)$. Using the cohomology ring structure, Münzner proved the following:

8.4.1. Theorem.   *If M is an isoparametric hypersurface of S^{n+1} with p distinct principal curvatures, then p must be $1, 2, 3, 4$ or 6.*

Next we state some restrictions on the possible multiplicities m_i. The first result of this type was proved by É. Cartan:

8.4.2. Theorem. *If M^n is an isoparametric hypersurface of S^{n+1} with three distinct principal curvatures, then $m_1 = m_2 = m \in \{1, 2, 4, 8\}$.*

Using delicate topological arguments, Münzner [Mü2] and Abresch [Ab] obtained restrictions on the m_i's for the case of $p = 4$ and $p = 6$. First we make a definition:

8.4.3. Definition. A pair of integers (m_1, m_2) is said to satisfy condition (*) if one of the following hold:
  (a) 2^k divides $(m_1 + m_2 + 1)$, where $2^k = \min\{2^\sigma \mid m_1 < 2^\sigma, \ \sigma \in N\}$,
  (b) if m_1 is a power of 2, then $2m_1$ divides $(m_2 + 1)$ or $3m_1 = 2(m_2 + 1)$.

8.4.4. Theorem. *Suppose M^n is isoparametric in S^{n+1} with p distinct principal curvatures.*
(i) If $p = 4$ and $m_1 \leq m_2$, then (m_1, m_2) must satisfy condition (*).
(ii) If $p = 6$, then $m_1 = m_2 \in \{1, 2\}$.

We will omit the difficult proof of these results and instead refer the reader to [Mü2] and [Ab].
  As consequence of Theorem 8.4.1, we have:

8.4.5. Theorem. *If M is a rank 2 isoparametric submanifold of Euclidean space, then the Coxeter group W associated to M is crystallographic, i.e., $W = A_1 \times A_1$, A_2, B_2, or G_2.*

8.4.6. Theorem. *If M is an irreducible rank 2 isoparametric submanifold of Euclidean space, then the marked Dynkin diagram associated to M must be one of the following:*

A_2 　　　　　$\underset{m \quad\ \ m}{\circ\!\!-\!\!\!-\!\!\!-\!\!\circ}$ 　　　　　$m \in \{1, 2, 4, 8\}$

B_2 　　　　　$\underset{m_1 \quad\ m_2}{\circ\!\!=\!\!\!=\!\!\circ}$ 　　　　　(m_1, m_2) satisfies (*)

G_2 　　　　　$\underset{m \quad\ \ m}{\circ\!\!\equiv\!\!\!\equiv\!\!\circ}$ 　　　　　$m \in \{1, 2\}$

8.5   Parallel foliations.

In section 7.2 we noted that most results of Chapter 6 work for infinite dimensional isoparametric submanifolds. Although the proof of the existence of parallel foliation for finite dimensional isoparametric submanifolds does not work in the infinite dimensional case, the topological results of section 8.3 lead to the existence of parallel foliation.

Let M be a PF submanifold of V with flat normal bundle, and Y the end point map of M. In general, the parallel set,

$$M_v = \{Y(v(x)) = x + v(x) \mid x \in M\},$$

defined by a parallel normal field v, may be a singular set, and $\mathcal{F} = \{M_v \mid v \text{ is a parallel normal field on } M\}$ need not foliate V. The main result of this section is that if M is isoparametric, then each M_v is an embedded submanifold of V and $\mathcal{F}$ gives an orbit-like singular foliation on V.

In what follows M is a rank k isoparametric submanifold of a Hilbert space V, $\nu_q = q + \nu(M)_q$ and $\triangle_q$ is the Weyl chamber of W on ν_q containing q.

8.5.1. Proposition. $M \cap \nu_q = W \cdot q$.

PROOF. It is easily seen that $W \cdot q \subseteq \nu_q$. Now suppose that $b \in M \cap \nu_q$. Then $b \in \nu_b \cap \nu_q$. But b is non-focal with respect to b, so it follows from Theorem 8.3.7 that we have $b \in W \cdot q$. ∎

8.5.2. Proposition.   *Suppose σ is a simplex of $\triangle_q$ and σ' is a simplex of $\triangle_{q'}$. If $\sigma \cap \sigma' \neq \emptyset$ then $\sigma = \sigma'$, and the slices $S_{q,\sigma}$ and $S_{q',\sigma'}$ are equal.*

PROOF.   Suppose $a \in \sigma \cap \sigma'$. Then q and q' are critical points of f_a with index zero, nullities $m_{q,\sigma}$, $m_{q',\sigma'}$, and critical submanifolds $S_{q,\sigma}$, $S_{q',\sigma'}$ of f_a at q and q' respectively. So it follows from Proposition 8.3.6 that $S_{q,\sigma} = S_{q',\sigma'}$. It then follows from the Slice Theorem 6.5.9 that we have $\sigma = \sigma'$. ∎

8.5.3. Proposition.   *Let σ be a simplex of a Weyl chamber in ν_q, $\varphi \in W$, and $S_{x,\sigma}$ the slice as in Theorem 6.5.9. Then $\varphi(S_{q,\sigma}) = S_{\varphi(q),\sigma}$.*

PROOF.   Using Theorem 6.5.9, we see that $S_{q,\sigma}$ is the leaf of the distribution $\bigoplus\{E_j \mid j \in I(q,\sigma)\}$ through q. But both $\varphi(S_{q,\sigma})$ and $S_{\varphi(q),\sigma}$ are the leaves of the distribution $\bigoplus\{E_j \mid j \in I(\varphi(q),\sigma)\}$ through $\varphi(q)$. So $\varphi(S_{q,\sigma}) = S_{\varphi(q),\sigma}$. ∎

8.5.4. Theorem.   *Let M be a rank k isoparametric submanifold of V, σ a simplex of $\triangle_q$ of dimension less than k, and $a \in \sigma$. Then f_a is nondegenerate in the sense of Bott, and the set $C(f_a)$ of critical points of f_a is $\bigcup\{S_{x,\sigma} | x \in W \cdot q\}$.*

PROOF.   Let $x \in W \cdot q$. Then x is a critical point of f_a with nullity $m_{x,\sigma}$ and $S_{x,\sigma}$ is the critical submanifold of f_a through x. Hence $S_{x,\sigma} \subseteq C(f_a)$. Conversely, if $y \in C(f_a)$ then $a \in \nu_y$. By Theorem 8.3.7, a is a focal point with respect to y. so there exists $\varphi \in W$ such that $\varphi^{-1}(y) = y_0$, and a simplex σ' in the Weyl chamber $\triangle_{y_0}$ on ν_{y_0} such that $a \in \sigma'$. Then it follows from Proposition 8.5.2 that $\sigma = \sigma'$ and $S_{q,\sigma} = S_{y_0,\sigma}$. Thus we have

$$\varphi(S_{q,\sigma}) = S_{\varphi(q),\sigma} = \varphi(S_{y_0,\sigma}) = S_{\varphi(y_0),\sigma} = S_{y,\sigma}. \quad \blacksquare$$

8.5.5. Theorem.   *Let M be an isoparametric submanifold in V, $q \in M$, and $\triangle_q$ the Weyl chamber of W on ν_q containing q. Let v be in $\nu(M)_q$, $\tilde{v}$ the parallel normal vector field on M determined by $\tilde{v}(q) = v$, and M_v the parallel submanifold $M_{\tilde{v}}$, i.e.,*

$$M_v = \{x + \tilde{v}(x) | \ x \in M\}.$$

Then:

(i) if $v \neq w$, and $q + v$ and $q + w$ are in $\triangle_q$, then M_v and M_w are disjoint,

(ii) given any $a \in V$ there exists a unique $v \in \nu(M)_q$ such that $q + v \in \triangle_q$ and $a \in M_v$,

(iii) each M_v is an embedded submanifold of V.

PROOF.   Suppose $(q + v)$, $(q + w)$ are in $\triangle_q$, and $M_v \cap M_w \neq \emptyset$. Let $a \in M_v \cap M_w$ then there exist $x, y \in M$ such that $a = x + \tilde{v}(x) = y + \tilde{w}(y)$. Since $a \in \triangle_q$ and $\tilde{v}$, $\tilde{w}$ are parallel, $\langle \tilde{v}, v_i \rangle$ and $\langle \tilde{w}, v_i \rangle$ are constant. So $a \in \triangle_x$ and $a \in \triangle_y$, which imply that x and y are critical points of f_a with index 0. If a is non-focal then $x = y$, so by Proposition 8.3.6 we have $v = w$. If a is focal (suppose a lies in the simplex σ of $\triangle_q$) then the two critical submanifolds $S_{x,\sigma}$ and $S_{y,\sigma}$ are equal. In particular, $y \in S_{x,\sigma}$. Using the same notation as in the Slice Theorem 6.5.9, we note that the slice $S_{x,\sigma}$ is a finite dimensional isoparametric submanifold in $x + \eta(\sigma) \subset a + \nu(M_v)_a$. Let $v = u_1 + u_2$, where u_1 is the orthogonal projection of v along $V(\sigma)$. Then $S_{x,\sigma}$ is contained in the sphere of radius $\|u_1\|$ and centered at $x + u_1$. So $y + \tilde{u}_1(y) = x + u_1$. Since $V(\sigma)$ is perpendicular to $S_{x,\sigma}$, $\tilde{u}_2(y) = u_2$. Therefore we have $y + \tilde{v}(y) = x + \tilde{v}(x) = a = y + \tilde{w}(y)$, which implies that $v = w$

To prove (ii), we note that since f_a is bounded from below and satisfies condition C, there exists $x_0 \in M$ such that $f_a(x_0)$ is the minimum. Then $a \in \Delta_{x_0}$, so there exists a parallel normal field $\tilde{v}$ such that $\tilde{v}(x_0) = a - x_0$.

If $x, y \in M_v$ and $x + \tilde{v}(x) = y + \tilde{v}(y) = b$, then both x and y are critical points of f_b with index 0. Then, by Proposition 8.3.6 (iii), $f_b(x) = f_b(y)$ is the absolute minimum of f_b and if b is non-focal then, by 8.3.6 (ii), $x = y$. If b is a focal point of M, then a is a focal point with respect to both x and y by Theorem 8.3.7. Suppose a lies in a simplex σ of Δ_x. Then by Proposition 8.3.6 again, $y \in N_{x,\sigma}$. Since $S_{x,\sigma}$ is isoparametric in $\eta(\sigma)$, it is an embedded submanifold, i.e., $x = y$.   ∎

8.5.6. Corollary.   Let M be an isoparametric submanifold of V and $q \in M$. Then $\mathcal{F} = \{M_v| \ q + v \in \Delta_q\}$ defines an orbit-like singular foliation on V, which will be called the isoparametric foliation of M. The leaf space of $\mathcal{F}$ is isomorphic to the orbit space ν_q/W.

8.5.7. Corollary.   If $a \in \sigma \subseteq \Delta_q$ and $a = q + v$, then the isoparametric foliation of $S_{q,\sigma}$ in $(a + \nu(M_v)_a)$ is $\{M_u \cap (a + \nu(M_u)_a)| \ M_u \in \mathcal{F}\}$.

8.6   Convexity theorem.

A well-known theorem of Schur ([Su]) can be stated as follows: Let M be the set of $n \times n$ Hermitian matrices with eigenvalues $a_1, \ldots, a_n$, and $u : M \to \mathbf{R}^n$ the map defined by $u((x_{ij})) = (x_{11}, \ldots, x_{nn})$. Then $u(M)$ is contained in the convex hull of $S_n \cdot a$, where S_n is the symmetric group acting on $\mathbf{R}^n$ by permuting the coordinates. Conversely, A. Horn ([Hr]) showed that the convex hull of $S_n \cdot a$ is contained in $u(M)$. Hence we have

8.6.1. Theorem. $u(M) = \text{cvx}(S_n \cdot a)$, the convex hull of $S_n \cdot a$.

Note that Theorem 8.6.1 can be viewed as a theorem about a certain symmetric space, because M is an orbit of the isotropy representation of the symmetric space $SL(n, C)/SU(n, C)$, and u is the orthogonal projection onto a maximal abelian subspace. B. Kostant ([Ks]) generalized this to any symmetric space; his result is :

8.6.2. Theorem.   Let G/K be a symmetric space, $\mathcal{G} = \mathcal{K} + \mathcal{P}$ the corresponding decomposition of the Lie algebra, $\mathcal{T}$ a maximal abelian subspace of $\mathcal{P}$, $W = N(T)/Z(T)$ the associated Weyl group of G/K acting on T, and $u : \mathcal{P} \to T$ the linear orthogonal projection onto T. Let M be an orbit of the isotropy representation of G/K through z, i.e., $M = Kz$. Then $u(M) = \operatorname{cvx}(W \cdot z)$.

The isotropy action of the compact symmetric space $G \times G/G$ is just the adjoint action of G on $\mathcal{G}$. Moreover, if we identify $\mathcal{G}$ to its dual $\mathcal{G}^*$ via the Killing form, then these orbits have a natural symplectic structure. In this case, the map u in Theorem 8.6.2 is the moment map. Recently, Theorem 8.6.2 has been generalized in the framework of symplectic geometry by Atiyah ([At]) and independently by Guillemin and Sternberg ([GS]) to the following:

8.6.3. Theorem. *Let M be a compact connected symplectic manifold with a symplectic action of a torus T, and $f : M \to T^*$ the moment map. Then $f(M)$ is a convex polyhedron.*

The orbits that occur in Kostant's theorem 8.6.2, are isoparametric. Moreover, as we shall now see, it turns out that the convexity result follows just from this geometric condition of being isoparametric. Since there are infinitely many families of rank 2 isoparametric submanifolds that are not orbits of any linear orthogonal representation, the Riemannian geometric proof of Theorem 8.6.2 gives a more general result [Te3].

8.6.4. Main Theorem. *Let $M^n \subseteq S^{n+k-1} \subseteq R^{n+k}$ be isoparametric, $q \in M$, and W the associated Weyl group of M. Let P denote the orthogonal projection of R^{n+k} onto the normal plane $\nu_q = q + \nu(M)_q$, and $u = P|M$ the restriction of P to M. Then $u(M) = \operatorname{cvx}(W \cdot q)$, the convex hull of $W \cdot q$.*

As we said above, our main tool for proving this is Riemannian geometry. However, the basic idea of the proof goes back to Atiyah ([At]), and Guillemin-Sternberg ([GS]). Although there is no symplectic torus action around, the height function of M plays the role of the Hamiltonian function in their symplectic proofs. In section 8.3, we showed that M is taut (Corollary 8.3.3), i.e., every non-degenerate Euclidean distance function f_a on M is perfect. Because $M \subseteq S^{n+k-1}$, the height function h_a and $-1/2f_a$ differ by a constant, i.e.,

$$f_a = -2h_a + (1 + \|a\|^2).$$

In particular f_a and h_a have the same critical point theory. Using our detailed knowledge of the Morse theory of these height functions, Theorem 8.6.4 can be proved rather easily. It seems that tautness and convexity are closely related, however, the precise relation is not yet clear.

Henceforth we assume that $M^n \subseteq S^{n+k-1} \subseteq R^{n+k}$ is isoparametric, W is its Weyl group, and we use the same notations as in Chapter 6. In particular, for $x \in M$, we let $\triangle_x$ denote the Weyl chamber on $\nu_x = x+\nu(M)_x$ containing x. First we recall following results concerning the height functions.

8.6.5. Theorem.   *Let $a \in R^{n+k}$ be a fixed non-zero vector, $h_a : M \to R$ the associated height function, i.e., $h_a(x) = \langle x, a \rangle$, and let $C(h_a)$ denote the set of all critical points of h_a.*

(i) $x \in C(h_a)$ if and only if $a \in \nu_x$.

(ii) If x_0 is an index 0 critical of h_a, then $b = h_a(x_0)$ is the absolute minimum value of h_a on M and $h_a^{-1}(b)$ is connected. Moreover,

(1) $a \in \triangle_{x_0}$,

(2) if $a \in \sigma$, a simplex of $\triangle_{x_0}$, then $h_a^{-1}(b) = S_{x_0,\sigma}$ (the slice through x_0 with respect to σ).

(iii) If $x \in C(h_a)$ and a is regular with respect to the W-action on ν_x, then h_a is non-degenerate and $C(h_a) = W \cdot x$.

(iv) If a is W-singular, then h_a is non-degenerate in the sense of Bott ([Bt]). More specifically, if $x^0 \in C(h_a)$ is a minimum and a lies on the simplex σ of $\triangle_{x^0}$, then

$$C(h_a) = \bigcup \{ S_{x,\sigma} \mid x \in W \cdot x^0 \}.$$

8.6.6. Lemma.   *We use the same notation as in Theorem 8.6.4. Let $u = P|M$, the restriction of P to M, and C the set of all singular points of u. Then C is the union of all slices $S_{x,\sigma}$ for x in $W \cdot q$ and σ a 1-simplex of some Weyl chamber of ν_q.*

PROOF.   We may assume that $\nu_q = R^k$. Let $t_1, \ldots, t_k$ be the standard base of R^k. Then $u = (u_1, \ldots, u_k)$, where $u_i(x) = h_{t_i}(x) = \langle x, t_i \rangle$. It is easy to see that the following statements are equivalent:

(1) $\mathrm{rank}(du_x) < k$.

(2) $du_1(x), \ldots, du_k(x)$ are linearly dependent.

(3) there exists a non-zero vector $a = (a_1, \ldots, a_k)$ such that

$$a_1 du_1(x) + \ldots + a_k du_k(x) = 0.$$

(4) x is a critical point of some height function h_a.

Then the lemma follows from Theorem 8.5.4.   ∎

8.6.7. Proof of the Main Theorem.   We will use induction on k to show that:

(*) $u(M) = \text{cvx}(W \cdot q)$, if M is isoparametric of rank k.

If $k = 1$, then M is a standard sphere of R^{n+1}, so $u(M)$ is the line segment joining q to $-q$. Suppose (*) is true when the codimension is less than k, and M^n is full and isoparametric in R^{n+k}. Then we want to show that $u(M) = D$, where $D = \text{cvx}(W \cdot q)$. We divide our proof into five steps.

(i) Let C denote the set of singular points of u. Then $u(C)$ is the union of finitely many $(k - 1)$-polyhedra, and $\partial D \subseteq u(C) \subseteq D$. So in particular, $D \setminus u(C)$ is open.

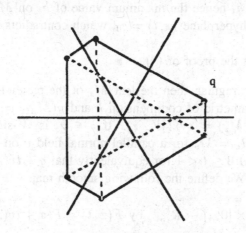

To see this, we note by Theorem 6.5.9 that if σ a 1-simplex on ν_q and $x \in W \cdot q$ then the slice $S_{x,\sigma}$ is a rank $k - 1$ isoparametric submanifold. So by the induction hypothesis, $u(S_{x,\sigma})$ is a $(k - 1)$-polyhedron. Then using Theorem 8.5.4, we have

$$u(C) = \bigcup \{P(S_{x,\sigma}) \mid x \in W \cdot q,$$

$$\sigma \text{ a } 1 - \text{simplex of some Weyl chamber of } \nu_q\}.$$

But it is also easy to see that

$$\partial D = \bigcup \{P(S_{x,\sigma}) \mid x \in W \cdot q, \ \sigma \text{ a } 1 - \text{simplex of } \Delta_x\}.$$

(ii) $\partial(u(M)) \subseteq u(C)$. This follows from the Inverse function theorem, because the image of a regular point of u is in the interior of $u(M)$.

(iii) $u(M) \subseteq D$. This follows from the fact that $u(C) \subseteq D$.

(iv) If U_i is a connected component of $D \setminus u(C)$, then either $U_i \subseteq u(M)$, or $U_i \cap u(M) = \emptyset$. To prove this, we proceed as follows: Suppose $U_i \cap u(M)^0$ is a non-empty proper subset of U_i, where $u(M)^0$ denotes the interior of $u(M)$. Then there is a sequence $y_n \in U_i \cap u(M)^0$ such that y_n converges to y, which is not in $U_i \cap u(M)^0$. Since $u(M)$ is compact, $y \in u(M)$. Using step (ii), we have $\partial(u(M)) \subseteq u(C)$. But by definition of U_i, $U_i \cap u(C) = \emptyset$, so we conclude that y is a regular value of u, hence $y \in U_i \cap u(M)^\circ$, a contradiction.

(v) $U_i \subseteq u(M)$ for all i. Suppose not, then we may assume $U_1 \cap u(M) = \emptyset$. Using step (i), we know that ∂U_1 is the union of $(k-1)$-polyhedra. Let μ be a $(k-1)$-face of ∂U_1, and t the outward unit normal of ∂U_1 at μ. Then by Euclidean geometry the height function h_t on M has local minimum value c_0 on μ, hence by Proposition 8.3.6, c_0 is the absolute minimum of h_t and $\mu \subset \partial D$. But by Euclidean geometry $\mu \subseteq \partial D$ implies that c_0 is also a local maximum value of h_t hence the maximum value of h_t on M, and hence M is contained in the hyperplane $\langle x, t \rangle = c_0$, which contradicts the fact that M is full.

This completes the proof of (*).   ∎

If $z \in \nu_q$ is W-regular, then the leaf M_z of the parallel foliation of M through z is isoparametric of codimension k and $\nu(M_z)_z = \nu(M)_q$. Hence (*) implies that $P(M_z) = \text{cvx}(W \cdot z)$. If $z \in \nu_q$ is W-singular, then we may assume that $M_z = M_v$ for a parallel normal field v on M, and M_{tv} is isoparametric for all $0 \leq t < 1$ (or equivalently that $q + tv(q)$ is W-regular for all $0 \leq t < 1$). We define the following smooth map

$$F : M \times [0,1] \to R^k, \quad \text{by } F(x,t) = P(x + tv(x)).$$

Let $u_t(x) = F(x,t)$, then $u_t(M) = P(M_{tv})$. By (*), $P(M_{tv}) = \text{cvx}(W \cdot (q + tv(q)))$ for all $0 \leq t < 1$. But $u_t \to u_1$ uniformly as $t \to 1$, so its image $u_t(M)$ converges in the Hausdorff topology to $u_1(M)$. But $q + tv(q)$ converges to $q + v(q) = z$, so $P(M_{tv})$ converges to the convex hull of $W \cdot (q + v(q)) = W \cdot z$, hence we obtain

8.6.8. Theorem.   *With the same assumption as in Theorem 8.6.4. Let $z \in \nu_q$, and M_z the leaf of the parallel foliation of M through z. Then $P(M_z) = \text{cvx}(W \cdot z)$.*

8.7 Marked Dynkin Diagrams for Isoparametric Submanifolds.

In this section we determine the possible marked Dynkin diagrams for both the finite and infinite dimensional isoparametric submanifolds.

Let M be a rank k irreducible isoparametric submanifold in a Hilbert space V, $\{\ell_i | i \in I\}$ the focal hyperplanes, $\{v_i | i \in I\}$ the curvature normals, m_i the corresponding multiplicities, and W the associated Coxeter group.

(1) If V is of finite dimension, then we may assume that $\ell_1, \ldots, \ell_k$ form a simple root system for W, and the marked Dynkin diagram has k vertices (one for each ℓ_i, $1 \leq i \leq k$) such that the ith vertex is marked with multiplicities m_i and there are $\alpha(g)$ edges joined the ith and j^{th} vertices if the angle between ℓ_i and ℓ_j is π/g, where $\alpha(g) = g - 2$ if $1 < g \leq 4$ and $\alpha(6) = 3$. So the possible marked Dynkin diagram for rank k finite dimensional irreducible isoparametric submanifolds are:

A_k

B_k

D_k

E_k

$k = 6, 7, 8$

F_4

G_2

(2) If V is an infinite dimensional Hilbert space, then we may assume that $\ell_1, \ldots, \ell_{k+1}$ form a simple root system for W, and the marked Dynkin diagram has $k + 1$ vertices (one for each ℓ_i, $1 \leq i \leq k + 1$); the ith vertex is marked with the multiplicity m_i. There are $\alpha(g)$ edges joining the ith and j^{th} vertices if the angle between ℓ_i and ℓ_j is π/g with $g > 1$, and there are infinitely many edges joining i^{th} and j^{th} vertices if ℓ_i is parallel to ℓ_j. So using the classification of the affine Weyl groups, we can easily write down the possible marked Dynkin diagrams for rank k infinite dimensional irreducible isoparametric submanifolds.

Let $q \in M$, and $\nu_q = q + \nu(M)_q$. Given $i \neq j \in I$, suppose ℓ_i is not parallel to ℓ_j and the angle between ℓ_i and ℓ_j is π/g. Then there exists a unique $(k-2)$-dimensional simplex σ of the chamber $\triangle$ on ν_q containing q such that $\sigma \subseteq \ell_i(q) \cap \ell_j(q)$. By the Slice Theorem, 6.5.9, the slice $S_{q,\sigma}$ is a finite dimensional rank 2 isoparametric submanifold with the Dihedral group of $2g$ elements as its Coxeter group, and m_i, m_j its multiplicities. So use the classification of Coxeter groups and the results in section 8.4 of Cartan, Münzner, and Abresch on rank 2 finite dimensional case, we obtain some immediate restrictions of the possible marked Dynkin diagrams for rank k isoparametric submanifolds of Hilbert spaces. In particular, we have

8.7.1. Theorem.   *If M^n is isoparametric in R^{n+k}, then the angle between any two focal hyperplanes ℓ_i and ℓ_j is π/g for some $g \in \{2,3,4,6\}$.*

8.7.2. Corollary.   *If M^n is an irreducible rank k isoparametric submanifold in R^{n+k}, then the associated Coxeter group W of M is an irreducible Weyl (or crystallographic) group.*

8.7.3. Proposition.   *There are at most two distinct multiplicities for an irreducible isoparametric submanifolds M of V.*

PROOF.   If the ith and $(i+1)^{th}$ vertices of the Dynkin diagram are joined by one edge, then by Theorem 8.4.2, $m_i = m_{i+1}$. But each irreducible Dynkin graph has at most one i_0 such that the i_0^{th} and $(i_0+1)^{th}$ vertices are joined by more than one edge. So the result follows.   ∎

8.7.4. Theorem.   *Let M^n be a rank k isoparametric submanifold in R^{n+k}. If all the multiplicities are even, then they are all equal to an integer m, where $m \in \{2,4,8\}$.*

PROOF.   If the ith and $(i+1)^{th}$ vertices of the Dynkin diagram are joined by two or four edges, then by Theorem 8.4.5, $m_i = m_{i+1} = 2$.   ∎

To obtain further such restrictions we need the more information on the cohomology ring of M. The details can be found in [HPT2]. Here we will only state the results without proof.

8.7.5. Theorem.   *The possible marked Dynkin diagrams of irreducible rank $k \geq 3$ finite dimensional isoparametric submanifolds are as follows:*

A_k $\underset{m}{\circ}\!\!-\!\!-\!\!-\!\!\underset{m}{\circ}\cdots\circ\!\!-\!\!-\!\!-\!\!\underset{m}{\circ}$ $m\in\{1,2,4\}$

B_k $\underset{m_1}{\circ}\!\!-\!\!-\!\!-\!\!\underset{m_1}{\circ}\cdots\underset{m_1}{\circ}\!\!=\!\!=\!\!\underset{m_2}{\circ}$ (m_1,m_2) satisfies $(*)$ below.

D_k $\circ\!\!-\!\!-\!\!-\!\!\circ \cdots \circ\!\!-\!\!-\!\!-\!\!\overset{\overset{\displaystyle m}{\circ}}{\underset{}{\circ}}\!\!-\!\!-\!\!-\!\!\circ$ $m\in\{1,2,4\}$

(with markings m, m, m)

E_k $\circ\!\!-\!\!-\!\!\circ \cdots \overset{\overset{\displaystyle m}{\circ}}{\underset{}{\circ}}\!\!-\!\!-\!\!\circ\!\!-\!\!-\!\!\circ$ $m\in\{1,2,4\}$ $k=6,7,8$

(with markings m, m, m)

F_4 $\circ\!\!-\!\!-\!\!\circ\!\!=\!\!\circ\!\!-\!\!-\!\!\circ$ $m_1=m_2=2$ or $m_1=1,\ m_2\in\{1,2,4,8\}$

(with markings m_1, m_1, m_2, m_2)

The pair (m_1,m_2) satisfies (*) if it satisfies one of the following conditions:

(1) $m_1=1$, m_2 is arbitrary,

(2) $m_1=2$, $m_2=2$ or $2r+1$,

(3) $m_1=4$, $m_2=1,5$, or $4r+3$,

(4) $k=3$, $m_1=8$, $m_2=1,3,7,11$, or $8r+7$.

As a consequence of Theorem 8.7.5, Theorem 8.4.4 and the Slice Theorem 6.5.9, we have:

8.7.6. Theorem. *The possible marked Dynkin diagrams of irreducible rank $k\geq 3$ infinite dimensional isoparametric submanifolds are as follows:*

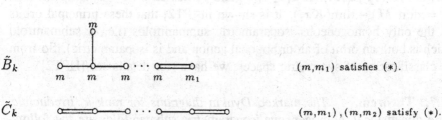

$\tilde{A}_1$ $\circ\!\!\overset{\infty}{-\!\!-}\!\!\circ$
        m_1 m_2

$\tilde{A}_k$ $\circ\!\!-\!\!-\!\!\circ \cdots \circ\!\!-\!\!-\!\!\overset{\overset{\displaystyle m}{\triangle}}{}\!\!-\!\!-\!\!\circ$ $m\in\{1,2,4\}$
        m m m

$\tilde{B}_2$ $\circ\!\!=\!\!\circ\!\!-\!\!-\!\!\circ$ $(m_1,m_2),(m_2,m_3)$ satisfy (*).
        m_1 m_2 m_3

$\tilde{B}_k$ $\circ\!\!-\!\!\overset{\overset{\displaystyle m}{\circ}}{}\!\!-\!\!\circ \cdots \circ\!\!=\!\!\circ$ (m,m_1) satisfies (*).
        m m m m m_1

$\tilde{C}_k$ $\circ\!\!=\!\!\circ\!\!-\!\!\circ \cdots \circ\!\!-\!\!\circ\!\!=\!\!\circ$ $(m,m_1),(m,m_2)$ satisfy (*).
        m_1 m m m m m_2

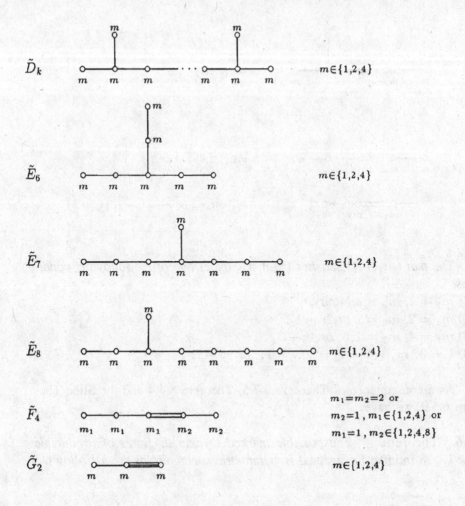

Let G/K be a rank k symmetric space, $\mathcal{G} = \mathcal{K} + \mathcal{P}$, $\mathcal{A}$ the maximal abelian subalgebra contained in $\mathcal{P}$, and $q \in \mathcal{A}$ a regular point with respect to the isotropy action K on $\mathcal{P}$. Then $M = Kq$ is a principal orbit, and is a rank k isoparametric submanifold of $\mathcal{P}$. The Weyl group associated to M as an isoparametric submanifold is the standard Weyl group associated to G/K, i.e., $W = N(\mathcal{A})/Z(\mathcal{A})$. If $x_i \in \ell_i(q)$ and x_i lies on a $(k-1)$-simplex, then $m_i = \dim(M) - \dim(Kx_i)$. It is shown in [PT2] that these principal orbits are the only homogeneous isoparametric submanifolds (i.e., a submanifold which is both an orbit of an orthogonal action and is isoparametric). So from the classification of symmetric spaces, we have (for details see [HPT2])

8.7.7. Theorem. *The marked Dynkin diagrams for rank k, irreducible, finite dimensional, homogeneous isoparametric submanifolds are the following:*

A_2 $\quad$ o—o $\qquad\qquad\qquad m\in\{1,2,4,8\}$
$\qquad\quad m \quad m$

A_k $\quad$ o—o $\cdots$ o—o $\qquad\qquad m\in\{1,2,4\}$
$\qquad\quad m \quad m \qquad\quad m$

B_k $\quad$ o—o $\cdots$ o⊜o $\qquad\qquad (m_1,m_2)$ satisfies (*)
$\qquad\quad m_1 \quad m_1 \quad m_1 \quad m_2$

D_k $\quad$ o—o $\cdots$ o—o—o $\qquad\qquad m\in\{1,2\}$
$\qquad\quad m \quad m \qquad\qquad m$
(with vertical branch labeled m above the junction)

E_k $\quad$ o—o $\cdots$ o—o—o $\qquad\qquad m\in\{1,2\}\quad k=6,7,8$
$\qquad\quad m \quad m \qquad\qquad m$
(with vertical branch labeled m above)

F_4 $\quad$ o—o⊜o—o $\qquad\qquad m_1=m_2=2$ or $m_1=1,\ m_2\in\{1,2,4,8\}$
$\qquad\quad m_1 \quad m_1 \quad m_2 \quad m_2$

G_2 $\quad$ o⊜o $\qquad\qquad\qquad m\in\{1,2\}$
$\qquad\quad m \quad m$

The pair (m_1,m_2) satisfies (*) in all of the following cases:
(1) $m_1 = 1$, m_2 is arbitrary,
(2) $m_1 = 2$, $m_2 = 2$ or $2m + 1$,
(3) $m_1 = 4$, $m_2 = 1,5,4m + 3$,
(4) $m_1 = 8$, $m_2 = 1$,
(5) $k = 2$, $m_1 = 6$, $m_2 = 9$.

8.7.8. Corollary. *The set of multiplicities* (m_1,m_2) *of homogeneous, isoparametric, finite dimensional submanifolds with* B_2 *as its Coxeter groups is*

$$\{(1,m),\ (2,2m+1),\ (4,4m+3),\ (9,6),\ (4,5),\ (2,2)\}.$$

8.7.9. Open problems. If we compare Theorem 8.7.5 and 8.7.7, it is natural to pose the following problems:

(1) Is it possible to have an irreducible, rank 3, finite dimensional isoparametric submanifold, whose marked Dynkin diagram is the following?

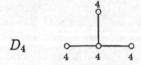

D_4

(It would be interesting if such an example does exist, however we expect that most likely it does not. Of course a negative answer to this problem would also imply the non-existence of marked Dynkin diagrams with uniform multiplicity 4 of D_k-type, $k > 5$ or E_k-type, $k = 6, 7, 8$.

(2) Is it possible to have an isoparametric submanifold whose marked Dynkin diagram is of the following type with $m > 1$:

B_3 ○———○═══○
           8   8 m

(3) Let $M^n \subseteq R^{n+k}$ be an irreducible isoparametric submanifold with uniform multiplicities. Is it necessarily homogeneous ?

If the answer to problem 3 is affirmative and if the answers to problem 1 and 2 are both negative, then the remaining fundamental problem would be:

(4) Are there examples of non-homogeneous irreducible isoparametric submanifolds of rank $k > 3$?

It follows from section 6.4 that if $n = 2(m_1 + m_2 + 1)$, and $f : R^n \to R$ is a homogeneous polynomial of degree 4 such that

$$\Delta f(x) = 8(m_2 - m_1) \|x\|^2, \quad \|\nabla f(x)\|^2 = 16 \|x\|^6, \qquad (8.7.1)$$

then the polynomial map $x \mapsto (|x|^2, f(x))$ is isoparametric and its regular levels are isoparametric submanifolds of R^{n+2} with

B_2 ○═══○
          m_1 . m_2

as its marked Dynkin diagram, i.e., B_2 is the associated Weyl group with (m_1, m_2) as multiplicities. Solving (8.7.1), Ozeki and Takeuchi found the first two families of non-homogeneous rank 2 examples. In fact, they constructed the isoparametric polynomial explicitly as follows:

8.7.10. Examples. (Ozeki-Takeuchi [OT1,2]) Let $(m_1, m_2) = (3, 4r)$ or $(7, 8r)$, $F = H$ or Ca (the quarternions or Cayley numbers) for $m_1 = 3$ or 7 respectively, and let $n = 2(m_1 + m_2 + 2)$. Let $u \mapsto \bar{u}$ denote the canonical involution of F. Then

$$(u, v) = \frac{1}{2}(u\bar{v} + v\bar{u})$$

defines an inner product on F, that gives an inner product on F^m. We let

$$f_0 : R^n = F^{2(r+1)} = F^{1+r} \times F^{1+r} \to R,$$

$$f_0(u, v) = 4(\|u\bar{v}^t\|^2 - (u, v)^2) + (\|u_1\|^2 - \|v_1\|^2 + 2(u_0, v_0))^2,$$

where $u = (u_0, u_1)$, $v = (v_0, v_1)$ and $u_0, v_0 \in F$, $u_1, v_1 \in F^r$. Then

$$f(u, v) = (\|u\|^2 + \|v\|^2)^2 - 2f_0(u, v)$$

satisfies (8.7.1). So the intersection of a regular level of f and S^{n-1} is isoparametric with B_2 as the associated Weyl group and $(3, 4r)$ or $(7, 8r)$ as multiplicities. These examples correspond to $(3, 4r)$ and $(7, 8r)$ are non-homogeneous. But there is also a homogeneous example with B_2 as its Weyl group and $(3, 4)$ as its multiplicities. So the marked Dynkin diagram does not characterize an isoparametric submanifold.

8.7.11. Examples.   Another family of non-homogeneous rank 2 isoparametric examples is constructed from the representations of the Clifford algebra $C\ell^{m+1}$ by Ferus, Karcher and Münzner (see [FKM] for detail). It is known from representation theory that every irreducible representation space of $C\ell^{m+1}$ is of even dimension, and it is given by a "Clifford system" $(P_0, \ldots, P_m)$ on R^{2r}, i.e., the $P_i's$ are in $SO(2r)$ and satisfy

$$P_i P_j + P_j P_i = 2\delta_{ij} Id.$$

Let $f : R^{2r} \to R$ be defined by

$$f(x) = \|x\|^4 - \sum_{i=0}^{m} \langle P_i(x), x \rangle^2.$$

Then f satisfies (8.7.1) with $m_1 = m$ and $m_2 = r - m_1 - 1$. If $m_1, m_2 > 0$, then the regular levels of the map $x \mapsto (\|x\|^2, f(x))$ are isoparametric with Coxeter group B_2 and multiplicities (m_1, m_2). Most of these examples are non-homogeneous.

8.7.12. Remark.   The classification of isoparametric submanifolds is still far from being solved. For example we do not know

(1) what the set of the marked Dynkin diagrams for rank 2 finite dimensional isoparametric submanifolds is,

(ii) what the rank k homogeneous infinite dimensional isoparametric submanifolds are.

Part II.   Critical Point Theory.

Chapter 9.
Elementary Critical Point Theory.

The essence of Morse Theory is a collection of theorems describing the intimate relationship between the topology of a manifold and the critical point structure of real valued functions on the manifold. This body of theorems has over and over again proved itself to be one of the most powerful and far-reaching tools available for advancing our understanding of differential topology and analysis. But a good mathematical theory is more than *just* a collection of theorems; in addition it consists of a tool box of related conceptualizations and techniques that have been gradually built up to help understand some circle of mathematical problems. Morse Theory is no exception, and its basic concepts and constructions have an unusual appeal derived from an underlying geometric naturality, simplicity, and elegance. In these lectures we will cover some of the more important theorems and applications of Morse Theory and, beyond that, try to give a feeling for and an ability to work with these beautiful and powerful techniques.

9.1 Preliminaries.

We will assume that the reader is familiar with the standard definitions and notational conventions introduced in the Appendix. We begin with some basic assumptions and further notational conventions. In all that follows $f : M \to R$ will denote a smooth real valued function on a smooth finite or infinite dimensional hilbert manifold M. We will make three basic assumptions about M and f:

(a) (Completeness). M is a complete Riemannian manifold.
(b) (Boundedness below) The function f is bounded below on M. We will let B denote the greatest lower bound of f, so our assumption is that $B > -\infty$.
(c) (Condition C) If $\{x_n\}$ is any sequence in M for which $|f(x_n)|$ is bounded and for which $\|df_{x_n}\| \to 0$, it follows that $\{x_n\}$ has a convergent subsequence, $x_{n_k} \to p$.
    (By continuity, $\|df_p\| = 0$, so that p is a critical point of f).
Of course if M is compact then with *any* choice of Riemannian metric for M all three conditions are automatically satisfied. In fact we recommend that a reader new to Morse theory develop intuition by always thinking of M as compact, and we will encourage this by using mainly compact surfaces

for our examples and diagrams. Nevertheless it is important to realize that in our formal proofs of theorems only (a), (b), and (c) will be used, and that as we shall see later these conditions do hold in important cases where M is not only non-compact, but even infinite dimensional.

Recall that p in M is called a *critical point* of f if $df_p = 0$. Other points of M are called *regular points* of f. Given a real number c we call $f^{-1}(c)$ the *c-level* of f, and we say it is a *critical level* (and that c is a *critical value* of f) if it contains at least one critical point of f. Other real numbers c (even those for which $f^{-1}(c)$ is empty!) are called *regular values* of f and the corresponding levels $f^{-1}(c)$ are called *regular levels*. We denote by M_c (or by $M_c(f)$ if there is any ambiguity) the "part of M below the level c", i.e., $f^{-1}((-\infty, c])$. It is immediate from the inverse function theorem that for a regular value c, $f^{-1}(c)$ is a (possibly empty) smooth, codimension one submanifold of M, that M_c is a smooth submanifold with boundary, and that $\partial M_c = f^{-1}(c)$. We will denote by $\mathcal{C}$ the set of all critical points of f, and by $\mathcal{C}_c$ the set $\mathcal{C} \cap f^{-1}(c)$ of critical points at the level c. Then we have the following lemma.

9.1.1. Lemma.   *The restriction of f to $\mathcal{C}$ is proper. In particular, for any $c \in R$, $\mathcal{C}_c$ is compact.*

PROOF.   We must show that $f^{-1}([a, b]) \cap \mathcal{C}$ is compact, i.e. if $\{x_n\}$ is a sequence of critical points with $a \le f(x_n) \le b$ then $\{x_n\}$ has a convergent subsequence. But since $\|\nabla f_{x_n}\| = 0$ this is immediate from Condition C.   ∎

Since proper maps are closed we have:

9.1.2. Corollary.   *The set $f(\mathcal{C})$ of critical values of f is a closed subset of R.*

Recall that the gradient of f is the smooth vector field ∇f on M dual to df, i.e., characterized by $Yf = \langle Y, \nabla f \rangle$ for any tangent vector Y to M. Of course if Y is tangent to a level $f^{-1}(c)$ then $Yf = 0$, so at each regular point x it follows that ∇f is orthogonal to the level through x. In fact it follows easily from the Schwarz inequality that, at a regular point, ∇f points in the direction of most rapid increase of f. We will denote by φ_t the maximal flow generated by $-\nabla f$. For each x in M $\varphi_t(x)$ is defined on an interval $\alpha(x) < t < \beta(x)$ and $t \mapsto \varphi_t(x)$ is the maximal solution curve of $-\nabla f$ with initial condition x. Thus $\frac{d}{dt}\varphi_t(x) = -\nabla f_{\varphi_t(x)}$ and so $\frac{d}{dt}f(\varphi_t(x)) = -\nabla f(f) = -\|\nabla f\|^2$, so $f(\varphi_t(x))$ is monotonically decreasing in t. Since f is bounded below by B it follows that $f(\varphi_t(x))$ has a limit as $t \to \beta(x)$. We shall now prove the important fact that $\{\varphi_t\}$ is a "positive semigroup", that is, for each x in M $\beta(x) = \infty$, so $\varphi_t(x)$ is defined for all $t > 0$.

9.1.3. Lemma.   *A C^1 curve $\sigma : (a, b) \to M$ of finite length has relatively compact image.*

PROOF.   Since M is complete it will suffice to show that the image of σ is totally bounded. Since $\int_a^b \|\sigma'(t)\| \, dt < \infty$, given $\epsilon > 0$ there exist $t_0 = a < t_1 < \ldots < t_n < t_{n+1} = b$ such that $\int_{t_i}^{t_{i+1}} \|\sigma'(t)\| \, dt < \epsilon$. Then by the definition of distance in M it is clear that the $x_i = \sigma(t_i)$ are ϵ-dense in the image of σ.   ∎

9.1.4. Proposition.   *Let X be a smooth vector field on M and let $\sigma : (a, b) \to M$ be a maximal solution curve of X. If $b < \infty$ then $\int_0^b \|X_{\sigma(t)}\| \, dt = \infty$. Similarly if $a > -\infty$ then $\int_a^0 \|X_{\sigma(t)}\| \, dt = \infty$.*

PROOF.   Since σ is maximal, if $b < \infty$ then $\sigma(t)$ has no limit point in M as $t \to b$. Thus, by the lemma, $\sigma : [0, b) \to M$ must have infinite length, and since $\sigma'(t) = X_{\sigma(t)}$, $\int_0^b \|X_{\sigma(t)}\| \, dt = \infty$.   ∎

9.1.5. Corollary.   *A smooth vector field X on M of bounded length, generates a one-parameter group of diffeomorphisms of M.*

PROOF.   Suppose $\|X\| \leq K < \infty$. If $b < \infty$ then $\int_0^b \|X_{\sigma(t)}\| \, dt \leq bK < \infty$, contradicting the Proposition. By a similar argument $a > -\infty$ is also impossible.   ∎

9.1.6. Theorem.   *The flow $\{\varphi_t\}$ generated by $-\nabla f$ is a positive semigroup; that is, for all $t > 0$ φ_t is defined on all of M. Moreover for any x in M $\varphi_t(x)$ has at least one critical point of f as a limit point as $t \to \infty$.*

PROOF.   Let $g(t) = f(\varphi_t(x))$ and note that $B \leq g(T) = g(0) + \int_0^T g'(t) \, dt = g(0) - \int_0^T \|\nabla f_{\varphi_t(x)}\|^2 \, dt$. Since this holds for all $T < \beta(x)$, by the Schwarz inequality

$$\int_0^{\beta(x)} \|\nabla f_{\varphi_t(x)}\| \, dt \leq \sqrt{\beta(x)} \left(\int_0^{\beta(x)} \|\nabla f_{\varphi_t(x)}\|^2 \, dt \right)^{\frac{1}{2}},$$

which is less than or equal to $\sqrt{\beta(x)}(g(0) - B)^{\frac{1}{2}}$, and hence would be finite if $\beta(x)$ were finite. It follows from the preceding proposition that $\beta(x)$ must be infinite and consequently $\|\nabla f_{\varphi_t(x)}\|$ cannot be bounded away from zero as $t \to \infty$, since otherwise $\int_0^\infty \|\nabla f_{\varphi_t(x)}\|^2 \, dt$ would be infinite, whereas we know it is less than $g(0) - B$. Finally, since $f(\varphi_t(x))$ is bounded, it now follows from Condition C that $\varphi_t(x)$ has a critical point of f as a limit point as $t \to \infty$. ∎

9.1.7. Remark.   An exactly parallel argument shows that as $t \to \alpha(x)$ either $f(\varphi_t(x)) \to \infty$ or else $\alpha(x)$ must be $-\infty$ and $\varphi_t(x)$ has a critical point of f as a limit point as $t \to -\infty$.

9.1.8. Corollary.   *If x in M is not a critical point of f then there is a critical point p of f with $f(p) < f(x)$.*

PROOF.   Choose any critical point of f that is a limit point of $\varphi_t(x)$ as $t \to \infty$. ∎

9.1.9. Theorem.   *The function f attains its infimum B. That is, there is a critical point p of f with $f(p) = B$.*

PROOF.   Choose a sequence $\{x_n\}$ with $f(x_n) \to B$. By the preceding corollary we can assume that each x_n is a critical point of f. Then by Condition C a subsequence of $\{x_n\}$ converges to a critical point p of f, and clearly $f(p) = B$. ∎

In order to understand and work effectively with a complex mathematical subject one must get behind its purely logical content and develop some intuitive picture of the key concepts. Normally these intuitions are imprecise and vary considerably from one individual to another, and this often can be a barrier to the easy communication of mathematical ideas. One of the pleasant and special features of Morse Theory is that it has a generally accepted metaphor for visualizing many of its basic concepts. Since much of the terminology and motivation of the theory is based on this metaphor we shall now explain it in some detail.

Starting with our smooth function $f : M \to R$ we build a "world" $W = M \times R$. We now identify M *not* with $M \times \{0\}$, (which we think of as "sea-level") but rather with the graph of f; that is we identify $x \in M$ with $(x, f(x)) \in W$.

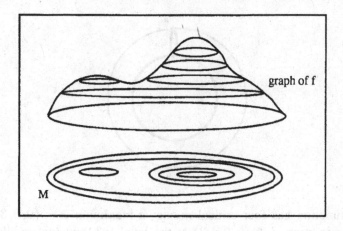

The projection $z : W \to R$, $(x,t) \mapsto z(x,t) = t$ we think of as "height above sea-level". Since $z(x, f(x)) = f(x)$ this means that our original function f represents altitude in our new realization of M. And this in turn means that the a-level of f becomes just that, it is the intersection of the graph of f with the altitude level-surface $z = a$ in W. The critical points of f are now the valleys, passes, and mountain summits of the graph of f, that is the points where the tangent hyperplane to M is horizontal. We think of the projection of W onto M as providing us with a "topo map" of our world; projecting the a-level of f in W into M gives us the old $f^{-1}(a)$ which we now think of as an isocline (surface of constant height) on this topographic map.

We give W the product Riemannian metric, and recall that the negative gradient vector field $-\nabla f$ represents the direction of "steepest descent" on the graph of f; pointing orthogonal to the level surfaces in the downhill direction. Thus (very roughly speaking) we may think of the flow φ_t we have been using as modelling the way a very syrupy liquid would flow down the graph of f under the influence of gravity. We shall return to this picture many times in the sequel to provide intuition, motivation, and terminology.

There is a particular Morse function that, while not completely trivial, is so intuitive and easy to analyze, that it is is everybody's favorite model, and we will use it frequently to illustrate various concepts and theorems. Informally it is the height above the floor on a tire standing in ready-to-roll position. More precisely, we take M to be the torus obtained by revolving the circle of radius 1 centered at $(0, 2)$ in the (x, y)-plane about the y-axis, and define $f : M \to R$ to be orthogonal projection on the z-axis.

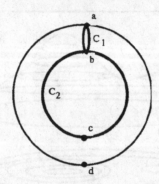

This function has four critical points: a maximum $a = (0,0,3)$ at the level 3, a minimum $d = (0,0,-3)$ at the level -3, and two saddle points $b = (0,0,1)$, and $c = (0,0,-1)$, at the levels 1 and -1 respectively. The reader should analyze the asymptotic properties of the flow $\varphi_t(x)$ of $-\nabla f$ in this case. Of course the four critical points are fixed. Other points on the circle $C_1 : x = 0, y^2 + (z-2)^2 = 1$ tend to b, other points on $C_2 : y = 0, x^2 + y^2 = 1$ tend to c, and all remaining points tend to the minimum, d. We shall refer to this function as the "height function on the torus".

The study of the flow $\{\varphi_t\}$ generated by $-\nabla f$ (or more generally of vector fields proportional to it) is one of the most important tools of Morse theory. We have seen a little of its power above and we shall see much more in what follows.

9.2   The First Deformation Theorem.

We shall now use the flow $\{\varphi_t\}$ generated by $-\nabla f$ to deform subsets of the manifold M, and see how this leads to a very general method (called "minimaxing") for locating critical points of f. We will then illustrate minimaxing with an introduction to Lusternik-Schnirelman theory.

9.2.1. Lemma.   *If O is a neighborhood of the set C_c of critical points of f at the level c, then there is an $\epsilon > 0$ such that $\|\nabla f\|$ is bounded away from zero on $f^{-1}(c - \epsilon, c + \epsilon) \setminus O$.*

PROOF.   Suppose not. Then for each positive integer n we could choose an x_n in $f^{-1}(c - \frac{1}{n}, c + \frac{1}{n}) \setminus O$ such that $\|\nabla f_{x_n}\| < \frac{1}{n}$. By Condition C, a subsequence of $\{x_n\}$ would converge to a critical point p of f with $f(p) = c$, so $p \in C_c$ and eventually the subsequence must get inside the neighborhood O of C_c, a contradiction.   ∎

Since C_c is compact,

9.2.2. Lemma.   *Any neighborhood of $\mathcal{C}_c$ includes the neighborhoods of the form $N_\delta(\mathcal{C}_c) = \{x \in M \mid \rho(x, \mathcal{C}_c) < \delta\}$ provided δ is sufficiently small.*

Now let U be any neighborhood of $\mathcal{C}_c$ in M, and choose a $\delta_1 > 0$ such that $N_{\delta_1}(\mathcal{C}_c) \subseteq U$. Since $\|\nabla f_p\| = 0$ on $\mathcal{C}_c$ we may also assume that $\|\nabla f_p\| \leq 1$ for $p \in N_{\delta_1}(\mathcal{C}_c)$.

If ϵ is small enough then, by 2.1, for any $\delta_2 > 0$ we can choose $\mu > 0$ such that $\|\nabla f_p\| \geq \mu$ for $p \in f^{-1}([c - \epsilon, c + \epsilon])$ and $\rho(p, \mathcal{C}_c) \geq \delta_2$ (i.e., $p \notin N_{\delta_2}(\mathcal{C}_c)$). In particular we can assume $\delta_2 < \delta_1$, so that $N_{\delta_2}(\mathcal{C}_c) \subseteq N_{\delta_1}(\mathcal{C}_c) \subseteq U$.

9.2.3. First Deformation Theorem.   *Let U be any neighborhood of $\mathcal{C}_c$ in M. Then for $\epsilon > 0$ sufficiently small $\varphi_1(M_{c+\epsilon} \backslash U) \subseteq M_{c-\epsilon}$.*

PROOF.   Let $\epsilon = \min(\frac{1}{2}\mu^2, \frac{1}{2}\mu^2(\delta_1 - \delta_2))$, where δ_1, δ_2, and μ are chosen as above. Let $p \in f^{-1}([c - \epsilon, c + \epsilon]) \backslash U$. We must show that $f(\varphi_1(p)) \leq c - \epsilon$, and since $f(\varphi_t(p))$ is monotonically decreasing we may assume that $\varphi_t(p) \in f^{-1}([c - \epsilon, c + \epsilon])$ for $0 \leq t < 1$. Thus by definition of δ_2 we can also assume that if $\rho(\varphi_t(p), \mathcal{C}_c) \geq \delta_2$ then $\|\nabla f_{\varphi_t(p)}\| \geq \mu$.

Since $\varphi_0(p) = p$ and $\frac{d}{dt}f(\varphi_t(p)) = -\|\nabla f_{\varphi_t(p)}\|^2$ we have:

$$f(\varphi_1(p)) = f(\varphi_0(p)) + \int_0^1 -\|\nabla f_{\varphi_t(p)}\|^2 \, dt$$

$$\leq c + \epsilon - \int_0^1 \|\nabla f_{\varphi_t(p)}\|^2 \, dt \, ,$$

so it will suffice to show that

$$\int_0^1 \|\nabla f_{\varphi_t(p)}\|^2 \, dt \geq 2\epsilon = \min(\mu^2, \mu^2(\delta_1 - \delta_2)) \, .$$

We will break the remainder of the proof into two cases.

Case 1. $\rho(\varphi_t(t), \mathcal{C}_c) > \delta_2$ for all $t \in [0, 1]$.

Then $\|\nabla f_{\varphi_t(p)}\| \geq \mu$ for $0 \leq t \leq 1$ and hence $\int_0^1 \|\nabla f_{\varphi_t(p)}\|^2 \, dt \geq \mu^2 \geq \min(\mu^2, \mu^2(\delta_1 - \delta_2))$.

Case 2. $\rho(\varphi_t(t), \mathcal{C}_c) \leq \delta_2$ for some $t \in [0, 1]$.

Let t_2 be the first such t. Since $p \notin U$, *a fortiori* $p \notin N_{\delta_1}(\mathcal{C}_c)$, i.e., $\rho(\varphi_0(p), \mathcal{C}_c) \geq \delta_1 > \delta_2$, so there is a last $t \in [0, 1]$ less than t_2 such that $\rho(\varphi_0(p), \mathcal{C}_c) \geq \delta_1$. We denote this value of t by t_1, so that $0 < t_1 < t_2 < 1$, and in the interval $[t_1, t_2]$ we have $\delta_1 \geq \rho(\varphi_t(p), \mathcal{C}_c) \geq \delta_2$. Note that $\rho(\varphi_{t_1}(p), \mathcal{C}_c) \geq \delta_1$ while $\rho(\varphi_{t_2}(p), \mathcal{C}_c) \leq \delta_2$ and hence by the triangle inequality $\rho(\varphi_{t_1}(p), \varphi_{t_2}(p)) \geq \delta_1 - \delta_2$. It follows that any curve joining

$\varphi_{t_1}(p)$ to $\varphi_{t_2}(p)$ has length greater or equal $\delta_1 - \delta_2$, and in particular this is so for $t \mapsto \varphi_t(p)$, $t_1 \leq t \leq t_2$. Since $\frac{d}{dt}\varphi_t(p) = -\nabla f_{\varphi_t(p)}$ this means:

$$\int_{t_1}^{t_2} \|\nabla f_{\varphi_t(p)}\| \, dt \geq \delta_1 - \delta_2 \,.$$

By our choice of δ_1, $\|\nabla f_{\varphi_t(p)}\| \leq 1$ for t in $[t_1, t_2]$, since $\rho(\varphi_t(p), \mathcal{C}_c) \leq \delta_1$ for such t. Thus

$$t_2 - t_1 = \int_{t_1}^{t_2} 1 \, dt \geq \int_{t_1}^{t_2} \|\nabla f_{\varphi_t(p)}\| \, dt \geq \delta_1 - \delta_2 \,.$$

On the other hand, by our choice of δ_2, for t in $[t_1, t_2]$ we also have $\|\nabla f_{\varphi_t(p)}\| \geq \mu$, since $\rho(\varphi_t(p), \mathcal{C}_c) \geq \delta_2$ for such t. Thus

$$\int_0^1 \|\nabla f_{\varphi_t(p)}\|^2 \, dt \geq \int_{t_1}^{t_2} \|\nabla f_{\varphi_t(p)}\|^2 \, dt$$

$$\geq \int_{t_1}^{t_2} \mu^2 \, dt = \mu^2(t_2 - t_1)$$

$$\geq \mu^2(\delta_2 - \delta_1)$$

$$\geq \min(\mu^2, \mu^2(\delta_1 - \delta_2)) \,. \quad \blacksquare$$

9.2.4. Corollary.   *If c is a regular value of f then, for some $\epsilon > 0$,* $\varphi_1(M_{c+\epsilon}) \subseteq M_{c-\epsilon}$.

PROOF.   Since $\mathcal{C}_c = \emptyset$ we can take $U = \emptyset$. $\blacksquare$

Let $\mathcal{F}$ denote a non-empty family of non-empty compact subsets of M. We define minimax$(f, \mathcal{F})$, the minimax of f over the family $\mathcal{F}$, to be the infimum over all F in $\mathcal{F}$ of the maximum of f on F. Now the maximum value of f on F is just the smallest c such that $F \subseteq M_c$. So minimax$(f, \mathcal{F})$, is the smallest c such that, for any positive ϵ, we can find an F in $\mathcal{F}$ with $F \subseteq M_{c+\epsilon}$. The family $\mathcal{F}$ is said to be invariant under the positive time flow of $-\nabla f$ if whenever $F \in \mathcal{F}$ and $t > 0$ it follows that $\varphi_t(F) \in \mathcal{F}$.

9.2.5. Minimax Principle.   *If $\mathcal{F}$ is a family of compact subsets of M invariant under the positive time flow of $-\nabla f$ then minimax$(f, \mathcal{F})$ is a critical value of M.*

PROOF.   By definition of minimax we can find an F in $\mathcal{F}$ with $F \subseteq M_{c+\epsilon}$. Suppose c were a regular value of f. Then by the above Corollary

$\varphi_1(M_{c+\epsilon}) \subseteq M_{c-\epsilon}$ and *a fortiori* $\varphi_1(F) \subseteq M_{c-\epsilon}$. But since $\mathcal{F}$ is invariant under the positive time flow of $-\nabla f$, $\varphi_1(F)$ is also in the family $\mathcal{F}$ and it follows that minimax$(f, \mathcal{F}) \leq c - \epsilon$, a contradiction. ∎

Of course any family $\mathcal{F}$ of compact subsets of M invariant under homotopy is *a fortiori* invariant under the positive time flow of $-\nabla f$. Here are a few important examples:

- If α is a homotopy class of maps of some compact space X into M take $\mathcal{F} = \{\text{im}(f) | f \in \alpha\}$.
- Let α be a homology class of M and let $\mathcal{F}$ be the set of compact subsets F of M such that α is in the image of $i_* : H_*(F) \to H_*(M)$.
- Let α be a cohomology class of M and let $\mathcal{F}$ be the set of compact subsets F of M that support α (i.e., such that α restricted to $M \backslash F$ is zero).

There are a number of related applications of the Minimax Principle that go under the generic name of "Mountain Pass Theorem". Here is a fairly general version.

9.2.6. Definition. Let M be connected. We will call a subset $\mathcal{R}$ of M a *mountain range* relative to f if it separates M and if, on each component of $M \backslash \mathcal{R}$, f assumes a value strictly less that $\inf(f|\mathcal{R})$.

9.2.7. Mountain Pass Theorem. *If M is connected and $\mathcal{R}$ is a mountain range relative to f then f has a critical value $c \geq \inf(f|\mathcal{R})$.*

PROOF. Set $\alpha = \inf(f|\mathcal{R})$ and let M^0 and M^1 be two different components of $M \backslash \mathcal{R}$. Define $M_\alpha^i = \{x \in M^i \mid f(x) < \alpha\}$. By assumption each M_α^i is non-empty, and since M is connected we can find a continuous path $\sigma : I \to M$ such that $\sigma(i) \in M_\alpha^i$. Let Γ denote the set of all such paths σ and let $\mathcal{F} = \{\text{im}(\sigma) \mid \sigma \in \Gamma\}$, so that $\mathcal{F}$ is a non-empty family of compact subsets of M. Since $\sigma(0)$ and $\sigma(1)$ are in different components of $M \backslash \mathcal{R}$ it follows that $\sigma(t_0) \in \mathcal{R}$ for some $t_0 \in I$, so $f(\sigma(t_0)) \geq \alpha$ and hence minimax$(f, \mathcal{F}) \geq \alpha$. Thus, by the Minimax Principle, it will suffice to show that if $\sigma \in \Gamma$ and $t > 0$ then $\varphi_t \circ \sigma \in \Gamma$, where φ_t is the positive time flow of $-\nabla f$. And for this it will clearly suffice to show that if x is in M_α^i then so is $\varphi_t(x)$. But since $f(\varphi_0(x)) = f(x) < \alpha$, and $f(\varphi_t(x))$ is a non-increasing function of t, it follows that $f(\varphi_t(x)) < \alpha$, so in particular $\varphi_t(x) \in M \backslash \mathcal{R}$, and hence x and $\varphi_t(x)$ are in the same component of $M \backslash \mathcal{R}$. ∎

In recent years Mountain Pass Theorems have had extensive applications in proving existence theorems for solutions to both ordinary and partial differential equations. For further details see [Ra].

We next consider Lusternik-Schnirelman Theory, an early and elegant application of the Minimax Principle. This material will not be used in the remainder of these notes and may be skipped without loss of continuity.

A subset A of a space X is said to be contractible in X if the inclusion map $i : A \to X$ is homotopic to a constant map of A into X. We say that A has category m in X (and write $\text{cat}(A, X) = m$)) if A can be covered by m (but no fewer) closed subsets of X, each of which is contractible in X. We define $\text{cat}(X) = \text{cat}(X, X)$. Here are some obvious properties of the set function cat that follow immediately from the definition.

(1) $\text{cat}(A, X) = 0$ if and only if $A = \emptyset$.
(2) $\text{cat}(A, X) = 1$ if and only if $\bar{A}$ is contractible in X.
(3) $\text{cat}(A, X) = \text{cat}(\bar{A}, X)$.
(4) If A is closed in X then $\text{cat}(A, X) = m$ if and only if A is the union of m (but not fewer) closed sets, each contractible in X.
(5) $\text{cat}(A, X)$ is monotone; i.e., if $A \subseteq B$ then $\text{cat}(A, X) \leq \text{cat}(B, X)$.
(6) $\text{cat}(A, X)$ is subadditive; i.e., $\text{cat}(A \cup B, X) \leq \text{cat}(A, X) + \text{cat}(B, X)$.
(7) If A and B are closed subsets of X and A is deformable into B in X (i.e., the inclusion $i : A \to X$ is homotopic as a map of A into X to a map with image in B), then $\text{cat}(A, X) \leq \text{cat}(B, X)$.
(8) If $h : X \to X$ is a homeomorphism then $\text{cat}(h(A), X) = \text{cat}(A, X)$.

To simplify our discussion of Lusternik-Schnirelman Theory we will temporarily assume that M is compact. For $m \leq \text{cat}(M)$ we define $\mathcal{F}_m$ to be the collection of all compact subsets F of M such that $\text{cat}(F, M) \geq m$. Note that $\mathcal{F}_m$ contains M itself and so is non-empty. We define $c_m(f) = \text{minimax}(f, \mathcal{F}_m)$. By the monotonicity of $\text{cat}(\ , M)$ we can equally well define $c_m(f)$ by the formula

$$c_m(f) = \inf\{a \in \boldsymbol{R} \mid \text{cat}(M_a(f), M) \geq m\} .$$

9.2.8. Proposition.   *For $m = 0, 1, \ldots, \text{cat}(M)$, $c_m(f)$ is a critical value of M.*

PROOF. This is immediate from The Minimax Principle, since by (7) above, $\mathcal{F}_m$ is homotopy invariant.   ∎

Now $\mathcal{F}_{m+1}$ is clearly a subset of $\mathcal{F}_m$, so $c_m(f) \leq c_{m+1}(f)$. But of course equality *can* occur (for example if f is constant). However as the next result shows, this will be compensated for by having more critical points at this level.

9.2.9. Lusternik-Schnirelman Multiplicity Theorem.
If $c_{n+1}(f) = c_{n+2}(f) = \cdots c_{n+k}(f) = c$ then there are at least k critical points at the level c. Hence if $1 \leq m \leq \text{cat}(M)$ then f has at least m critical points at or below the level $c_m(f)$. In particular every smooth function $f : M \to \boldsymbol{R}$ has at least $\text{cat}(M)$ critical points altogether.

PROOF.   Suppose that there are only a finite number r of critical points $x_1, \ldots, x_r$ at the level c and choose open neighborhoods O_i of the x_i whose closures are disjoint closed disks (hence in particular contractible). Putting $O = O_1 \cup \ldots \cup O_r$, clearly $\text{cat}(O, M) \leq r$. By the First Deformation Theorem, for some $\epsilon > 0$ $M_{c+\epsilon} \setminus O$ can be deformed into $M_{c-\epsilon}$. Since $c - \epsilon < c = c_{n+1}$, $\text{cat}(M_{c-\epsilon}, M) < n+1$, and so by (7) above $\text{cat}(M_{c+\epsilon} \setminus O, M) \leq n$. Thus, by subadditivity and monotonicity of cat,

$$\text{cat}(M_{c+\epsilon}, M) \leq \text{cat}((M_{c+\epsilon} \setminus O) \cup O, M) \leq n + r$$

and hence

$$c < c + \epsilon < \inf\{a \in R \mid \text{cat}(M_a, M) > n + r + 1\} = c_{n+r+1}(f).$$

Since on the other hand $c = c_{n+k}(f)$, (and $c_m(f) \leq c_{m+1}(f)$) it follows that $n + r + 1 > n + k$, so $r \geq k$.   ∎

Taken together the following two propositions make it easy to compute exactly the category of some spaces.

9.2.10. Proposition.   *If M is connected, and A is a closed subset of M, then* $\text{cat}(A, M) \leq \dim(A) + 1$.

PROOF.   (Cf. [Pa5]). Let $\{O_\alpha\}$ be a cover of A by A-open sets, each contractible in M. Letting $n = \dim(A)$, by a lemma of J. Milnor (cf. [Pa4, Lemma 2.4]), there is a an open cover $\{G_{i\beta}\}$, $i = 0, 1, \ldots, n$, $\beta \in B_i$ of A, refining the covering by the O_α, such that $G_{i\beta} \cap G_{i\beta'} = \emptyset$ for $\beta \neq \beta'$. Since each $G_{i\beta}$ is contractible in M, and M is connected, it follows that $G_i = \bigcup\{G_{i\beta} \mid \beta \in B_i\}$ is contractible in M for $i = 0, 1, \ldots, n$. Let $\{U_{i\beta}\}$, $\beta \in B_i$ be a cover of A by A-open sets with $\overline{U}_{i\beta} \subseteq G_{i\beta}$. Then for $i = 0, 1, \ldots, n$, $A_i \overset{\text{def}}{=} \bigcup\{\overline{U}_{i\beta} \mid \beta \in B_i\}$ is a subset of G_i and hence contractible in M, and $A = \cup A_i$. Finally, since the $\overline{U}_{i\beta}$ are closed in A and locally finite, each A_i is closed in A and hence in M, so $\text{cat}(A, M) \leq n + 1$.   ∎

9.2.11. Proposition. $\text{cat}(M) \geq \text{cuplong}(M) + 1$, *provided M is connected.*

PROOF.   Cf. [BG].
The topological invariant $\text{cuplong}(M)$ is defined as the largest integer n such that, for some field F, there exist cohomology classes $\gamma_i \in H^{k_i}(M, F)$, $i = 1, \ldots, n$, with positive degrees k_i, such that $\gamma_1 \cup \ldots \cup \gamma_n \neq 0$. Thus

9.2.12. Proposition.   *If M is an n-dimensional manifold and for some field F their is a cohomology class $\gamma \in H^1(M, F)$ such that $\gamma^n \neq 0$, then* $\text{cat}(M) = n + 1$.

9.2.13. Corollary.   *The n-dimensional torus T^n and the n-dimensional projective space RP^n both have category $n + 1$.*

Recall that RP^n is the quotient space obtained by identifying pairs of antipodal points, x and $-x$, of the unit sphere S^n in R^{n+1}. Thus a function on RP^n is the same as a function on S^n that is "even", in the sense that it takes the same value at antipodal points x and $-x$.

9.2.14. Proposition.   *Any smooth even function on S^n has at least $n + 1$ pairs of antipodal critical points.*

An important and interesting application of the latter proposition is an existence theorem for certain so-called "non-linear eigenvalue problems". Let $\Phi : R^n \to R^n$ be a smooth map. If $\lambda \in R$ and $0 \neq x \in R^n$ satisfy $\Phi(x) = \lambda x$, then x is called an eigenvector and λ an eigenvalue of Φ. In applications Φ is often of the form ∇F for some smooth real-valued function $F : R^n \to R$, and moreover F is usually even. For example if A is a self-adjoint linear operator on R^n and we define $F(x) = \frac{1}{2}\langle Ax, x \rangle$, then F is even, $\nabla F = A$, and we are led to the standard linear eigenvalue problem. Usually we look for eigenvectors on $S_r = \{x \in R^n \mid \|x\| = r\}$, $r > 0$.

9.2.15. Proposition.   *A point x of S_r is an eigenvector of ∇F if and only if x is a critical point of $F|S_r$. In particular if F is even then each S_r contains at least n pairs of antipodal eigenvectors for ∇F.*

PROOF.   Define $G : R^n \to R$ by $G(x) = \frac{1}{2}\|x\|^2$, so $\nabla G_x = x$ and hence all positive real numbers are regular values of G. In particular $S_r = G^{-1}(\frac{1}{2}r^2)$ is a regular level of G. By the Lagrange Multiplier Theorem (cf. Appendix A) x in S_r is a critical point of $F|S_r$ if and only if $\nabla F_x = \lambda \nabla G_x = \lambda x$ for some real λ.   ∎

9.3   The Second Deformation Theorem.

We will call a closed interval $[a, b]$ of real numbers *non-critical* with respect to f if it contains no critical values of f. Recalling that the set $f(\mathcal{C})$ of critical values of f is closed in R it follows that for some $\epsilon > 0$ the interval $[a - \epsilon, b + \epsilon]$ is also non-critical. If $[a, b]$ is non-critical then the set $\mathcal{N} = f^{-1}([a, b])$ will be called a *non-critical neck* of M with respect to f. We will now prove the important fact that $\mathcal{N}$ has a very simple structure: namely it is diffeomorphic to $\mathcal{W} \times [a, b]$ where $\mathcal{W} = f^{-1}(b)$.

Since $(\nabla f)f = \|\nabla f\|^2$, on the set $M \setminus \mathcal{C}$ of regular points, where $\|\nabla f\| \neq 0$, the smooth vector field $Y = -\frac{1}{\|\nabla f\|^2}\nabla f$ satisfies $Yf = -1$. More generally if $F : R \rightarrow R$ is any smooth function vanishing in a neighborhood of $f(\mathcal{C})$, then $X = (F \circ f)Y$ is a smooth vector field on M that vanishes in a neighborhood of $\mathcal{C}$, and $Xf = -(F \circ f)$. We denote by Φ_t the flow on M generated by X. Let us choose $F : R \rightarrow R$ to be a smooth, non-negative function that is identically one on a neighborhood of $[a, b]$ and zero outside $[a - \epsilon/2, b + \epsilon/2]$.

9.3.1. Proposition.   *With the above choice of F, the vector field X on M has bounded length and hence the flow Φ_t it generates is a one-parameter group of diffeomorphisms of M.*

PROOF.   From the definition of Y it is clear that $\|Y\| = \frac{1}{\|\nabla f\|}$ so that $\|X\| = \frac{1}{\|\nabla f\|}|F \circ f|$. Since F has compact support it is bounded, and since $|F \circ f|$ vanishes outside $f^{-1}([a - \epsilon/2, b + \epsilon/2])$, it will suffice to show that $\frac{1}{\|\nabla f\|}$ is bounded on $f^{-1}([a - \epsilon/2, b + \epsilon/2])$, or equivalently that $\|\nabla f\|$ is bounded away from zero on $f^{-1}([a - \epsilon/2, b + \epsilon/2])$. But if not, then by Condition C we could find a sequence $\{x_n\}$ in $f^{-1}([a - \epsilon/2, b + \epsilon/2])$ converging to a critical point p of f. Then $f(p) \in [a - \epsilon/2, b + \epsilon/2]$, contrary to our assumption that the interval $[a - \epsilon, b + \epsilon]$ contains no critical values of f. ∎

Denote by $\gamma(t, c)$ the solution of the ordinary differential equation $\frac{d\gamma}{dt} = -F(\gamma)$ with initial value c. Since $\frac{d}{dt}(f \circ \Phi_t(x)) = X_{\varphi_t(x)}f = -F(f \circ \Phi_t(x))$, it follows that $f(\Phi_t(x)) = \gamma(t, f(x))$, and hence that $\Phi_t(f^{-1}(c)) = f^{-1}(\gamma(t, c))$. In particular *the flow Φ_t permutes the level sets of f*. From the definition of $\gamma(t, c)$ it follows that $\gamma(t, c) = c - t$ for $c \in [a, b]$ and $c - t \geq a$, while $\gamma(t, c) = c$ if $c > b + \epsilon$ or $c < a - \epsilon$. Since $\Phi_t(f^{-1}(c)) = f^{-1}(\gamma(t, c))$, it follows that if we write W for the b level of f, then Φ_{b-c} maps W diffeomorphically onto $f^{-1}(c)$ for all c in $[a, b]$ while, for all t, Φ_t is the identity outside the non-critical neck $f^{-1}([a - \epsilon/2, b + \epsilon/2])$.

In all that follows we shall denote by I the unit interval $[0, 1]$, and if $G : X \times I \rightarrow Y$ is any map, then for t in I we shall write $G_t : X \rightarrow Y$ for the map $G_t(x) = G(x, t)$. Recall that an *isotopy* of a smooth manifold M is a smooth map $G : M \times I \rightarrow M$ such that G_t is a diffeomorphism of M for all t in I and G_0 is the identity map of M. If A and B are subsets of M with $B \subseteq A$ then we say G *deforms* A onto B if $G_t(A) \subseteq A$ for all t and $G_1(A) = B$. And we say that G *fixes* a subset S of M if $G_t(x) = x$ for all (x, t) in $S \times I$. Finally if $f : M \rightarrow R$ then we shall say G *pushes down the levels* of f if for all $c \in R$ and $t \in I$ we have $G_t(f^{-1}(c)) = f^{-1}(c')$, where $c' \leq c$.

9.3.2. Second Deformation Theorem.   *If the interval $[a, b]$ is non-critical for the smooth function $f : M \to R$ then there is a deformation G of M that pushes down the levels of f and deforms M_b onto M_a. If $\epsilon > 0$ then we can assume G fixes the complement of $f^{-1}(a - \epsilon, b + \epsilon)$.*

PROOF.   Using the above notation we can define the deformation G by $G(x, t) = \Phi_{(b-a)t}(x)$.   ∎

9.3.3. Non-Critical Neck Principle.   *If $[a, b]$ is a non-critical interval of a smooth function $f : M \to R$ and W is the b-level of f, then there is a diffeomorphism of the non-critical neck $\mathcal{N} = f^{-1}([a, b])$ with $W \times [a, b]$, under which the restriction of f to $\mathcal{N}$ corresponds to the projection of $W \times [a, b]$ onto $[a, b]$.*

PROOF.   We define the map G of $W \times [a, b]$ into $\mathcal{N}$ by $G(x, t) = \Phi_{(b-t)}(x)$. Since $x \in W$, $f(x) = b$ and hence $f(G(x,t)) = (b - (b - t)) = t$. If $v \in TW_x$ then $DG(v, \frac{\partial}{\partial t}) = D\Phi_t(v) + X$. Now Φ_t maps W diffeomorphically onto $\tilde{W} = f^{-1}(t)$ and $TM_{\Phi_t(x)}$ is clearly spanned by the direct sum of $T\tilde{W}_{\Phi_t(x)}$ and $X_{\Phi_t(x)}$. It now follows easily from the Inverse Function Theorem that Φ is a diffeomorphism.   ∎

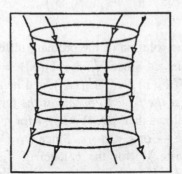

A Non-Critical Neck.
The ellipses represent the level surfaces, and the vertical curves represent flowlines of the gradient flow.

The intuitive content of the above results deserves being emphasized. As a ranges over a non-critical interval the diffeomorphism type of the a-level of f, the diffeomorphism type of M_a, and even the diffeomorphism type of the the pair (M, M_a) is *constant*, that is it is independent of a. Now, as we shall see shortly, if we assume that our function f satisfies a certain simple, natural, and generic non-degeneracy assumption (namely, that it is what is called a Morse function) then the set of critical points of f is discrete. For simplicity let us assume for the moment that M is compact. Then the set of critical points is finite and of course the set of critical values of f is then *a fortiori* finite. Let us denote them, in increasing order, by $c_1, c_2, \ldots, c_k$, and let us choose real numbers $a_0, a_1, \ldots, a_k$ with $a_0 < c_1 < a_1 < c_2 < \ldots < a_{k-1} < c_k < a_k$. Notice that c_1 must be the minimum of f, so that M_{a_0} is empty. And similarly c_k is the maximum of f so that M_{a_k} is all of M. More generally, by the above remark, the diffeomorphism type of M_{a_i} does not depend on the choice of a_i in the interval (c_i, c_{i+1}), so **we can think of a Morse function f as providing us with a specific method for "building up" our manifold M inductively in a finite number of discrete stages,** starting with the empty M_{a_0} and then, step by step, creating $M_{a_{i+1}}$ out of M_{a_i} by some "process" that takes place at the critical level c_{i+1}, finally ending up with M. Moreover the "process" that gives rise to the sudden changes in the topology of $f^{-1}(a)$ and of M_a as a crosses a critical value is not at all mysterious. From the point of view of M_a it is called "adding a handle", while from the point of view of the level $f^{-1}(a)$ it is just a "cobordism". From either point of view it can be analyzed fairly completely and is the basis for almost all classification theorems for manifolds.

9.4 Morse Functions.

An elementary corollary of the Implicit Function Theorem is an important local canonical form theorem for a smooth function $f : M \to R$ in the neighborhood of a regular point p; namely $f - f(p)$ is linear in a suitable coordinate chart centered at p. Equivalently, in this chart f coincides near p with its first order Taylor polynomial: $f(p) + df_p$.

But what if p is a critical point of f? Of course f will not necessarily be locally constant near p, but a natural conjecture is that, under some "generic" non-degeneracy assumption, we should again have a local canonical form for f near p, namely in a suitable local chart, (called a Morse Chart), f should coincide with its *second* order Taylor polynomial near p. That such a canonical form does exist generically is called The Morse Lemma and plays a fundamental role in Morse Theory. Before stating it precisely we review some standard linear algebra, adding some necessary infinite dimensional touches.

Let V be the model hilbert space for M, and let $\mathcal{A} : V \times V \to R$ be a continuous, symmetric, bilinear form on V. We denote by $f_{\mathcal{A}} : V \to R$ the associated homogeneous quadratic polynomial; $f_{\mathcal{A}}(x) = \frac{1}{2}\mathcal{A}(x, x)$. Now $\mathcal{A}$ defines a bounded linear map $\hat{\mathcal{A}} : V \to V^*$ by $\hat{\mathcal{A}}(x)(v) = \mathcal{A}(x, v)$. Using the canonical identification of V with V^* we can interpret $\hat{\mathcal{A}}$ as a bounded linear map $A : V \to V$, characterized by $\hat{\mathcal{A}}(x, v) = \langle Ax, v \rangle$, so that $f_{\mathcal{A}}(x) = \frac{1}{2}\langle Ax, x \rangle$. Since $\mathcal{A}$ is symmetric, A is self-adjoint. The bilinear form $\mathcal{A}$ is called non-degenerate if $\hat{A} : V \to V^*$ (or $A : V \to V$) is a linear isomorphism, i.e., if 0 does not belong to $\text{Spec}(A)$, the spectrum of A. While we will be concerned primarily with the non-degenerate case, for now we make a milder restriction. Let $V^0 = \ker(A)$. The dimension of V^0 is called the *nullity* of the quadratic form f_A. There is a densely-defined self-adjoint linear map $A^{-1} : (V^0)^\perp \to (V^0)^\perp$. But of course A^{-1} may be unbounded. Since $\|A\| = \sup\{|\lambda| \mid \lambda \in \text{Spec}(A)\}$ and $\text{Spec}(A^{-1}) = (\text{Spec}(A))^{-1}$, equivalently $\text{Spec}(A)$ might have 0 as a limit point. It is *this* that we assume does not happen.

9.4.1. Assumption.   *Zero is not a limit point of the Spectrum of A. Equivalently, if A does not have a bounded inverse then $V^0 = \ker(A)$ has positive dimension and A has a bounded inverse on $(V^0)^\perp$.*

(Of course in finite dimensions this is a vacuous assumption).

Choose $\epsilon > 0$ so that $(-\epsilon, \epsilon) \cap \text{Spec}(A)$ contains at most zero. Let $p^+ : R \to R$ be a continuous function such that $p^+(x) = 1$ for $x \geq \epsilon$ and $p^+(x) = 0$ for $x \leq \frac{\epsilon}{2}$. And define $p^- : R \to R$ by $p^-(x) = p^+(-x)$. Finally let $p^0 : R \to R$ be continuous with $p^0(0) = 1$ and $p^0(x) = 0$ for $|x| \geq \frac{\epsilon}{2}$. Then using the functional calculus for self-adjoint operators [Lang], we can define three commuting orthogonal projections $P^+ = p^+(A)$, $P^0 = p^0(A)$, and $P^- = p^-(A)$ such that $P^+ + P^0 + P^-$ is the identity map of V. Clearly $V^0 = \text{im}(P^0)$ and we define $V^+ = \text{im}(P^+)$ and $V^- = \text{im}(P^-)$, so that V is the orthogonal direct sum $V^+ \oplus V^0 \oplus V^-$. (In the finite dimensional case V^+ and V^- are respectively the direct sums of the positive and of the negative eigenspaces of A). The dimension of V^- is called the *index* of the quadratic form f_A and the dimension of V^+ is called its *coindex*.

Let $\varphi : R \to R$ be a continuous strictly positive function with $\varphi(\lambda) = \sqrt{\frac{2}{|\lambda|}}$ for $|\lambda| \geq \epsilon$, and $\varphi(0) = 1$. Then $\Phi = \varphi(A)$ is a self-adjoint linear diffeomorphism of V with itself. Since $\frac{1}{2}\varphi(\lambda)\lambda\varphi(\lambda) = \text{sgn}(\lambda) = p^+(\lambda) - p^-(\lambda)$ for all λ in $\text{Spec}(A)$, it follows that $\frac{1}{2}\Phi A \Phi = P^+ - P^-$, so that

$$f_A(\Phi(x)) = \tfrac{1}{2}\langle A\Phi x, \Phi x\rangle$$
$$= \langle \tfrac{1}{2}\Phi A \Phi x, x\rangle$$
$$= \langle P^+ x, x\rangle - \langle P^- x, x\rangle$$
$$= \|P^+ x\|^2 - \|P^- x\|^2.$$

9.4.2. Proposition.   *Let $A : V \to V$ be a bounded self-adjoint operator and $f_A : V \to R$ the homogeneous quadratic polynomial $f_A(x) = \frac{1}{2}\langle Ax, x\rangle$. If 0 is not a limit point of $\text{Spec}(A)$ then V has an orthogonal decomposition $V =$*

$V^+ \oplus V^0 \oplus V^-$ *(with $V^0 = \ker(A)$)* *and a self-adjoint linear diffeomorphism* $\Phi : V \approx V$ *such that*

$$f_A(\Phi(x)) = \|P^+(x)\|^2 - \|P^-(x)\|^2 \, ,$$

where P^+ and P^- are the orthogonal projections of V on V^+ and V^- respectively.

We now return to our smooth function $f : M \to R$. (For the moment we do not need the Riemannian structure on M.)

We associate to each pair of smooth vector fields X and Y on M, a smooth real valued function $B(X,Y) = X(Yf)$. We note that $B(X,Y)(p)$ is just the directional derivative of Yf at p in the direction X_p, so in particular its value depends on X only through its value, X_p, at p. Now if p is a critical point of f then $B(X,Y)(p) - B(Y,X)(p) = X_p(Yf) - Y_p(Xf) = [X,Y]_p(f) = df_p([X,Y]) = 0$. It follows that in *this* case $B(X,Y)(p) = B(Y,X)(p)$ also only depends on Y through its value, Y_p, at p. This proves:

9.4.3. Hessian Theorem. *If p is a critical point of a smooth real valued function $f : M \to R$ then there is a uniquely determined symmetric bilinear form $\mathrm{Hess}(f)_p$ on TM_p such that, for any two smooth vector fields X and Y on M, $\mathrm{Hess}(f)_p(X_p, Y_p) = X_p(Yf)$.*

We call $\mathrm{Hess}(f)_p$ the *Hessian bilinear form* associated to f at the critical point p, and we will also denote the related Hessian quadratic form by $\mathrm{Hess}(f)_p$ (i.e., $\mathrm{Hess}(f)_p(v) = \frac{1}{2}\mathrm{Hess}(f)_p(v,v)$). (Given a local coordinate system $x_1, \ldots, x_n$ for M at p, evaluating $\mathrm{Hess}(f)_p(\frac{\partial}{\partial x_i}, \frac{\partial}{\partial x_j})$ we see that the matrix of $\mathrm{Hess}(f)_p$ is just the classical "Hessian matrix" of second partial derivatives of f.)

We shall say that the critical point p is *non-degenerate* if $\mathrm{Hess}(f)_p$ is non-degenerate, and we define the nullity, index, and coindex of p to be respectively the nullity, index, and coindex of $\mathrm{Hess}(f)_p$. Finally, f is called a *Morse Function* if all of its critical points are non-degenerate.

Using the Riemannian structure of M we have a self-adjoint operator $\mathrm{hess}(f)_p$, defined on TM_p, and characterized by $\langle \mathrm{hess}(f)_p(X), Y \rangle = \mathrm{Hess}(f)_p(X,Y)$. Then the nullity of p is the dimension of the kernel of $\mathrm{hess}(f)_p$, p is a non-degenerate critical point of f when $\mathrm{hess}(f)_p$ has a bounded inverse, and, in finite dimensions, the index of p is the sum of the dimensions of eigenspaces of $\mathrm{hess}(f)_p$ corresponding to negative eigenvalues.

Let ∇ denote any connection on TM (*not* necessarily the Levi-Civita connection). Then ∇ induces a family of associated connections on all the tensor bundle over M, characterized by the fact that covariant differentiation commutes with contraction and the "product rule" holds. The latter means that, for example given vector fields X and Y on M,

$$\nabla_X(Y \otimes df) = \nabla_X(Y) \otimes df + Y \otimes \nabla_X(df).$$

Contracting the latter gives:

$$X(Yf) = df(\nabla_X Y) + i_Y i_X(\nabla df).$$

If we define $\text{Hess}^\nabla(f)$ to be ∇df then we can rewrite this equation as

$$\text{Hess}^\nabla(f)(X,Y) = X(Yf) - df(\nabla_X Y).$$

This has two interesting consequences. First, interchanging X and Y and subtracting gives:

$$\text{Hess}^\nabla(f)(X,Y) - \text{Hess}^\nabla(f)(Y,X) = df(\tau^\nabla(X,Y)),$$

where τ^∇ is the torsion tensor of ∇. Thus if ∇ is a symmetric connection (i.e. $\tau^\nabla = 0$), as is the Levi-Civita connection, then $\text{Hess}^\nabla(f)$ is a *symmetric* covariant two-tensor field on M. And in any case, at a critical point p of f, where $df_p = 0$, we have:

$$\text{Hess}^\nabla(f)(X_p, Y_p) = X_p(Yf) = \text{Hess}(f)_p(X_p, Y_p).$$

9.4.4. Proposition.   *If ∇ denotes the Levi-Civita connection for M, then* $\text{Hess}^\nabla(f) \overset{\text{def}}{\equiv} \nabla df$ *is a symmetric two-tensor field on M that at each critical point p of f agrees with* $\text{Hess}(f)_p$.

9.4.5. Corollary. $\text{hess}^\nabla(f) \overset{\text{def}}{\equiv} \nabla(\nabla f)$ *is a field of self adjoint operators on M that at each critical point p of f agrees with* $\text{hess}(f)_p$.

There is yet another interpretation of $\text{Hess}(f)_p$ that is often useful. The differential df of f is a section of T^*M that vanishes at p, so *its* differential, $D(df)_p$, is a linear map of TM_p into $T(T^*M)_{0_p}$ (where 0_p denotes the zero element of T^*M_p). Now $T(T^*M)_{0_p}$ is canonically the direct sum of two subspaces; the "vertical" subspace, tangent to the fiber T^*M_p, which we identify with T^*M_p, and the "horizontal" space, tangent to the zero section, which we identify with TM_p. If we compose $D(df)_p$ with the projection onto the vertical space we get a linear map $TM_p \to T^*M_p$ that, under the natural isomorphism of bilinear maps $V \times V \to R$ with linear maps $V \to V^*$, is easily seen to correspond to $\text{Hess}(f)_p$. With this alternate definition of

$\mathrm{Hess}(f)_p$, the condition for p to be non-degenerate is that $\mathrm{Hess}(f)_p$ map TM_p isomorphically onto T^*M_p.

It is clear that at the critical point p of f, $\mathrm{Hess}(f)_p$ determines the second order Taylor polynomial of f at p. But what is less obvious is that, at least in the non-degenerate case, f "looks like" its second order Taylor polynomial near p, a fact known as the Morse Lemma.

Let us put $V = T^*M_p$, $A = \mathrm{hess}(f)_p$, and let V^+, V^0, and V^- be as above, i.e., the maximal subspaces of V on which A is positive, zero, and negative. Recall that a chart for M centered at p is a diffeomorphism Φ of a neighborhood O of 0 in V onto a neighborhood U of p in M with $\Phi(0) = p$. We call Φ a *Morse chart of the first kind at p* if $f(\Phi(v)) - f(p) = \mathrm{Hess}(v) = \frac{1}{2}\langle Av, v\rangle$. And Φ is called a *Morse chart of the second kind at p* (or simply a *Morse chart at p*) if $f(\Phi(v)) - f(p) = \|P^+v\|^2 - \|P^-v\|^2$, where P^+ and P^- are the orthogonal projections on V^+ and V^-. It is clear that a Morse chart of the second kind *is* a Morse chart of the first kind. Moreover, by the proposition at the beginning of this section, if a Morse chart of the first kind exists at p, then so does a Morse chart of the second kind. In this case we shall say simply that Morse charts exist at p.

9.4.6. Morse Lemma.   *If p is a non-degenerate critical point of a smooth function $f : M \to R$ then Morse charts exist at p.*

PROOF.   Since the theorem is local we can take M to be V and assume p is the origin 0. Also without loss of generality we can assume $f(0) = 0$. We must show that, after a smooth change of coordinates φ, f has the form $f(x) = \frac{1}{2}\langle Ax, x\rangle$ in a neighborhood O of 0. Since $df_p = 0$, by Taylor's Theorem with remainder we can write f near 0 in the form $f(x) = \frac{1}{2}\langle A(x)x, x\rangle$, where $x \mapsto A(x)$ is a smooth map of O into the self-adjoint operators on V. Since $A(0) = A = \mathrm{hess}(f)_0$ is non-singular, $A(x)$ is also non-singular in a neighborhood of 0, which we can assume is O. We define a smooth map B of O into the group $GL(V)$ of invertible operators on V by $B(x) = A(x)^{-1}A(0)$, and note that $B(0)$ is I, the identity map of V. Now a square root function is defined in the neighborhood of I by a convergent power series with real coefficients, so we can define a smooth map C of O into $GL(V)$ by $C(x) = \sqrt{B(x)}$. Since $A(0)$ and $A(x)$ are self-adjoint it is immediate from the definition of B that $B(x)^*A(x) = A(x)B(x)$. This same relation then holds if we replace $B(x)$ by any polynomial in $B(x)$, and hence if we replace $B(x)$ by $C(x)$ which is a limit of such polynomials. Thus

$$C(x)^*A(x)C(x) = A(x)C(x)^2 = A(x)B(x) = A(0)$$

or $A(x) = C_1(x)^*AC_1(x)$, where we have put $C_1(x) = C(x)^{-1}$. If we define a smooth map φ of O into V by $\varphi(x) = C_1(x)x$, then $f(x) = \langle C_1(x)^*AC_1(x)x, x\rangle = \langle A\varphi(x), \varphi(x)\rangle$, so it remains only to check that φ is a valid change of coordinates at 0, i.e., that $D\varphi_0$ is invertible. But $D\varphi_x = C_1(x) + D(C_1)_x(x)$, so in particular $D\varphi_0 = C_1(0) = I$. ∎

9.4.7. Corollary 1.   *A non-degenerate critical point of a smooth function $f : M \to R$ is isolated in the set C of all critical points of f. In particular if f is a Morse function then C is a discrete subset of M.*

PROOF.   Maintaining the assumptions and notations introduced in the proof of the Morse Lemma we have $f(x) = \frac{1}{2}\langle Ax, x \rangle$ in a neighborhood $\mathcal{O}$ of 0, and hence $df_x = Ax$ for x in $\mathcal{O}$. Since A is invertible, df_x does not vanish in $\mathcal{O}$ except at 0. ∎

9.4.8. Corollary 2.   *If a Morse function $f : M \to R$ satisfies Condition C then for any finite interval $[a, b]$ of real numbers there are only a finite number of critical points p of f with $f(p) \in [a, b]$. In particular the set C of critical values of f is a discrete subset of R.*

PROOF.   We saw earlier that Condition C implies that f restricted to C is proper, so the set of critical points p of f with $f(p) \in [a, b]$ is compact. But by Corollary 1 it is also discrete. ∎

Since we are going to be focusing our attention on Morse functions, a basic question to answer is, whether they necessarily exist, and if so how rare or common are they. Fortunately, at least in the finite dimensional case this question has an easy and satisfactory answer; Morse functions form an open, dense subspace in the C^2 topology of the space $C^2(M, R)$ of all C^2 real valued functions on M. The easiest, but not the most elementary, approach to this problem is through Thom's transversality theory. Let ξ be a smooth vector bundle of fiber dimension m over a smooth n-manifold M . Recall that if s_1 and s_2 are two C^1 sections of ξ with $s_1(p) = s_2(p) = v$, then we say that these sections have *transversal intersection* (or are transversal) at p if, when considered as submanifolds of M, their tangent spaces at v span the entire tangent space to ξ at v. We say s_1 and s_2 are transversal if they have transversal intersection wherever they meet. Since each section has dimension n, and ξ has dimension $m + n$ the condition for transversality is that the intersection of their tangent spaces at v should have dimension $(n+n)-(n+m) = n-m$. So if ξ has fiber dimension n then this intersection should have dimension zero and, since Ds_i maps TM_p isomorphically onto the tangent space to s_i at v, this just means that $Ds_1(u) \neq Ds_2(u)$ for $u \neq 0$ in TM_p. In particular for ξ the cotangent bundle T^*M, a section s vanishing at p is transversal to the zero section at p if and only if $\text{im}(Ds)$ is disjoint from the horizontal space at p, or equivalently if and only if the composition of Ds with projection onto the vertical subspace, T^*M_p is an isomorphism. Recalling our alternate interpretation of $\text{Hess}(f)_p$ above we see:

9.4.9. Lemma. *The critical point p of $f : M \to R$ is non-degenerate if and only if df is transversal to the zero section of T^*M at p. Thus f is a Morse function if and only if df is transversal to the zero section.*

Thom's k-jet transversality theorem [Hi, p.80] states that if s_0 is a C^{k+1} section of a smooth vector bundle ξ over a compact manifold M and $J^k\xi$ is the corresponding bundle of k-jets of sections of ξ, then in the space $C^{k+1}(\xi)$ of C^{k+1} sections of ξ with the C^{k+1} topology, the set of sections s whose k-jet extension $j_k s$ is transversal to $j_k s_0$ is open and dense. If we take for ξ the trivial bundle $M \times R$ then a section becomes just a real valued function, and we can identify $J^1\xi$ with T^*M so that $j_1 f$ is just df. Finally, taking $k = 1$ and letting s_0 be the zero section, Thom's theorem together with the above lemma gives the desired conclusion, that Morse functions are open and dense in $C^2(M, R)$.

As a by-product of the section on the Morse Theory of submanifolds of Euclidean space, we will find a much more elementary approach to this question, that gives almost as complete an answer.

9.5 Passing a Critical Level.

We now return to our basic problem of Morse Theory; reconstructing the manifold M from knowledge about the critical point structure of the function $f : M \to R$.

To get a satisfactory theory we will supplement the assumptions (a), (b), and (c) of the Introduction with the following additional assumption:

(d) f is a Morse function.

As we saw in the preceding section this implies that for any finite interval $[a, b]$ their are only a finite number of critical points p of f with $f(p)$ in $[a, b]$, and hence only a finite number of critical values of f in $[a, b]$.

Our goal is to describe how M_α changes as α changes from one non-critical value a to another b. Now, by the Second Deformation Theorem, the diffeomorphism type of M_α is constant for α in a non-critical interval of f, hence we can easily reduce our problem to the case that their is a single critical value c in (a, b), and without loss of generality we can assume that $c = 0$. So what we want to see is how to build M_ϵ out of $M_{-\epsilon}$ when 0 is the unique critical value of f in $[-\epsilon, \epsilon]$. In general there could be a finite number of critical points $p_1, \ldots, p_k$ at the level 0, and eventually we shall consider that case explicitly. But the discussion will be greatly simplified (with no essential loss of generality) by assuming at first that there is a *unique* critical point p at the level 0. We will let k and l denote the index and coindex of f at p and $n = k + l$ the dimension of M. If $n = \infty$ then one or both of k and l will also be infinite; nevertheless we shall write R^k, and R^l for the Hilbert spaces of dimension k and l, and $R^n = R^l \times R^k$.

As in all good construction projects we will proceed in stages, and start with some blueprints before filling in the precise mathematical details.

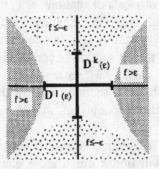

- We will denote by $D^k(\epsilon)$, and $D^l(\epsilon)$ the disks of radius $\sqrt{\epsilon}$ centered at the origin in R^k and R^l respectively. We will write D^k and D^l for the unit disks. The product $D^l \times D^k$, attached in a certain way to $M_{-\epsilon}$ will be called a *handle of index* k.

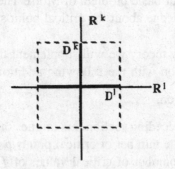

- We will construct a smooth submanifold N of M with $M_{-\epsilon} \subseteq N \subseteq M_\epsilon$. Namely, $N = M_{-\epsilon}(g) = \{x \in M \mid g(x) \le -\epsilon\}$, where $g : M \to R$ is a certain smooth function that agrees with f where f is greater than ϵ (so that $M_\epsilon = M_\epsilon(f) = M_\epsilon(g)$). Moreover the interval $[-\epsilon, \epsilon]$ is non-critical for g, so by the Second Deformation Theorem there is an isotopy of M that deforms $M_\epsilon = M_\epsilon(g)$ onto $M_{-\epsilon}(g) = N$.

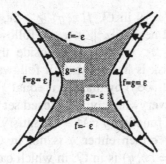

- The manifold N has a second description. Namely, N is an adjunction space that consists of $M_{-\epsilon}$ together with a subset $\mathcal{H}$, (called the "handle") that is diffeomorphic to the above product of disks and is glued onto $\partial M_{-\epsilon}$, the boundary of $M_{-\epsilon}$, by a diffeomorphism of $\partial D^l \times D^k$ onto $\mathcal{H} \cap \partial M_{-\epsilon}$.

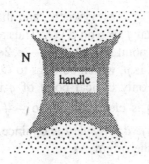

Thus **when we pass a critical level $f^{-1}(c)$ of f that contains a single non-degenerate critical point of index k, $M_{c+\epsilon}$ is obtained from $M_{c-\epsilon}$ by attaching to the latter a handle of index k.**

Now for the details. We identify a neighborhood $\mathcal{O}$ of p in M with a neighborhood of the origin in $R^n = R^l \times R^k$, using a Morse chart (of the second kind). We will regard a point of $\mathcal{O}$ as a pair (x, y), where $x \in R^l$ and $y \in R^k$. We suppose ϵ is chosen small enough that 0 is the only critical level of f in $[-2\epsilon, 2\epsilon]$, or equivalently so that p is the only critical point of f with $|f(p)| \leq 2\epsilon$. We can also assume ϵ so small that the closed disk of radius $2\sqrt{\epsilon}$ in R^{k+l} is included in $\mathcal{O}$. Thus f is given in $\mathcal{O}$ by $f(x, y) = \|x\|^2 - \|y\|^2$. Choose a smooth, non-increasing function $\lambda : R \to R$ that is identically 1 on $t \leq \frac{1}{2}$, positive on $t < 1$, and zero for $t \geq 1$. Then the function g is defined in $\mathcal{O}$ by $g(x, y) = f(x, y) - \frac{3\epsilon}{2}\lambda(\|x^2\|/\epsilon)$.

9.5.1. Lemma.   *The function g can be extended to be a smooth function $g : M \to R$ that is everywhere less than f and agrees with f wherever $f \geq \epsilon$ and also, outside $\mathcal{O}$, wherever $f \geq -2\epsilon$. In particular $M_\epsilon(g) = M_\epsilon(f)$.*

PROOF. Suppose (x, y) in $\mathcal{O}$, $f(x, y) \geq -2\epsilon$, and $g(x, y) \neq f(x, y)$. Then $\lambda(\|x^2\|/\epsilon) \neq 0$ and hence $\|x^2\| < \epsilon$. It follows that $\|x^2\| + \|y^2\| = 2\|x^2\| - f(x, y) < 2\epsilon + 2\epsilon$, i.e., (x, y) is inside the disk of radius $2\sqrt{\epsilon}$. Recalling that the latter disk is interior to $\mathcal{O}$ it follows that if we extend g to the remainder of $f^{-1}([-2\epsilon, \infty))$ by making it equal f outside $\mathcal{O}$, then it will be smooth. Since $g \leq f$ everywhere on the closed set $f^{-1}([-2\epsilon, \infty))$ we can now further extend it to a function $g : M \rightarrow R$ satisfying the same inequality on all of M. If $f(q) \geq \epsilon$ then either q is not in $\mathcal{O}$, so $g(q) = f(q)$ by definition of g, or else $q = (x, y)$ is in $\mathcal{O}$, in which case $\|x\|^2 \geq f(x, y) \geq \epsilon$, so $\lambda(\|x^2\|/\epsilon) = 0$, and again $g(q) = f(q)$. ∎

9.5.2. Lemma.   *For the function g, extended as above, the interval $[-\epsilon, \epsilon]$ is a non-critical interval. (In fact p is the only critical point of g in $S = g^{-1}([-2\epsilon, \epsilon])$, and $g(p) = -\frac{3\epsilon}{2}$).*

PROOF. Recalling that $f \geq g$ everywhere, and that, outside $\mathcal{O}$, $f = g$ wherever $f \geq -2\epsilon$, it follows that $f = g$ on $S \backslash \mathcal{O}$. Thus any critical point of g in $S \backslash \mathcal{O}$ would also be a critical point of f in $f^{-1}[-2\epsilon, 2\epsilon]$. But by our choice of ϵ, the only such critical point is p, which belongs to $\mathcal{O}$. Thus it will suffice to show that, inside of $\mathcal{O}$, the only critical point of g is $p = (0, 0)$, where $g(x, y) = f(x, y) - \frac{3\epsilon}{2}\lambda(\frac{\|x^2\|}{\epsilon})$ is clearly equal to $-\frac{3\epsilon}{2}\lambda(0) = -\frac{3\epsilon}{2} < -\epsilon$. But in $\mathcal{O}$, $dg = (2 - 3\lambda'(\frac{\|x\|^2}{\epsilon}))x\,dx + 2y\,dy$ and, since λ' is a non-positive function, this vanishes only at the origin. ∎

Now it is time to make the concept of "attaching a handle" mathematically precise.

9.5.3. Definition. Let P and N be smooth manifolds with boundary, having the same dimension $n = k + l$, and with P a smooth submanifold of N. Let α be a homeomorphism of $D^l \times D^k$ onto a closed subset $\mathcal{H}$ of N. We shall say that N *arises from P by attaching a handle of index k and coindex l (or a handle of type (k, l)) with attaching map α* if:

(1) $N = P \cup \mathcal{H}$,

(2) $\alpha|(D^l \times S^{k-1})$ is a diffeomorphism onto $\mathcal{H} \cap \partial P$,

(3) $\alpha|(D^l \times \mathring{D}^k)$ is a diffeomorphism onto $N \backslash P$.

Here $\mathring{D}^k$ denotes the interior of the k-disk. Of course $D^l \times D^k$ is not a *smooth* manifold (it has a "corner" along $\partial D^l \times \partial D^k$), but both $D^l \times S^{k-1}$ and $D^l \times \mathring{D}^k$) *are* smooth manifolds with boundary.

Note that if $k < \infty$, (so, in particular, if $n < \infty$) then $l = n - k$ is determined by k, so in this case it is common to speak simply of attaching a handle of index k.

The following example (with $k = l = 1$) is a good one to keep in mind:

P is the lower hemisphere of the standard S^2 in R^3, (think of it as a basket), and $\mathcal{H}$, the handle of the basket, is a tubular neighborhood of that part of a great circle lying in the upper hemisphere. Of course, where the handle and basket meet, the sharp corner should be smoothed.

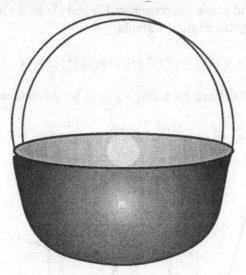

Another example that can be easily visualized ($k = 1, l = 2$) is the "solid torus" formed by gluing a 1-handle $D^2 \times D^1$ to the unit disk in R^3 (a bowling ball with a carrying handle).

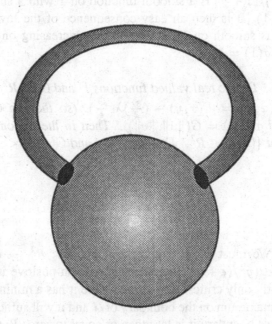

Recall that, in the case of interest to us, $P = M_{-\epsilon}(f)$, $N = M_{-\epsilon}(g)$, and we define the handle $\mathcal{H}$ to be the closure of the set of $(x, y) \in \mathcal{O}$ such that $f(x, y) > -\epsilon$ and $g(x, y) < -\epsilon$. Then recalling that, outside of $\mathcal{O}$, f and g agree where $f \geq -\epsilon$, it follows from the definition of N as $M_{-\epsilon}(g)$ that $N = M_{-\epsilon}(f) \cup \mathcal{H}$. What remains then is to define the homeomorphism α of $D^l \times D^k$ onto $\mathcal{H}$, and prove the properties (2) and (3) of the above definition.

We define α by the explicit formula:

$$\alpha(x, y) = (\epsilon\sigma(\|y\|^2))^{\frac{1}{2}} x + (\epsilon\sigma(\|y\|^2)\|x\|^2 + \epsilon)^{\frac{1}{2}} y$$

where $\sigma : I \to I$ is defined by taking $\sigma(s)$ to be the unique solution of the equation

$$\frac{\lambda(\sigma)}{(1 + \sigma)} = \frac{2}{3}(1 - s).$$

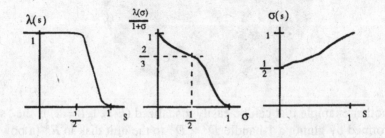

Clearly $\lambda(\sigma)/(1 + \sigma)$ is a smooth function on I with a strictly negative derivative on $[0, 1)$. It is then an easy consequence of the Inverse Function Theorem that σ is smooth on $[0, 1)$ and strictly increasing on I. Moreover $\sigma(0) = 1/2$ and $\sigma(1) = 1$.

9.5.4. Lemma.   *Define real valued functions F and G on $\mathbf{R}^2$ by $F(x, y) = x^2 - y^2$ and $G(x, y) = F(x, y) - (\frac{3\epsilon}{2})\lambda(\frac{x^2}{\epsilon})$. (so that, in $\mathcal{O}$, $f(u, v) = F(\|u\|, \|v\|)$ and $g(u, v) = G(\|u\|, \|v\|)$). Then in the region $\mathcal{U}$ that is the closure of the set $\{(x, y) \in \mathbf{R}^2 \mid F(x, y) > -\epsilon \text{ and } G(x, y) < -\epsilon\}$ we have*

$$x^2 \leq \epsilon\sigma\left(\frac{y^2}{\epsilon + x^2}\right).$$

PROOF.   We must show that the function $h : \mathbf{R}^2 \to \mathbf{R}$, defined by $h(x, y) = x^2 - \epsilon\sigma(y^2/(\epsilon + x^2))$, is everywhere non-positive in $\mathcal{U}$. Now for fixed y, h is clearly only critical for $x = 0$, where it has a minimum. Hence h must assume its maximum on the boundary of $\mathcal{U}$ and it will suffice to show that everywhere on this boundary it is less than or equal to zero. But the boundary of $\mathcal{U}$ is the closure of the union of the two curves $\partial_1 = \{(x, y) | F(x, y) = $

$-\epsilon,\quad G(x,y) < -\epsilon\}$ and $\partial_2 = \{(x,y)|F(x,y) > -\epsilon,\quad G(x,y) = -\epsilon\}$ and we will show that $h \leq 0$ both on ∂_1 and on ∂_2.

Indeed on ∂_1, since $G < F$, $(-3\epsilon/2)\lambda(x^2/\epsilon) < 0$ so $\lambda(x^2/\epsilon) > 0$, which implies $x^2/\epsilon < 1$ or $x^2 < \epsilon$. On the other hand, since $x^2 - y^2 = F(x,y) = -\epsilon$, $y^2/(\epsilon+x^2) = 1$ so $\sigma(y^2/(\epsilon+x^2)) = 1$ and hence $h(x,y) = x^2 - \epsilon < 0$.

On ∂_2 we again have $G < F$, so as above $x^2/\epsilon < 1$. The equality $G(x,y) = -\epsilon$ gives

$$\frac{y^2}{\epsilon+x^2} = 1 - \left(\frac{3}{2}\right)\frac{\lambda(x^2/\epsilon)}{(1+x^2/\epsilon)}.$$

Now $x^2/\epsilon < 1/2$ would imply both $\lambda(x^2/\epsilon) = 1$ and $1 + x^2/\epsilon < \frac{3}{2}$, so the displayed inequality would give the impossible $y^2/(\epsilon + x^2) < 0$. Thus $1/2 \leq x^2/\epsilon < 1$, so x^2/ϵ is in the range of σ, say $x^2/\epsilon = \sigma(\rho)$. Then by definition of σ,

$$\frac{y^2}{\epsilon+x^2} = 1 - \left(\frac{3}{2}\right)\frac{\lambda(\sigma(\rho))}{(1+\sigma(\rho))} = 1 - \left(\frac{3}{2}\right)\left(\frac{2}{3}\right)(1-\rho) = \rho,$$

and hence

$$h(x,y) = x^2 - \epsilon\sigma\left(\frac{y^2}{\epsilon+x^2}\right) = \epsilon\sigma(\rho) - \epsilon\sigma(\rho) = 0,$$

so h vanishes on ∂_2 as well. ∎

The remainder of the proof is now straightforward. We will leave to the reader the easy verifications that if $(u,v) = \alpha(x,y)$ then $f(u,v) \geq -\epsilon$ and $g(u,v) \leq -\epsilon$, so that α maps $D^l \times D^k$ into $\mathcal{H}$.

Conversely, suppose that (u,v) belongs to $\mathcal{H}$. Then $F(\|u\|, \|v\|) = \|u\|^2 - \|v\|^2 \geq -\epsilon$ and $G(\|u\|, \|v\|) \leq -\epsilon$. Thus $\|v\|^2/(\epsilon + \|u\|^2) \leq 1$, so $y = (\epsilon + \|u\|^2)^{-1/2}v \in D^k$. Also $\sigma(\|v\|^2/(\epsilon + \|u\|^2))$ is well defined, and by the preceding Lemma $\|u\|^2/\epsilon\sigma(\|v\|^2/(\epsilon + \|u\|^2)) \leq 1$ so that $x = (\epsilon\sigma(\|v\|^2/(\epsilon + \|u\|^2)))^{-1/2}u \in D^l$. It follows that $\beta(u,v) = (x,y)$ defines a map $\beta : \mathcal{H} \to D^l \times D^k$, and it is elementary to check that α and β are mutually inverse maps, so that α is a homeomorphism of $D^l \times D^k$ onto $\mathcal{H}$. Since σ is smooth and has positive derivative in $[0,1)$ it follows that α is a diffeomorphism on $D^l \times \mathring{D}^k$. On $D^l \times S^{k-1}$ the map α reduces to

$$\alpha(x,y) = \epsilon^{1/2}x + (\epsilon(\|y\|^2 + 1))^{1/2}y$$

which is clearly a diffeomorphism onto $\mathcal{H} \cap \partial M_{-\epsilon}$. This completes the proof that $M_{c+\epsilon}$ is diffeomorphic to $M_{c-\epsilon}$ with a handle of index k attached.

Finally, let us see what modifications are necessary when we pass a critical level that contains more than one critical point. First note that the whole process of adjoining a handle to $M_{-\epsilon}$ took place in a small neighborhood of p (the domain of a Morse chart at p). Thus if we have several critical points at the same level then we can carry out the same attaching process independently in disjoint neighborhoods of these various critical points.

9.5.5. Definition. Suppose we have a sequence of smooth manifolds $N = N_0, N_1, \ldots, N_s = M$ such that N_{i+1} arises from N_i by attaching a handle of type (k_i, l_i) with attaching map α_i. If the images of the α_i are disjoint then we shall say that M *arises from N by the disjoint attachment of handles of type* $((k_1, l_1), \ldots, (k_s, l_s))$ *with attaching maps* $(\alpha_1, \ldots, \alpha_s)$.

9.5.6. Theorem.   *Let f be a Morse function that is bounded below and satisfies Condition C on a complete Riemannian manifold M. Suppose $c \in (a, b)$ is the only critical value of f in the interval $[a, b]$, and that $p_1, \ldots, p_s$ are all the critical points of f at the level c. Let p_i have index k_i and coindex l_i. Then M_b arises from M_a by the disjoint attachment of handles of type* $((k_1, l_1), \ldots, (k_s, l_s))$.

Let us return to our example of the height function on the torus. That is, we take M to be the surface of revolution in $\mathbf{R}^3$, formed by rotating the circle $x^2 + (y - 2)^2 = 1$ about the x-axis. The function $f : M \rightarrow \mathbf{R}$ defined by $f(x, y, z) = z$ is a Morse function with critical points at $(0, 0, -3), (0, 0, -1), (0, 0, 1)$, and $(0, 0, 3)$, and with respective indices 0,1,1,2. Here is a diagram showing the sequence of steps in the gradual building up of this torus, starting with a disk (or 0-handle), adding two consecutive 1-handles, and finally completing the torus with a 2-handle.

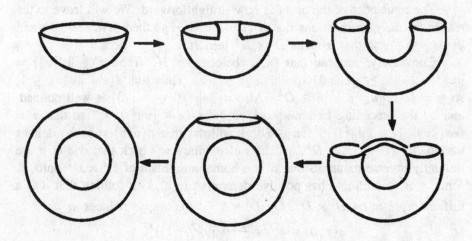

9.6 Morse Theory of Submanifolds.

As we shall now see, there is a more detailed Morse theory for submanifolds of a Euclidean space. In this section proofs of theorems will often be merely sketched or omitted entirely, since details can be found in the first two sections of Chapter 4.

We assume in what follows that M is a compact, smooth n-manifold smoothly embedded in R^N, and we let k denote the codimension of the embedding. (We recall that, by a classical theorem of H. Whitney, any abstractly given compact (or even second countable) n-manifold can *always* be embedded as a closed submanifold of R^{2n+1}, so for $k > n$ we are not assuming anything special about M. We will consider M as a *Riemannian* submanifold of R^N, i.e., we give it the Riemannian metric induced from R^N.

Let $L(R^N, R^N)$ denote the vector space of linear operators from R^N to itself and $L^s(R^N, R^N)$ the linear subspace of self-adjoint operators. We define a map $P : M \to L^s(R^N, R^N)$, called the *Gauss map* of M by $P_x = $ orthogonal projection of R^N onto TM_x. We denote the kernel of P_x (that is the normal space to M at x) by ν_x. We will write $P_x^\perp$ for the orthogonal projection $I - P_x$ of R^N onto ν_x. Since the Gauss map is a map of M into a vector space, at each point x of M it has a well-defined differential $(DP)_x : TM_x \to L^s(R^N, R^N)$.

9.6.1. Definition. For each normal vector v to M at x we define a linear map $A_v : TM_x \to R^N$, called the *shape operator of M at x in the direction v*, by $A_v(u) = -(DP)_x(u)(v)$.

Since the tangent bundle TM and normal bundle $\nu(M)$ are both subbundles of the trivial bundle $M \times R^N$, the flat connection on the latter induces connections ∇^T and ∇^ν on TM and on $\nu(M)$, Explicitly, given $u \in TM_x$, a smooth curve $\sigma : (-\epsilon, \epsilon) \to M$ with $\sigma'(0) = u$, and a smooth section $s(t)$ of TM (resp. $\nu(M)$) along σ, we define $\nabla_u^T(s)$ (resp. $\nabla_u^\nu(s)$) by $P_x(s'(0))$ (resp. $P_x^\perp(s'(0))$. Clearly ∇^T is just the Levi-Civita connection for M.

The following is an easy computation.

9.6.2. Proposition. *Given u in TM_x and e in $\nu(M)_x$ let $\sigma : (-\epsilon, \epsilon) \to M$ be a smooth curve with $\sigma'(0) = u$ and let $s(t)$ and $v(t)$ be respectively tangent and normal vector fields along σ with $v(0) = e$. Let Pe denote the section $x \mapsto P_x(e)$ of $T(M)$. Then:*

(i) $A_e(u) = -P_x v'(0)$; hence each A_v maps TM_x to itself,

(ii) $A_e(u) = \nabla_u^T(Pe)$,

(iii) $\langle A_e(u), s(0) \rangle = \langle e, s'(0) \rangle$.

Suppose $F : R^N \to R$ is a smooth real valued function on R^N and $f = F|M$ is its restriction to M. Since $df = dF|TM_x$, it follows immediately from the definition of the gradient of a function that for x in M we have $\nabla f_x = P_x(\nabla F_x)$, and as a consequence we see that *the critical points of f are just the points of M where ∇F is orthogonal to M.* We will use this fact in what follows without further mention. Also, as we saw in the section on Morse functions, at a critical point x of f $\mathrm{Hess}(f)_x = \nabla^T(\nabla f)$.

We define a smooth map $H : S^{N-1} \times R^N \to R$ by $H(a, x) = \langle a, x \rangle$ and, for each $a \in S^{N-1}$, we define $H_a : R^N \to R$ and $h_a : M \to R$ by $H_a(x) = H(a, x)$ and $h_a = H_a|M$. Each of the functions h_a is called a "height" function. Intuitively, if we think of a as the unit vector in the "vertical" direction, so $\langle a, x \rangle = 0$ defines the sea-level surface, then $h_a(x)$ represents the height of a point $x \in M$ above sea-level. Similarly we define $F : R^N \times M \to R$ by $F(a, x) = \frac{1}{2}\|x - a\|^2$, and for $a \in R^N$ we define $F_a : R^N \to R$ and $f_a : M \to R$ by $F_a(x) = F(a, x)$ and $f_a = F_a|M$. Somewhat illogically we will call each f_a a "distance" function.

For certain purposes the height functions have nicer properties, while for others the distance functions behave better. Fortunately there is one situation when there is almost no difference between the height function h_a and the distance function f_a.

9.6.3. Proposition.   *If M is included in some sphere centered at the origin, then h_a and f_{-a} differ by a constant; hence they have the same critical points and the same Hessians at each critical point.*

PROOF.   Suppose that M is included in the sphere of radius ρ, i.e., for x in M we have $\|x\|^2 = \rho^2$. Then

$$f_{-a}(x) = \frac{1}{2}\|x + a\|^2$$
$$= \frac{1}{2}(\|x\|^2 + \|a\|^2) + \langle x, a \rangle$$
$$= \frac{1}{2}(\rho^2 + \|a\|^2) + h_a(x). \quad \blacksquare$$

Thus if the particular embedding of M in Euclidean space is not important we can always use stereographic projection to embed M in the unit sphere in one higher dimension and get both the good properties of height functions and of distance functions at the same time.

9.6.4. Proposition.   *The gradient of h_a at a point x of M is $P_x a$, the projection of a on TM_x, so the critical points of h_a are just those points x of M where a lies in the space ν_x, normal to M at x. Similarly the gradient of f_a at x is $P_x(x - a)$, so the critical points of f_a are the points x of M where the line segment from a to x meets M orthogonally.*

PROOF.   Since H_a is linear, $d(H_a)_x(v) = H_a(v) = \langle a, v \rangle$, so that $(\nabla H_a)_x = a$. Similarly, since F_a is quadratic we compute easily that $d(F_a)_x(v) = \langle x - a, v \rangle$ so $(\nabla F_a)_x = x - a$. $\quad \blacksquare$

By another easy computation we find:

9.6.5. Proposition.   *At a critical point x of h_a, $\text{hess}(h_a)_x = A_a$. Similarly at a critical point x of f_a, $\text{hess}(f_a)_x = I + A_{x-a}$.*

Thus, because the hessian of h_v is self-adjoint we see

9.6.6. Corollary.   *For each v in $\nu(M)$, A_v is a self-adjoint operator on TM_x.*

We recall that for v in $\nu(M)_x$, the second fundamental form of M at x in the direction v is the quadratic form II_v on TM_x defined by A_v, i.e.,

$$II_v(u_1, u_2) = \langle A_v u_1, u_2 \rangle,$$

and the eigenvalues of A_v are called the principal curvatures of M at x in the normal direction v.

9.6.7. Proposition.   *Given e in $\nu(M)_x$, let $v(t) = x + te$. Then for all real t, x is a critical point of $f_{v(t)}$ with hessian $I - tA_e$. Thus the nullity of $f_{v(t)}$ at x is just the multiplicity of t^{-1} as a principal curvature of M at x in the direction e. In particular, x is a degenerate critical point of $f_{v(t)}$ if and only if t^{-1} is a principal curvature of M at x in the direction e. If 1 is not such a principal curvature then x is a non-degenerate critical point of f_{x+e}, and its index is*

$$\sum_{0 < t < 1} \text{nullity of } f_{v(t)} \text{ at } x.$$

PROOF.   The first statement follows directly from the above propositions by taking $a = x + te$, and it is then immediate that the nullity of $f_{v(t)}$ is $\mu(t^{-1})$, where $\mu(\lambda)$ denotes the multiplicity of λ as an eigenvalue of A_e. On the other hand, the multiplicity of λ as an eigenvalue of $\text{hess}(f_{x+te})_x = 1 - A_e$ is clearly $\mu(1 - \lambda)$. Since $\lambda < 0$ if and only if $1 - \lambda$ equals t^{-1} for some t in $(0, 1)$, the formula for the index of f_{x+e} at x follows. ∎

We will denote by $Y : \nu(M) \to R^N$ the "exponential" or "endpoint" map $(x, v) \mapsto x + v$ of the normal bundle to M into the ambient R^N.

9.6.8. Definition.   If $a = Y(x, e)$ then a is called *non-focal* for M with respect to x if $DY_{(x,e)}$ is a linear isomorphism. If on the contrary $DY_{(x,e)}$ has a kernel of positive dimension m then a is called a *focal point* of multiplicity m for M with respect to x. A point a of R^N is called a *focal point* of M if, for some $x \in M$, a is focal for M with respect to x.

9.6.9. Proposition.   *The point $a = Y(x, e)$ is a focal point of multiplicity m for M with respect to x if and only if x is a degenerate critical point of f_a of nullity m.*

PROOF.   Let $\gamma(t) = (\sigma(t), v(t))$ be a smooth normal field to M along a smooth curve $\sigma(t)$, with $\sigma(0) = x$ and $v(0) = e$. Then:

$$
\begin{aligned}
DY_{(x,e)}(\gamma'(0)) &= \left(\frac{d}{dt}\right)_{t=0} Y(\sigma(t), v(t)) \\
&= \left(\frac{d}{dt}\right)_{t=0} (\sigma(t) + v(t)) \\
&= \sigma'(0) + v'(0) \\
&= \sigma'(0) + P_x v'(0) + P_x^\perp v'(0) \\
&= (I - A_e)\sigma'(0) + P_x^\perp v'(0).
\end{aligned}
$$

since by a proposition above $A_e \sigma'(0) = -P_x v'(0)$. Now taking $\sigma(t) \equiv x$ and $v(t) = e + tv$ gives the geometrically obvious fact that $DY_{(x,e)}$ reduces to the identity on the subspace $\nu(M)_x$. It then follows by elementary linear algebra that $\ker(DY_{(x,e)})$ and $\ker(I - A_e)$ have the same dimension. Since we have seen that $\mathrm{hess}(f_a) = I - A_e$ the final statement follows. ∎

9.6.10. Corollary.   *If $a \in R^N$ is not a focal point of M then the distance function f_a is a Morse function on M.*

9.6.11. Morse Index Theorem.   *If M is a compact, smooth submanifold of R^N, $x \in M$, $e \in \nu(M)_x$, and $a = x + e$ is non-focal for M with respect to x, then x is a non-degenerate critical point of the "distance function" $f_a : M \to R$, $v \mapsto \left(\frac{1}{2}\right)\|v - a\|^2$, and the index of x as a critical point of f_a is just equal to the number of focal points for M with respect to x along the segment joining x to a, each counted with its multiplicity.*

PROOF.   Immediate from the above. ∎

Next recall Sard's Theorem. Suppose X and Y are smooth, second countable manifolds of the same dimension and $F : X \to Y$ is a C^1 map. A point p of X is called a *regular point* of F if $DF_p : TX_p \to TY_{f(p)}$ is a linear isomorphism, or equivalently if F is a local diffeomorphism at p. A point q of Y is called a *regular value* of F if all points of $F^{-1}(q)$ are regular points of F; other points of N are called *critical values* of F. Then Sard's Theorem [DR, p.10] states that **the set of critical values of F has measure zero**, so that in particular regular values are dense. Taking $X = \nu(M)$, $Y = R^N$, and $F = Y$, the critical values are those points of R^N which are focal points of M. Thus, by the above Corollary, the distance function f_a is a Morse function for almost all $a \in R^N$. In particular if f_a is not itself a Morse function, that is if a is a focal point of M, we can nevertheless choose a sequence a_n of non-focal points converging to a, and then f_{a_n} will be a sequence of Morse functions converging to f_a in the C^∞ topology.

As an easy application of this fact we can now give a simple proof that any smooth real valued function on M, $G : M \to R$, can be approximated in the C^∞ topology by Morse functions. From the above remark it will suffice to show that G can be realized as a distance function, and of course it does no harm to change G by adding a constant. Define an embedding of M in the sphere of radius r in R^{N+2} by $x \mapsto \left(x, G(x), \sqrt{r^2 - \|x\|^2 - G(x)^2} \right)$, where of course r is chosen greater than the maximum of $\sqrt{\|x\|^2 + G(x)^2}$. Then, looked at in R^{N+2}, G is clearly the height function h_a, where $a = (0, 1, 0)$. So, by an earlier remark, G differs by a constant from the distance function f_{-a}.

9.7 The Morse Inequalities.

First we review some terminology.

We will be dealing with categories of pairs of spaces (X, A). We assume the reader is familiar with the usual notions of maps $(X, A) \to (Y, B)$, homotopies between such maps, etc. As usual we identify the pair $(X, \emptyset)$ with X. Homology groups $H_*(X, A)$ will always be with respect to some fixed principal ideal domain $\mathcal{R}$. In our applications $\mathcal{R}$ will usually be either Z or Z_2.

Let X be a space and A a closed subspace of X. A *retraction* of X onto A is a map $r : X \to A$ that is the identity on A. If such a map exists we call A a *retract* of X. If their is a homotopy $\rho : X \times I \to X$ such that ρ_0 is the identity map of X and $\rho_1 = r$ then we call ρ a *deformation retraction* of X onto A and call A a *deformation retract* of X. And finally if in addition $\rho_t|A$ is the identity map of A for all t in I then we call ρ a *strong deformation retraction*, and call A a *strong deformation retract* of A.

9.7.1. Lemma. *Let X be a convex subset of R^n, and A a closed subset of X. If r is a retraction of X onto A then $\rho(x, t) = (1 - t)x + tr(x)$ is a strong deformation retraction of X onto A.*

PROOF. Trivial. ∎

9.7.2. Proposition. $(0 \times D^k) \cup (D^l \times S^{k-1})$ *is a strong deformation retract of $D^l \times D^k$.*

PROOF.   Since $D^l \times D^k$ is convex in $\mathbf{R}^{l+k}$ we need only define a retraction

$$r : D^l \times D^k \to (0 \times D^k) \cup (D^l \times S^{k-1}).$$

Of course $r(0, y) = (0, y)$ and, for $x \neq 0$,

$$r(x, y) = \begin{cases} \left(0, \frac{2\|y\|}{2 - \|x\|}\right) & \text{if } \|y\| \leq 1 - \frac{\|x\|}{2}; \\ \left(\|x\| + 2\|y\| - 2)\frac{x}{\|x\|}, \frac{y}{\|y\|}\right) & \text{otherwise.} \end{cases} \qquad \blacksquare$$

Here is a diagram of the retraction r.

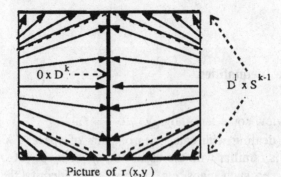

$0 \times D^k$

$D^l \times S^{k-1}$

Picture of r (x,y)

We next recall the concept of attaching a k–cell to a space. Let Y be closed a subspace of a space X, and $G : D^k \to X$ a continuous map of the k–disk onto another closed subspace, e^k, of X. We will write $X = Y \cup_g e^k$ and say X is obtained from Y by attaching a k–cell with attaching map $g \overset{\text{def}}{=} G|S^{k-1}$ if:

(1) $X = Y \cup e^k$,

(2) G maps $\overset{\circ}{D}{}^k = D^k \setminus S^{k-1}$ homeomorphically onto $e^k \setminus Y$, and

(3) g maps S^{k-1} onto $\partial e^k \overset{\text{def}}{=} e^k \cap Y$.

G is called the *characteristic map* of the attaching. In our applications G will actually be a homeomorphism of D^k onto e^k.

Note that X can be reconstructed from Y and the attaching map $g :$ $S^{k-1} \to Y$ by taking the topological sum of D^k and Y and identifying x in $S^{k-1} = \partial D^k$ with $g(x)$ in Y.

Since by (2) we have a relative homeomorphism of the pairs of spaces, (D^k, S^{k-1}) and $(e^k, \partial e^k)$, it follows that the homology groups $H_l(D^k, S^{k-1})$ and $H_l(e^k, \partial e^k)$ are isomorphic. On the other hand we have an excision isomorphism between $H_l(D^k, S^{k-1})$ and $H_l(X, Y)$. Hence:

9.7.3. Proposition.   *If X is obtained from Y by attaching a k−cell then:*

$$H_l(X,Y) \approx H_l(e^k, \partial e^k)$$

$$\approx H_l(D^k, S^{k-1}) = \begin{cases} \mathcal{R} & \text{if } l=k; \\ 0 & \text{otherwise.} \end{cases}$$

If, for $i = 1,2$, X_i is a space and A_i is a subspace of X_i, then a map $f_1 : (X_1, A_1) \to (X_2, A_2)$ is called a *homotopy equivalence* of these pairs if there exists a map $f_2 : (X_2, A_2) \to (X_1, A_1)$ such that $f_1 \circ f_2$ and $f_2 \circ f_1$ are homotopic (as maps of pairs) to the respective identity maps. (f_2 is then called a homotopy inverse for f_1). If there is a homotopy equivalence $f_1 : (X_1, A_1) \to (X_2, A_2)$ then we say (X_1, A_1) and (X_2, A_2) are *homotopy equivalent* or *have the same homotopy type*. In this case $H_*(f_1) : H_*(X_1, A_1) \to H_*(X_2, A_2)$ is an isomorphism with inverse $H_*(f_2)$.

Suppose in particular X_2 is a subspace of X_1 and r is a strong deformation retraction of X_1 onto X_2. Then if $A_1 \subseteq A_2$, $r : (X_1, A_1) \to (X_2, X_2 \cap A_1)$ is a homotopy equivalence. (The inclusion $i : (X_2, X_2 \cap A_1) \to (X_1, A_1)$ is a homotopy inverse).

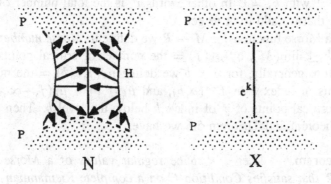

9.7.4. Theorem.   *Let N and P be smooth manifolds with boundary. If N arises from P by attaching a handle of type (k,l) then N has as a strong deformation retract a closed subspace $X = P \cup_g e^k$, obtained from P by attaching a k−cell e^k. In particular (N, P) has the homotopy type of P with a k−cell attached, so if $k < \infty$ then*

$$H_l(N, P) = \begin{cases} \mathcal{R} & \text{if } l=k; \\ 0 & \text{otherwise.} \end{cases}$$

PROOF. Let $\alpha : D^l \times D^k \approx \mathcal{H}$ be the map attaching the handle $\mathcal{H}$ to P to get N. Define $G : D^k \approx e^k$ by $G \overset{\text{def}}{=} \alpha|(0 \times D^k)$. The deformation retraction of $N = P \cup \mathcal{H}$ onto $P \cup e^k$ is of course the identity on P and equals $\alpha \circ r \circ \alpha^{-1}$ on $\mathcal{H}$, where r is the strong deformation retraction $r : D^l \times D^k \to (0 \times D^k) \cup (D^l \times S^{k-1})$ of the above proposition. ∎

9.7.5. Remark. Of course, more generally, If N arises from P by disjointly attaching handles of type $((k_1, l_1), \ldots, (k_s, l_s))$ then N has as a strong deformation retract a closed subspace $X = P \cup_{g_1} e^{k_1} \ldots \cup_{g_s} e^{k_s}$, obtained from P by disjointly attaching cells $e^{k_1}, \ldots, e^{k_s}$.

Suppose we have a sequence of closed subspaces X_i of X, $i = 0 \ldots n$, with

$$A = X_0 \subseteq X_1 \ldots \subseteq X_n = X$$

and maps $g_i : S^{k_i - 1} \to X_i$, $i = 0, \ldots, n-1$, such that $X_{i+1} \approx X_i \cup_{g_i} e^{k_i}$, i.e., X_{i+1} is homeomorphic to X_i with a k_i–cell attached by the attaching map g_i. In this case we call the pair (X, A) a (relative) *spherical complex*, and the sequence of attaching maps is called a *cell decomposition* for (X, A). If we only have a homotopy equivalence of X_{i+1} with $X_i \cup_{g_i} e^{k_i}$ then we shall call (X, A) a *homotopy spherical complex,* and call the sequence of g_i's a *homotopy cell decomposition.* In either case, for a given cell decomposition or homotopy cell decomposition we will denote by ν_i the number of cells $e^{k_0}, \ldots, e^{k_{n-1}}$ with $k_j = i$. In other words ν_i is the total number of cells of dimension i that we add to A to get X.

Given a Morse function $f : M \to R$ we define its *Morse numbers* $\mu_k(f)$, $0 \leq \ldots \leq k \leq \dim(M)$, by $\mu_k(f) =$ the number of critical points of f of index k. More generally, for $a < b$ we define $\mu_k(f, a, b) =$ the number of critical points of index k in $f^{-1}(a, b)$, and $\mu_k(f, b) = \mu_k(f, -\infty, b) =$ the number of critical points of f of index k below the level b. Then from the preceding theorem and Theorem 5.6 we have.

9.7.6. Theorem. *Let $a < b$ be regular values of a Morse function $f : M \to R$ that satisfies Condition C on a complete Riemannian manifold M. Then (M_b, M_a) is a homotopy spherical complex. In fact it has a homotopy cell decomposition with the number ν_k of cells of dimension k equal to $\mu_k(f, a, b)$.*

9.7.7. Corollary. *Any compact, smooth manifold M is a homotopy spherical complex, and in fact for any Morse function $f : M \to R$ there is a homotopy cell decomposition for M with $\nu_k = \mu_k(f)$.*

But wait! In the above theorem we have apparently ignored critical points of infinite index. Is this really legitimate? Yes, for the next proposition implies that attaching a handle of infinite index to a hilbert manifold does not change its homotopy type; so insofar as their effect on homotopy type is concerned we can simply ignore critical points of infinite index. (There is a beautiful result of N. Kuiper that infinite dimensional hilbert manifolds of the same homotopy type are diffeomorphic, so passing a critical point of infinite index does not even change diffeomorphism type.)

9.7.8. Proposition.   *If D^∞ is the closed unit disk in an infinite dimensional Hilbert space V, and $S^\infty = \partial D^\infty$ is the unit sphere in V, then there is a deformation retraction of D^∞ onto S^∞. Hence if A is any space and $g : S^\infty \to A$ is any continuous map then there is a deformation retraction of the adjunction space $X = A \cup_g D^\infty$ onto A, and in particular X has the same homotopy type as A.*

PROOF.   Since D^∞ is convex, it will suffice to show that there is a retraction of D^∞ onto S^∞. Now recall the standard proof of the Brouwer Fixed Point Theorem. If there were a fixed point free map $h : D^n \to D^n$ it would imply the existence of a deformation retraction r of D^n onto S^{n-1}; namely $r(x)$ is the point where the ray from $h(x)$ to x meets S^{n-1}. If $n < \infty$ this would contradict the fact that $H_n(D^n, S^{n-1}) = Z$, so there can be no such retraction and hence no such fix point free map. But when $n = \infty$ we will see that such a fixed point free map *does* exist, and hence so does the retraction r. This will be a consequence of two simple lemmas.

9.7.9. Lemma.   *D^∞ has a closed subspace homeomorphic to R.*

PROOF.   Let $\{e_n\}$ be an orthonormal basis for V indexed by Z, and define $F : R \to D^\infty$ by $F(t) = \cos(\frac{1}{2}(t - n)\pi)e_n + \sin(\frac{1}{2}(t - n)\pi)e_{n+1}$ for $n \leq t \leq n + 1$. It is easily checked that F is a homeomorphism of R into D^∞ with closed image.   ∎

9.7.10. Lemma.   *If a normal space X has a closed subspace A homeomorphic to R then it admits a fixed point free map $H : X \to X$.*

PROOF.   Since A is homeomorphic to R it admits a fixed point free map $h : A \to A$, corresponding to say translation by 1 in R. Since A is closed in X and X is normal, by the Tietze Extension Theorem h can be extended to a continuous map $H : X \to A$, and we may regard H as a map $H : X \to X$. If $x \in A$ then $x \neq h(x) = H(x)$, while if $x \in X \setminus A$ then, since $H(x) \in A$, again $H(x) \neq x$.   ∎

While the number ν_k of cells of dimension k in a cell decomposition for a spherical complex (X, A) is clearly *not* in general a topological invariant, there are important relations between the ν_k and topological invariants of (X, A). In particular there are the famous "Morse inequalities", relating certain alternating sums of the ν_k to corresponding alternating sums of betti numbers. We consider these next.

In what follows all pairs of spaces (X, A) considered are assumed "admissible", that is homotopy spherical complexes. We fix a field F, and for each admissible pair (X, A) and non-negative integer k we define $b_k(X, A)$, the kth betti number of (X, A) with respect to F, to be the dimension of $H_k(X, A; F)$, and we recall that the Euler characteristic of (X, A), $\chi(X, A)$, is defined to be the alternating sum, $\sum_k (-1)^k b_k(X, A)$, of the betti numbers. (We shall see that it is independent of F).

For each non-negative integer k we define another topological invariant,

$$S_k(X, A) = \sum_{m=0}^{k} (-1)^{k-m} b_m(X, A).$$

Thus:

$$S_0 = b_0,$$
$$S_1 = b_1 - b_0 = b_1 - S_0,$$
$$\cdots\cdots\cdots$$
$$S_k = b_k - b_{k-1} + \ldots \pm b_0 = b_k - S_{k-1},$$
$$\chi = b_0 - b_1 + b_2 - \ldots$$

9.7.11. Proposition.   *The Euler characteristic χ is additive and each S_k is subadditive. That is, given*

$$X_0 \subseteq X_1 \subseteq \ldots \subseteq X_n,$$

with all the pairs (X_i, X_{i-1}) admissible, we have:

$$S_k(X_n, X_0) \leq \sum_{i=1}^{n} S_k(X_i, X_{i-1}),$$

$$\chi(X_n, X_0) = \sum_{i=1}^{n} \chi(X_i, X_{i-1}).$$

PROOF.   By induction it suffices to show that for an admissible triple (X, Y, Z) we have $S_k(X, Z) \leq S_k(X, Y) + S_k(Y, Z)$, and $\chi(X, Z) = \chi(X, Y) + \chi(Y, Z)$. The long exact homology sequence for this triple:

$$\xrightarrow{\partial_{m+1}} H_m(Y, Z) \xrightarrow{i_m} H_m(X, Z) \xrightarrow{\partial_m} H_{m-1}(Y, Z) \longrightarrow$$

gives the short exact sequences:

$$0 \longrightarrow \mathrm{im}\,(\partial_{m+1}) \longrightarrow H_m(Y,Z) \longrightarrow \mathrm{im}\,(i_m) \longrightarrow 0,$$

$$0 \longrightarrow \mathrm{im}\,(i_m) \longrightarrow H_m(X,Z) \longrightarrow \mathrm{im}\,(j_m) \longrightarrow 0,$$

$$0 \longrightarrow \mathrm{im}\,(j_m) \longrightarrow H_m(X,Y) \longrightarrow \mathrm{im}\,(\partial_m) \longrightarrow 0,$$

and these in turn imply the identities:

$$b_m(Y,Z) = \dim \mathrm{im}\,(\partial_{m+1}) + \dim \mathrm{im}\,(i_m)$$
$$b_m(X,Z) = \dim \mathrm{im}\,(i_m) + \dim \mathrm{im}\,(j_m)$$
$$b_m(X,Y) = \dim \mathrm{im}\,(j_m) + \dim \mathrm{im}\,(\partial_m).$$

Subtracting the first and third equation from the second,

$$b_m(X,Z) - b_m(X,Y) - b_m(Y,Z) = -(\dim \mathrm{im}\,(\partial_m) + \dim \mathrm{im}\,(\partial_{m+1}))$$

so multiplying by $(-1)^{k-m}$, summing from $m = 0$ to $m = k$, and using that $\partial_0 = 0$ we get

$$S_k(X,Z) - S_k(X,Z) - S_k(X,Z) = -\dim \mathrm{im}\,(\partial_{k+1}) \leq 0.$$

Similarly, multiplying instead by $(-1)^m$, summing, and using that eventually $\partial_k = 0$ gives the additivity of χ. ∎

9.7.12.   Theorem.   *Let (X,A) be a homotopy spherical complex admitting a homotopy cell decomposition with ν_k cells of dimension k. If $b_k = b_k(X,A)$ denotes the kth betti number of (X,A) with respect to some fixed field F, then:*

$$b_0 \leq \nu_0,$$
$$b_1 - b_0 \leq \nu_1 - \nu_0,$$

$$\cdots\cdots\cdots\cdots\cdots$$

$$b_k - b_{k-1} + \ldots \pm b_0 \leq \nu_k - \nu_{k-1} + \ldots \pm \nu_0.$$

Moreover

$$\chi(X,A) \stackrel{\mathrm{def}}{\equiv} \sum_i (-1)^i b_i = \sum_i (-1)^i \nu_i.$$

PROOF. Let

$$A = X_0 \subseteq X_1 \subseteq \ldots \subseteq X_n = X$$

with $X_{i+1} = X_i \cup_{g_i} e^{k_i}$ be the cell decomposition for (X, A). Note that since $b_m(X_{i+1}, X_i) = \delta_{m\,k_i}$, it follows that $\sum_{i=0}^{n-1} b_m(X_{i+1}, X_i) = \nu_m$. Hence

$$\sum_{i=0}^{n-1} S_k(X_{i+1}, X_i) = \sum_{i=0}^{n-1} \sum_{m=0}^{k} (-1)^{k-m} b_m(X_{i+1}, X_i) = \sum_{m=0}^{k} (-1)^{k-m} \nu_m,$$

and

$$\sum_{i=0}^{n-1} \chi_k(X_{i+1}, X_i) = \sum_{i=0}^{n-1} \sum_{m=0}^{k} (-1)^m b_m(X_{i+1}, X_i) = \sum_{m=0}^{k} (-1)^m \nu_m,$$

so the theorem is immediate from the additivity of χ and the subadditivity of the S_k. ∎

9.7.13. Corollary.   *Let $a < b$ be regular values of a Morse function $f : M \to R$ that satisfies Condition C on a complete Riemannian manifold M. Let $\mu_k = \mu_k(f, a, b)$ denote the number of critical points of index k of f in $f^{-1}(a, b)$, and let $b_k = b_k(M_b, M_a)$ denote the kth betti number of (M_b, M_a) over some field F. Then:*
(Morse Inequalities)

$$b_0 \leq \mu_0,$$
$$b_1 - b_0 \leq \mu_1 - \mu_0,$$
$$\dotsb\dotsb\dotsb\dotsb$$
$$b_k - b_{k-1} + \dots \pm b_0 \leq \mu_k - \mu_{k-1} + \dots \pm \mu_0.$$

Moreover
(Euler Formula)

$$\chi(X, A) \stackrel{def}{\equiv} \sum_i (-1)^i b_i = \sum_i (-1)^i \mu_i.$$

Finally:
(Weak Morse Inequalities)

$$b_k \leq \mu_k.$$

PROOF.   The Morse Inequalities and Euler Formula are immediate from the theorem and Theorem 7.6. The Weak Morse Inequalities follow by adding two adjacent Morse Inequalities. ∎

9.7.14. Definition. A Morse function $f : M \to R$ on a compact manifold is called a *perfect* Morse function if all the Morse inequalities are equalities, or equivalently if $\mu_k(f) = b_k(M)$ for $k = 0, 1, \ldots, \dim(M)$.

Consider again our basic example of the height function on the torus T^2. Recall that $\mu_0 = 1$, $\mu_1 = 2$, and $\mu_2 = 1$. Since the torus is connected $b_0 = 1$, and since it is oriented $b_2 = 1$. Then by the Euler Formula we must have $b_1 = \mu_1 = 2$. In particular this is an example of a perfect Morse function.

More generally let Σ be an oriented surface of genus g, i.e., a sphere with g handles. There is a Morse function on Σ (an obvious generalization of the height function on the torus) that has one maximum, one minimum, and $2g$ saddles. The same argument as above shows that $b_0 = b_2 = 1$, $b_1 = 2g$, and that this is a perfect Morse function.

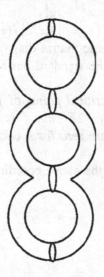

Now let $f : \Sigma \to R$ be *any* Morse function on Σ. We can rewrite the Euler Formula as a formula for the number of mountain passes on Σ, μ_1, in terms of the number of mountain peaks, μ_2, the number of valleys, μ_0, and the number of handles, g; namely

$$\mu_1 = (\mu_2 - 1) + (\mu_0 - 1) + 2g.$$

So, for a compact oriented surface, a Morse function is perfect precisely when it has a unique minimum and a unique maximum.

9.7.15. Theorem. *Suppose $f : M \to R$ is a Morse function on a compact manifold such that all the odd Morse numbers μ_{2k+1} are zero. Then all the odd betti number b_{2k+1} also vanish, and for the even betti numbers we have $b_{2k} = \mu_{2k}$. In particular f is a perfect Morse function.*

PROOF.   That the odd betti numbers are zero is immediate from the weak Morse inequalities. The Euler Formula then becomes

$$\chi(M) = b_0 + b_2 + \ldots + b_{2m} = \mu_0 + \mu_2 + \ldots + \mu_{2m},$$

so the weak Morse inequalities $b_{2k} \leq \mu_{2k}$ must in fact all be equalities.   ∎

As a typical application of the above result we will compute the betti numbers of n dimensional complex projective space, CP^n. Recall that CP^n is the quotient space of $C^{n+1} \setminus \{0\}$ under the equivalence relation $z \sim \lambda z$ for some non-zero $\lambda \in C$. For z in C^{n+1} we put $z = (z_0, \ldots, z_n)$ and if $z \neq 0$ then $[z]$ is its class in CP^n. The open sets $\mathcal{O}_k = \{z \in C^{n+1} \mid z_k \neq 0\}$, $k = 0, 1, \ldots n$ cover CP^n and in $\mathcal{O}_k$ we have coordinates $\{x_j^k, y_j^k\}$ $1 \leq j \leq n+1$ $j \neq k$ defined by $\frac{z_j}{z_k} = x_j^k + iy_j^k$.

Define $f : CP^n \to R$ by $f(z) = \langle Az, z \rangle / \langle z, z \rangle$, where $\langle w, z \rangle = \sum_i w_i \bar{z}_i$ and A is the hermitian symmetric matrix diag$(\lambda_0, \ldots, \lambda_n)$ with $(\lambda_0 < \lambda_1 < \ldots < \lambda_n)$. Let $e_0, \ldots, e_n$ be the standard basis for C^{n+1}.

9.7.16. Proposition.   *The critical points of f are the $[e_k]$. Moreover $[e_k]$ is non-degenerate and has index $2k$ Thus f is a perfect Morse function and the betti numbers b_k of CP^n are zero for k odd and 1 for $k = 0, 2, \ldots, 2n$.*

PROOF.   Exercise. Use the above coordinates to compute the differential and Hessian of f.   ∎

Chapter 10.

Advanced Critical Point Theory.

10.1 Refined Minimaxing.

Our original Minimax Principle located critical *levels*. Now we will look for more refined results that locate critical *points*.

In all that follows we assume that f is a smooth real valued function bounded below and satisfying Condition C on a complete Riemannian manifold M, and that M_0 is a closed subspace of M that is invariant under the positive time flow φ_t generated by $-\nabla f$. (In our applications M_0 will either be empty, or of the form M_c, or a subset of the set $\mathcal{C}$ of critical points of f.)

Let Y be a compact space and Y_0 a closed subspace of Y. We denote by $[(Y, Y_0), (M, M_0)]$ the set of homotopy classes of maps $h : Y \to M$ such that $h(Y_0) \subseteq M_0$. Given $\alpha \in [(Y, Y_0), (M, M_0)]$ we define $\mathcal{F}_\alpha = \{\text{im}\,(h) \mid h \in \alpha\}$, but we will use α and $\mathcal{F}_\alpha$ almost interchangeably, as in $\text{minimax}(f, \alpha) \overset{\text{def}}{\equiv} \text{minimax}(f, \mathcal{F}_\alpha)$. Clearly $\mathcal{F}_\alpha$ is invariant under the positive time flow φ_t, so by the Minimax Principle $\text{minimax}(f, \alpha)$ is a critical value of f.

In general given a family $\mathcal{F}$ of closed subsets F of M invariant under φ_t for $t > 0$, we shall say that "$\mathcal{F}$ hangs up at the level c" to indicate that $c = \text{minimax}(f, \mathcal{F})$. If further $S \subseteq f^{-1}(c)$ then we will say "$\mathcal{F}$ hangs up on S" if given any neighborhood U of S in M there is an $\epsilon > 0$ such that some F in $\mathcal{F}$ is included in $M_{c-\epsilon} \cup U$.

10.1.1. Refined Minimax Principle.
Let $\mathcal{F}$ be a family of closed subsets of M that is invariant under the positive time flow φ_t generated by $-\nabla f$. If $\mathcal{F}$ hangs up at the level c, then in fact it hangs up on $\mathcal{C}_c$.

PROOF. Since $\mathcal{C}$ is pointwise invariant under φ_t, given any neighborhood U of $\mathcal{C}_c$ there is a neighborhood O of $\mathcal{C}_c$ with $\varphi_1(O) \subseteq U$. By the First Deformation Theorem we may choose an $\epsilon > 0$ so that $\varphi_1(M_{c+\epsilon} \backslash O) \subseteq M_{c-\epsilon}$. Choose F in $\mathcal{F}$ with $F \subseteq M_{c+\epsilon}$. Then $\varphi_1(F) \in \mathcal{F}$, and since F is the union of $F \cap (M_{c+\epsilon} \backslash O)$ and $F \cap O$ it follows that

$$\varphi_1(F) \subseteq \varphi_1(M_{c+\epsilon} \backslash O) \cup \varphi(O) \subseteq M_{c-\epsilon} \cup U. \quad \blacksquare$$

To get still more precise results we assume f is a Morse function.

10.1.2. Theorem.   *Let $f : M \to R$ be a Morse function and assume that $\alpha \in [(Y, Y_0), (M, M_0)]$ hangs up at the level c of f, where $c > \max(f|M_0)$. Assume that f has a single critical point p at the level c, having index k, and let e^k denote the descending cell of radius $\sqrt{\epsilon}$ in some Morse chart of the second kind at p. Then for ϵ sufficiently small α has a representative h with $im(h) \subseteq M_{c-\epsilon} \cup e^k$.*

PROOF.   Since $c > \max(f|M_0)$, for ϵ small $M_0 \subseteq M_{c-\epsilon}$ and we can choose a neighborhood U of p with $U \subseteq M_{c+\epsilon}$. Since by the preceding proposition α hangs up on $C_c = \{p\}$ we can find a representative g of α with $im(g) \subseteq M_{c-\epsilon} \cup U \subseteq M_{c+\epsilon}$. But by an earlier result there is a deformation retraction ρ of $M_{c+\epsilon}$ onto $M_{c-\epsilon} \cup e^k$. Then $h = \rho \circ g$ also represents α and has its image in $M_{c-\epsilon} \cup e^k$.   ∎

10.1.3. Remark.   Of course if there are several critical points $p_1, \ldots, p_s$ at the level c having indices $k_1, \ldots, k_s$, then by a similar argument we can find a representative of α with its image in $M_{c-\epsilon} \cup e^{k_1} \cup \ldots \cup e^{k_s}$.

We will call (Y, Y_0) a smooth relative m–manifold if $Y \setminus Y_0$ is a smooth m–dimensional manifold. In that case, by standard approximation theory, we can approximate h by a map $\tilde{h}$ that agrees with h on $h^{-1}(M_{c-\epsilon})$ (and in particular on Y_0), and is a smooth map of $Y \setminus h^{-1}(M_{c-\epsilon})$ into e^k. Since e^k is convex this approximating map $\tilde{h}$ is clearly homotopic to h rel Y_0. In other words, when (Y, Y_0) a smooth relative m–manifold we can assume that the map h of the above theorem is smooth on $h^{-1}(e^k)$.

10.1.4. Corollary.   *If (Y, Y_0) a smooth relative m–manifold then $m \geq k$.*

PROOF.   We can assume the h of the theorem is smooth on $h^{-1}(e^k)$. Then by Sard's Theorem [DR,p.10] if $m < k$ the image of h could not cover $\overset{\circ}{e}{}^k$ and we could choose a z in $\overset{\circ}{e}{}^k$ not in the image of h. Since $\overset{\circ}{e}{}^k - \{z\}$ deformation retracts onto $\partial e^k \subseteq M_{c-\epsilon}$, α would have a representative with image in $M_{c-\epsilon}$, contradicting the assumption that α hangs up at the level c. Thus $m < k$ is impossible.   ∎

10.1.5. Corollary.   *If (Y, Y_0) is a smooth connected 1–manifold and α is non-trivial (i.e., no representative is a constant map), then $k = 1$.*

PROOF.   By the preceding corollary we have only to rule out the possibility that $k = 0$. But if $k = 0$ then $e^k = \{p\}$, so $M_{c-\epsilon} \cup e^k$ is the disjoint union of $M_{c-\epsilon}$ and $\{p\}$. Since Y is connected either $im(h) \subseteq \{p\}$

or else $\operatorname{im}(h) \subseteq M_{c-\epsilon}$. But the first alternative contradicts the non-triviality of α and the second contradicts that α hangs up at the level c. ∎

10.1.6. Corollary.   *If f has two distinct relative minima, x_0 and x_1, in the same component of M then it also has a critical point of index 1 in that component.*

PROOF. Take $Y = I$, $Y_0 = \{0, 1\}$, $M_0 = \{x_0, x_1\}$. We can assume that $f(x_0) \leq f(x_1)$. By the Morse Lemma there is a neighborhood U of x_1 not containing x_0 such that $f(x) > f(x_1) + \epsilon$ for all x in ∂U. Since any path from x_0 to x_1 must meet ∂U, it follows that minimax$(f, \alpha) > f(x_1) = \max(f|M_0)$ and we can apply the previous corollary. ∎

10.1.7. Remark.   Here is another proof: the second Morse inequality can be rewritten as $\mu_1 \geq b_1 + (\mu_0 - b_0)$. If M is connected then $b_0 = 1$, so if $\mu_0 > 1$ then $\mu_1 \geq 1$.

10.1.8. Corollary.   *If M is not simply connected then f has at least one critical point of index 1.*

PROOF. Take $Y = S^1$, Y_0 and M_0 empty, and choose any non-trivial free homotopy class α of maps $h : S^1 \to M$. Or, let x_0 be a minimum point of f in a non simply connected component of M, $Y = I$, $Y_0 = \{0, 1\}$, and let α be a non-trivial element of $\Pi_1(M, x_0)$. ∎

10.1.9. Remark.   This does not quite follow from the Morse inequality $\mu_1 \geq b_1$. The trouble is that $H_1(M)$ is the "abelianized" fundamental group, i.e., $\Pi_1(M)$ modulo its commutator subgroup. So if the fundamental group is non-trivial but perfect (e.g., the Poincaré Icosohedral Space) then $b_1 = 0$.

10.1.10. Corollary.   *If M is connected and f has no critical points with index k in the range $1 \leq k \leq m$ then $\Pi_i(M)$ is trivial for $i = 1, \dots m$.*

PROOF. If α is a non-trivial element of $\Pi_j(M) = [(S^j, \emptyset), (M, \emptyset)]$ then α hangs up on a critical point of index k, where $1 \leq k \leq j$. ∎

10.2 Linking Type.

Recall that, under our basic assumptions (a), (b), and (c) of Section 9.1, a Morse function $f : M \to R$ gives us a homotopy cell decomposition for the M_a. Each time we pass a critical level c with a single critical point of index k, $M_{c+\epsilon}$ has as a deformation retract $M_{c-\epsilon}$ with a k-cell attached. We would like to use this to compute inductively the the homology of the M_a, and hence eventually of M which is the limit of the M_a.

Let us review the general method involved. Let A be a homotopy spherical complex, $g : S^{k-1} \to A$ an attaching map, and $X \sim A \cup_g e^k$ (by which we mean X has $A \cup_g e^k$ as a deformation retract). We would like to compute the homology of X from that of A. We write $G : (D^k, S^{k-1}) \to (X, A)$ for the characteristic map of the attaching, so $g = G|S^{k-1}$. Now G induces a commutative diagram for the exact homology sequences of the pairs (D^k, S^{k-1}) and (X, A),

$$
\begin{array}{ccccccc}
H_m(D^k, S^{k-1}) & \xrightarrow{\partial} & H_{m-1}(S^{k-1}) & \longrightarrow & H_{m-1}(D^k) & \longrightarrow \\
\downarrow{\scriptstyle H_m(G)} & & \downarrow{\scriptstyle H_{m-1}(g)} & & \downarrow{\scriptstyle H_{m-1}(G)} & \\
H_m(X, A) & \xrightarrow{\partial} & H_{m-1}(A) & \longrightarrow & H_{m-1}(X) & \longrightarrow
\end{array}
$$

Since G is a relative homeomorphism, $H_m(G) : H_m(D^k, S^{k-1}) \to H_m(X, A)$ is an isomorphism. On the other hand D^k is contractible and hence all the $H_m(D^k)$ are zero, and it follows that the boundary maps $\partial : H_m(D^k, S^{k-1}) \to H_{m-1}(S^{k-1})$ are also isomorphisms. Thus in the exact sequence for (X, A) we can replace $H_m(D^k, S^{k-1})$ by $H_{m-1}(S^{k-1})$ and $\partial : H_m(D^k, S^{k-1}) \to H_{m-1}(A)$ by $H^{m-1}(g) : H_{m-1}(S^{k-1}) \to H_{m-1}(A)$, getting the exact sequence

$$\longrightarrow H_m(S^{k-1}) \xrightarrow{H_m(g)} H_m(A) \xrightarrow{i_m} H_m(X) \xrightarrow{j_m} H_{m-1}(S^{k-1}) \xrightarrow{H_{m-1}(g)} $$

When $m \neq k, k-1$ then $H_m(S^{k-1})$ and $H_{m-1}(S^{k-1})$, are both zero, so $H_m(X) \approx H_m(A)$. On the other hand for the two special values of m we get two short exact sequences

$$0 \longrightarrow H_k(A) \longrightarrow H_k(X) \longrightarrow \mathrm{Ker}(H_{k-1}(g)) \longrightarrow 0$$

and

$$0 \longrightarrow \mathrm{Im}(H_{k-1}(g)) \longrightarrow H_{k-1}(A) \longrightarrow H_{k-1}(X) \longrightarrow 0$$

from which we can in principle compute $H_k(X)$ and $H_{k-1}(X)$ **if we know** $H_{k-1}(g)$.

Unfortunately, in the Morse theoretic framework, there is no good algorithm for deriving the information needed to convert the above two exact sequences into a general tool for computing the homology of X. As a result we will now restrict our attention to what seems at first to be a very special case (called "linking type") where the computation of $H_*(X)$ becomes trivial. Fortunately, it is a case that is met surprisingly often in practice.

A choice of orientation for R^k is equivalent to a choice of generator $[D^k, S^{k-1}]$ for $H_k(D^k, S^{k-1}; \mathbf{Z})$, and we will denote by $[e^k, \partial e^k]$ the corresponding generator for $H_k(e^k, \partial e^k; \mathbf{Z})$ and for $H_k(X, A; \mathbf{Z})$. For a general coefficient ring $\mathcal{R}$ we may regard $H_k(X, A; \mathcal{R})$ as a free $\mathcal{R}$ module with basis $[e^k, \partial e^k]$. The following definition is due to M. Morse.

10.2.1. Definition. We say that (X, A) is of *linking type* over $\mathcal{R}$ if $[e^k, \partial e^k]$ is in the kernel of $\partial_k : H_k(X, A; \mathcal{R}) \to H_{k-1}(A; \mathcal{R})$, (so that in fact $\partial_k \equiv 0$), or equivalently if $[e^k, \partial e^k]$ is in the image of $j_k : H_k(X, A; \mathcal{R}) \to H_k(X, A; \mathcal{R})$ In this case we call any $\mu \in \mathbf{Z}_k(X; \mathcal{R})$ (or $[\mu] \in H_k(X; \mathcal{R})$) such that $j_k([\mu]) = [e^k, \partial e^k]$ a *linking cycle* for (X, A) over $\mathcal{R}$.

10.2.2. Remark. Clearly another equivalent condition for (X, A) to be of linking type is that the fundamental class $[S^{k-1}]$ of S^{k-1} be in the kernel of $H_{k-1}(g) : H_{k-1}(S^{k-1}) \to H_{k-1}(A)$

10.2.3. Theorem. *If (X, A) is of linking type over $\mathcal{R}$ and $[\mu] \in H_k(X; \mathcal{R})$ is a linking cycle for (X, A) then $H_*(X; \mathcal{R}) = H_*(A; \mathcal{R}) \oplus \mathcal{R}[\mu]$.*

PROOF. From the exact sequence for (X, A),

$$\to H_{m+1}(X, A) \xrightarrow{\partial_{m+1}} H_m(A) \xrightarrow{i_m} H_m(X) \xrightarrow{j_m} H_m(X, A) \to$$

since $\partial_k = 0$ and all $H_m(X, A) = 0$, except perhaps for $m = k, k - 1$, we have $H_m(X) = H_m(A)$ except for $m = k$. Taking $m = k$ and using $\partial_k = 0$ and $H_k(X, A) = \mathcal{R}[e^k, \partial e^k]$, we have the short exact sequence

$$0 \to H_k(A) \xrightarrow{i_k} H_k(X) \xrightarrow{j_k} \mathcal{R}[e^k, \partial e^k] \to 0,$$

and this is clearly split by the map $r[e^k, \partial e^k] \mapsto r[\mu]$ of $H_k(X, A)$ to $H_k(X)$. ∎

Now let $A = X_0 \subseteq X_1 \subseteq \ldots \subseteq X_n = X$ be a homotopy cell decomposition for (X, A); say X_i has as a deformation retract $X \cup_{g_i} e^{k_i}$. We shall say that this is a *cell decomposition of linking type* if each pair (X_i, X_{i-1}) is of linking type. By an easy induction from the previous theorem we see that the inclusions $i_\ell : X_\ell \to X$ induce injections $H_*(i_\ell) : H_*(X_\ell) \to H_*(X)$. So for such a homotopy cell decomposition we will identify each $H_*(X_\ell)$ with a sub-module of $H_*(X)$, and therefore identify a linking cycle $[\mu_\ell] \in H_{k_\ell}(X_\ell)$ for the pair $(X_\ell, X_{\ell-1})$ with an element of $H_{k_\ell}(X)$. With these conventions we define *a set of linking cycles* for the above homotopy cell decomposition of linking type to be a sequence of homology classes $\mu_1, \ldots, \mu_n$ such that $[\mu_\ell]$ is in the submodule $H_{k_\ell}(X_\ell)$ of $H_{k_\ell}(X)$ and $H_*(j_\ell)([\mu_\ell]) = [e^{k_\ell}, \partial e^{k_\ell}]$, where $H_*(j_\ell)$ denotes the projection $H_{k_\ell}(X_\ell) \rightarrow H_{k_\ell}(X_\ell, X_{\ell-1})$. Then by induction from the preceding theorem,

10.2.4. Theorem.   *With the above assumptions and notation:*

$$H_*(X) = H_*(A) \oplus \bigoplus_{\ell=1}^{n} \mathcal{R}[\mu_\ell].$$

Now let us specialize to the homotopy cell decompositions associated to a Morse function $f : M \rightarrow R$ that is bounded below and satisfies Condition C on a complete Riemannian manifold. Let a be a non-critical value of f and let $p_1 \ldots, p_n$ be all the critical points of finite index of f below the level a, ordered so that $c_i = f(p_i) \leq c_{i+1}$. Assume p_i has index k_i, and let e^{k_i} denote the descending cell in some Morse coordinate system at p_i. We have seen that M_a has a homotopy cell decomposition $\emptyset = X_0 \subseteq X_1 \subseteq \ldots \subseteq X_n = M_a$ with X_{i+1} having as a deformation retract $X_i \cup_{g_i} e^{k_i}$. (In the generic case that p_i is the unique critical point at its level c_i we may take $X_i = M_{c_{i+1}-\epsilon}$, $i = 0 \ldots n-1$.) We say that *the critical point p_i is of linking type over $\mathcal{R}$ if* (X_{i+1}, X_i) is of linking type over $\mathcal{R}$. And we say that *the Morse function f is of linking type over $\mathcal{R}$ if* all its critical points are of linking type over $\mathcal{R}$. In this case we let $[\mu_i] \in H_{k_i}(M_a)$ denote a linking cycle for (X_{i+1}, X_i), and we call $[\mu_i]$ a *linking cycle for the critical point p_i.*

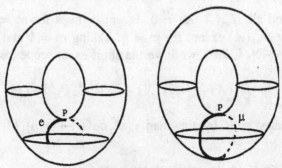

The descending cell e
and the linking cycle μ
at a critical point p.

By the previous theorem we have: $H_*(M_a) = \bigoplus_{i=1}^{n} \mathcal{R}[\mu_i]$. Note that if $a < b$ are two regular values of f then in particular it follows that $H_*(M_a)$ injects into $H_*(M_b)$. Let $\{a_n\}$ be a sequence of regular values of f tending to infinity. Then clearly M is the inductive limit of the subspaces M_{a_n}. Hence

10.2.5.   Theorem.   *Let $f : M \rightarrow R$ be a Morse function of linking type over $\mathcal{R}$ that is bounded below and satisfies Condition C on a complete*

Riemannian manifold. For each critical point p of f let $k(p)$ denote the index of p and let $\mu_p \in H_{k(p)}(M; \mathcal{R})$ be a a linking cycle for p over $\mathcal{R}$. Then $H_*(M_a; \mathcal{R})$ is a free $\mathcal{R}$ module generated by these $[\mu_p]$.

10.2.6. Corollary.   *If a Morse function on a compact manifold is of linking type over a field then it is a perfect Morse function.*

It is clearly important to have a good method for constructing linking cycles. In the next section we will study a very beautiful criterion, that goes back to Bott and Samelson, for recognizing when certain geometric cycles are linking cycles.

10.3 Bott-Samelson Type.

In this section our coefficient ring for homology, $\mathcal{R}$, is for simplicity assumed to be either $\mathbf{Z}$ or $\mathbf{Z}_2$. Y will denote a compact, connected, smooth k−manifold with (possibly empty) boundary ∂Y. We recall that $H_k(Y, \partial Y; \mathbf{Z}_2) \approx \mathbf{Z}_2$. The non-zero element of $H_k(Y, \partial Y; \mathbf{Z}_2)$ is denoted by $[Y, \partial Y]$ and is called the *fundamental class* of of $(Y, \partial Y)$ (over $\mathbf{Z}_2$). We say that "Y is oriented over $\mathbf{Z}_2$". Over $\mathbf{Z}$ there are two possibilities. Recall that Y is called orientable if it has an atlas of coordinate charts such that the Jacobians of all the coordinate changes are positive functions, otherwise non-orientable. If Y is non-orientable then $H_k(Y, \partial Y; \mathbf{Z}) = 0$, while if Y is orientable then $H_k(Y, \partial Y; \mathbf{Z}) \approx \mathbf{Z}$. In the latter case, a choice of one of the two possible generators is called an orientation of Y, and Y together with an orientation is called an oriented k−manifold. The chosen generator for $H_k(Y, \partial Y; \mathbf{Z})$ is again denoted by $[Y, \partial Y]$ and is called the fundamental class of the oriented manifold over $\mathbf{Z}$ (its reduction modulo 2 is clearly the fundamental class over $\mathbf{Z}_2$). Either over $\mathbf{Z}$ or $\mathbf{Z}_2$ the fundamental class $[Y, \partial Y]$ has the following characteristic property. If Δ is a k−disk embedded in the interior of Y, then the inclusion $(Y, \partial Y) \hookrightarrow (Y, Y \setminus \overset{\circ}{\Delta})$ induces a map $H_k(Y, \partial Y) \to H_k(Y, Y \setminus \overset{\circ}{\Delta})$. On the other hand we have an excision isomorphism $H_k(Y, Y \setminus \overset{\circ}{\Delta}) \approx H_k(\Delta, \partial \Delta)$. Then, under the composition of these two maps, the fundamental class $[Y, \partial Y]$ is mapped onto $\pm [\Delta, \partial \Delta]$.

10.3.1. Proposition.   *Let p be a non-degenerate critical point of index k and co-index l, lying on the level c of $f : M \to \mathbf{R}$, and let e^k and e^l be the descending and ascending cells of radius ϵ in a Morse chart for f at p. Let Y be a compact, smooth k−manifold with boundary, oriented over $\mathcal{R}$ and $\varphi : (Y, \partial Y) \to (M_c, M_{c-\epsilon})$ a smooth map, such that:*

(1) $im(\varphi) \cap f^{-1}(c) = \{p\}$,

(2) $\varphi^{-1}(p) = \{y_0\}$, *and*

(3) φ *is transversal to* e^l *at* y_0.

Then. $H_k(\varphi) : H_k(Y, Y_0) \to H_k(M_c, M_{c-\epsilon})$ *maps* $[Y, Y_0]$ *to* $\pm[e^k, \partial e^k]$. *(Here, as in the preceding section,* $[e^k, \partial e^k]$ *denotes the image of the fundamental class of* $(e^k, \partial e^k)$ *in* $H_k(M_c, M_{c-\epsilon})$ *under inclusion.)*

PROOF. By (1) and (2), if $y \neq y_0$ then $f(\varphi(y)) < c$; hence if N is any neighborhood of y_0 then, for ϵ small enough, φ maps $(Y, Y \setminus N)$ into $(M_c, M_{c-\epsilon})$. In the given Morse coordinates at p let P denote the projection onto the descending space R^k along the ascending space R^l. Then (3) says that $P \circ (D\varphi)_{y_0}$ maps TY_{y_0} isomorphically onto $T(e^k)_p$, so by the Inverse Function Theorem, for ϵ sufficiently small, $P \circ \varphi$ maps a closed disk neighborhood Δ of y_0 in Y diffeomorphically onto the neighborhood e^k of p in R^k. And by the first remark in the proof we can assume further that φ maps $(Y, Y \setminus \overset{\circ}{\Delta})$ into $(M_c, M_{c-\epsilon})$. Taking ϵ small enough we can suppose that both φ and $P \circ \varphi$ map Δ into some convex neighborhood of p, so by a standard interpolation argument we can find a smooth map $\tilde{\varphi} : (Y, \partial Y) \to (M_c, M_{c-\epsilon})$ that agrees with $P \circ \varphi$ in Δ, agrees with φ outside a slightly larger neighborhood of y_0, and is homotopic to φ rel Y_0. We now have the commutative diagram:

$$
\begin{array}{ccc}
(Y, \partial Y) & \xrightarrow{\tilde{\varphi}} & (M_c, M_{c-\epsilon}) \\
{\scriptstyle inc}\downarrow & & \uparrow{\scriptstyle inc} \\
(Y, Y \setminus \overset{\circ}{\Delta}) & \xrightarrow{\tilde{\varphi}} & (M_{c-\epsilon} \cup e^k, M_{c-\epsilon}) \\
{\scriptstyle exc}\downarrow & & \uparrow{\scriptstyle inc} \\
(\Delta, \partial\Delta) & \xrightarrow{\tilde{\varphi}} & (e^k, \partial e^k)
\end{array}
$$

where *inc* indicates an inclusion, and *exc* an excision.

Now, since $\tilde{\varphi}$ is a diffeomorphism of $(\Delta, \partial\Delta)$ onto $(e^k, \partial e^k)$, it follows that $H_k(\tilde{\varphi})([\Delta, \partial\Delta]) = \pm[e^k, \partial e^k]$. But, since $\tilde{\varphi}$ and φ are homotopic, $H_k(\varphi) = H_k(\tilde{\varphi})$, and the conclusion follows from the diagram and the characteristic property of fundamental classes stated above. ∎

10.3.2. Definition. Let Y be a compact, smooth, connected k-manifold that is oriented over $\mathcal{R}$, and $\varphi : Y \to M$ a smooth map. If p is a nondegenerate critical point of index k of $f : M \to R$, then we call (Y, φ) a *Bott-Samelson cycle for* f *at* p (over $\mathcal{R}$) if $f \circ \varphi : Y \to R$ has a unique nondegenerate maximum that is located at $y_0 = \varphi^{-1}(p)$. We say that the critical point p is of *Bott-Samelson type* (over $\mathcal{R}$) if such a pair (Y, φ) exists, and we say a Morse function f is of Bott-Samelson type over $\mathcal{R}$ if all of its critical points are of Bott-Samelson type.

10.3.3. Theorem.   If (Y, φ) is a Bott-Samelson cycle for f at p then $H_*(\varphi)([Y])$ is a linking cycle for f at p.

PROOF.   Immediate from the preceding proposition and the definition of linking cycle. Conditions (1) and (2) of the proposition are obviously satisfied, and (3) is an easy consequence of the non-degeneracy of $f \circ \varphi$ at $\varphi^{-1}(p)$. ∎

10.3.4. Corollary.   Let $f : M \to R$ be a Morse function that satisfies Condition C and is bounded below on a complete Riemannian manifold M. If f is of Bott-Samelson type over $\mathcal{R}$ then it also of linking type over $\mathcal{R}$. If for each critical point p of f, (Y_p, φ_p) is a Bott-Samelson cycle for f at p over $\mathcal{R}$ then, for a regular value a of f, $H_*(M_a; \mathcal{R})$ is freely generated as an $\mathcal{R}$-module by the $H_*(\varphi_p)([Y_p])$ with $f(p) < a$, and $H_*(M_a; \mathcal{R})$ is freely generated by all the $H_*(\varphi_p)([Y_p])$.

10.3.5. Remark.   Suppose all the critical points of index less than or equal to k of a Morse function $f : M \to R$ are of Bott-Samelson type. Does it follow that, for $l \le k$, $b_l(M) = \mu_l(M)$? By the following proposition the example of a function on S^1 with two local minima and two local maxima shows this is already false for $k = 0$.

10.3.6. Proposition.   If $f : M \to R$ is a Morse function then every local minimum (i.e., critical point of index zero) is of Bott-Samelson type. If M is compact then a local maximum $\{p\}$ of f is of Bott-Samelson type over $\mathcal{R}$ provided that the component M_0 of $\{p\}$ in M is oriented over $\mathcal{R}$ and p is the unique global maximum of $f|M_0$.

PROOF.   If x ia local minimum then $Y = \{x\}$ is an oriented, connected 0−manifold and if φ is the inclusion of Y into M then (Y, φ) is a Bott-Samelson cycle for f at x. Similarly, in the local maximum case, provided M_0 is oriented over $\mathcal{R}$ and p is the unique global maximum of $f|M_0$ then the inclusion of M_0 into M is a Bott-Samelson cycle for f at p. ∎

10.3.7. Corollary.   A smooth function on the circle S^1 is a Morse function of Bott-Samelson type provided that its only critical points are one non-degenerate local minimum and one non-degenerate local maximum.

10.3.8. Corollary.   Let M be a smooth, compact, connected surface, oriented over $\mathcal{R}$, and let $f : M \to R$ be a Morse function with a unique local maximum. A necessary and sufficient condition for f to be of Bott-Samelson *type is that for each saddle point p of f there exist a circle S^1 immersed in M that is tangent to the descending direction at p and everywhere else lies below the level $f(p)$. In this case the fundamental classes of these circles will generate $H_1(M)$.*

10.3.9. Remark.   This gives another proof that a surface of genus g has first betti number $2g$. There is a standard embedding of the surface in R^3 with the height function having $2g$ saddles and for which the corresponding circles are obvious (if you look at the illustration of the case $g = 3$ in Section 9.7, the circles stare out of the page).

Chapter 11.
The Calculus of Variations.

Let X be a compact Riemannian manifold and let $\mathcal{M}_0$ denote some space of smooth mappings of X into a manifold Y, or more generally some space of smooth sections of a fiber bundle E over X with fiber Y (the case $E = X \times Y$ gives the maps $X \to Y$).

A *Lagrangian function* L for $\mathcal{M}_0$ of order k is a function $L : \mathcal{M}_0 \to C^\infty(X, \mathbf{R})$ that is a partial differential operator of order k. This means that $L(\varphi) : X \to \mathbf{R}$ can be written as a function of the partial derivatives of φ up to order k with respect to local coordinates in X and E. (More precisely, but more technically, L should be of the form $F \circ j_k$, where $j_k : C^\infty(E) \to C^\infty(J^k E)$ is the k-jet extension map and F is a smooth map $J^k(E) \to \mathbf{R}$.)

Given such a Lagrangian function L we can associate to it a real valued function $\mathcal{L} : \mathcal{M}_0 \to \mathbf{R}$, called *the associated action integral*, (or action functional) by $\mathcal{L}(\varphi) = \int_X L(\varphi) \, d\mu(x)$, where $d\mu$ is the Riemannian volume element.

The general problem of the Calculus of Variations is to study the "critical points" of such action integrals in the following sense. Let $\varphi \in \mathcal{M}_0$. Given a smooth path φ_t in $\mathcal{M}_0$ (in the sense that $(t, x) \mapsto \varphi_t(x)$ is smooth) we can compute $\left(\frac{d}{dt}\right)_{t=0} \mathcal{L}(\varphi_t)$. If this is zero for all smooth paths φ_t with $\varphi_0 = \varphi$ then φ is called a critical point of the functional $\mathcal{L}$. We shall see below that the condition for φ to be a critical point of $\mathcal{L}$ can be written as a system of partial differential equations of order $2k$ for φ, called the Euler-Lagrange equations corresponding to the Lagrangian L. Of course if we can interpret $\mathcal{M}_0$ as a smooth manifold and $\mathcal{L} : \mathcal{M}_0 \to \mathbf{R}$ as a smooth function on this manifold, then "critical point" in the above sense will be equivalent to critical point in the sense we have been using it previously, namely that $d\mathcal{L}_\varphi = 0$. Moreover in this case the Euler-Lagrange equation is equivalent to $\nabla \mathcal{L}(\varphi) = 0$.

To see what the Euler-Lagrange equation of a k-th order Lagrangian L looks like we consider the following simple example: Let $I = (0, 1)$ and $\Omega = I^n \subset \mathbf{R}^n$, $\mathcal{M}_0 = C_o^\infty(\Omega, \mathbf{R})$, the space of smooth functions with compact support in Ω, and $L(u) = L(j_k(u)) = L(u, D^\alpha u)$, i.e., L is a function of u and its partial derivatives $D^\alpha u$ up to order k. Here $\alpha = (\alpha_1, \ldots, \alpha_n)$ is an n-tuple of non-negative integers, $|\alpha| = \alpha_1 + \ldots + \alpha_n$, and

$$D^\alpha u = \frac{\partial^{|\alpha|} u}{\partial x_{\alpha_1} \cdots \partial x_{\alpha_n}}.$$

If $u, h \in \mathcal{M}_0$, then $u_t = u + th \in \mathcal{M}_0$, and

$$L(u + th) = L(u + th, D^\alpha u + tD^\alpha h).$$

Since h and all its partial derivatives vanish near $\partial\Omega$, there are no boundary terms when we integrate by parts in the following:

$$
\begin{aligned}
d\mathcal{L}_u(h) &= \left(\frac{d}{dt}\right)_{t=0}(\mathcal{L}(u+th)) \\
&= \int_\Omega \left\{\frac{\partial L}{\partial u}h + \sum_\alpha \frac{\partial L}{\partial(D^\alpha u)}D^\alpha h\right\}dx \\
&= \int_\Omega \left\{\frac{\partial L}{\partial u} + \sum_\alpha (-1)^{|\alpha|}D^\alpha\left(\frac{\partial L}{\partial(D^\alpha u)}\right)\right\}h\,dx
\end{aligned}
$$

So the Euler-Lagrange equation is

$$
\frac{\partial L}{\partial u} + \sum_\alpha (-1)^{|\alpha|}D^\alpha\left(\frac{\partial L}{\partial(D^\alpha u)}\right) = 0.
$$

In the same way, if $\mathcal{M}_0 = C_o^\infty(\Omega, \mathbf{R}^m)$, then the Euler-Lagrange equations for $u = (u_1, \ldots, u_m)$ are

$$
\frac{\partial L}{\partial u_j} + \sum_\alpha (-1)^{|\alpha|}D^\alpha\left(\frac{\partial L}{\partial(D^\alpha u_j)}\right) = 0 \qquad 1 \leq j \leq m,
$$

which is a determined system of m PDE of order $2k$ for the m functions $u_1, \ldots, u_m$.

In general $\mathcal{M}_0$ is the space of smooth sections of a fiber bundle E on a compact Riemannian manifold X. To compute the first variation, $(\frac{d}{dt})_{t=0}\mathcal{L}(u_t)$, it suffices to compute it for deformation u_t having "small support", i.e., a smooth curve $u_t \in \mathcal{M}_0$ such that $u_0 = u$ on all of X, and $u_t = u$ outside a compact subset of a coordinate neighborhood U. One standard method for computing the first variation is to choose a trivialization of E over U so that, locally, smooth sections of E are represented by $\mathbf{R}^m$-valued maps defined on an open neighborhood of $\mathbf{R}^n$. Here $n = \dim(X)$ and m is the fiber dimension of E. Then the Euler-Lagrange equation of $\mathcal{L}$ can be computed locally just as above. If the Lagrangian L is natural then it is usually easy to interpret the local formula this leads to in an invariant manner. A second standard method to get the Euler-Lagrange equations is to use covariant rather than ordinary derivatives to get an invariant expression for $(\frac{d}{dt})_{t=0}\mathcal{L}(u_t)$ directly. Both methods will be illustrated below.

Many important objects in geometry, analysis, and mathematical physics are critical point of variational problems. For example, geodesics, harmonic maps, minimal submanifolds, Einstein metrics, solutions of the Yamabe equation, Yang-Mills fields, and periodic solutions of a Hamiltonian vector field.

Now the Euler-Lagrange equations of a variational problem are usually a highly non-linear system of PDE, and there is no good general theory for solving them—*except* going back to the variational principle itself. Often we would be happy just to prove an existence theorem, i.e., prove that the set C_0 of critical points of $\mathcal{L} : \mathcal{M}_0 \to R$ is non-empty. Can we apply our general theory? Sometimes we can. Here are the steps involved:

(1) Complete $\mathcal{M}_0$ to a complete Riemannian manifold $\mathcal{M}$ of sections of E. Usually this is some Sobolev completion of $\mathcal{M}_0$. Choosing the correct one is an art! For (2) to work the Sobolev norm used must be "strong enough", while for (5) to work this norm cannot be "too strong". In order that both work the choice must be *just* right.

(2) Extend $\mathcal{L}$ to a smooth map $\tilde{\mathcal{L}} : \mathcal{M} \to R$. (If the correct choice is made in step (1) this is usually easy.)

(3) Boundedness from below. Show that $\tilde{\mathcal{L}}$ is bounded below. (Usually easy.)

(4) Verify Condition C for $\tilde{\mathcal{L}}$. (Usually a difficult step.)

(5) (Regularity) Show that a solution φ in $\mathcal{M}$ of $d\tilde{\mathcal{L}}_\varphi = 0$ is actually in $\mathcal{M}_0$, and hence is a critical point of $\mathcal{L}$. This is usually a difficult step. In the simplest case L is a homogeneous quadratic form, so that the Euler-Lagrange equations are linear. Then technically it comes down to proving the ellipticity of these equations. In the general non-linear case we again usually must show some sort of ellipticity for the Euler-Lagrange operator and then prove a regularity theorem for a class of elliptic non-linear equations that includes the Euler-Lagrange equations.

In general, for all but the simplest Calculus of Variations problems, carrying out this program turns out to be a technical and difficult process if it can be done at all. Many research papers have consisted in verifying all the details under particular assumptions about the nature of L. Often some special tricks must be used, such as dividing out some symmetry group of the problem or solving a "perturbed" problem with Lagrangian L^ϵ and letting $\epsilon \to 0$ ([U],[SU]).

In section 1 we will discuss Sobolev manifolds of sections of fiber bundles over compact n-dimensional manifolds needed for step (1) of the program. We will only give the full details for the case $n = 1$, needed in section 2 where we work out in complete detail the above five steps for the the geodesic problem. But the geodesic problem is misleadingly easy. To give some of the flavor and complexity of the analysis that comes into carrying out the program for more general Calculus of Variations problems we study a second model problem in section 3; namely the functional

$$J(u) = \int_X \|\nabla u\|^2 + f u^2 \, dv(g)$$

with constraint $\int_X |u|^p \, dv(g) = 1$ on a compact Riemannian manifold (X, g).

The corresponding Euler-Lagrange equation is

$$\triangle u + fu = \lambda u^{p-1}$$

for some constant λ, an equation that has important applications to the problem of prescribing scalar curvature. But this equation can also be viewed as an excellent model equation for studying the feasibility of the above general program. For example, as we shall see, it turns out that whether or not Condition C is satisfied depends on the value of the exponent p.

11.1   Sobolev manifolds of fiber bundle sections.

If M is a compact n-dimensional manifold and ξ is a smooth vector bundle over M then we can associate to ξ a sequence of hilbert spaces $H_k(\xi)$ of sections of ξ. Somewhat roughly we can say that a section σ of ξ is in $H_k(\xi)$ if (with respect to local coordinates in M and a local trivialization for ξ) all its partial derivatives of order less than or equal to k are locally square summable. Moreover these so-called Sobolev spaces are functorial in the following sense: if η is a second smooth vector bundle over M and $\varphi : \xi \to \eta$ is a smooth vector bundle morphism, then $\sigma \mapsto \varphi \circ \sigma$ is a continuous linear map $H_k(\varphi) : H_k(\xi) \to H_k(\eta)$. When $k > n/2$ it turns out that $H_k(\xi)$ is a dense linear subspace of the Banach space $C^0(\xi)$ of continuous sections of ξ and moreover that the inclusion map $H_k(\xi) \hookrightarrow C^0(\xi)$ is a continuous (and in fact compact) linear map. In this case H_k is also functorial in a larger sense; namely, if $\varphi : \xi \to \eta$ is a smooth *fiber* bundle morphism then $\sigma \mapsto \varphi \circ \sigma$ is of course *not* necessarily linear, but it is a *smooth* map $H_k(\varphi) : H_k(\xi) \to H_k(\eta)$. It follows easily from this that, for a fiber bundle E over M, when $k > n/2$ we can in a natural way define a hilbert *manifold* $H_k(E)$ of sections of E. $H_k(E)$ is characterized by the property that if a vector bundle ξ is an open sub-bundle of E, then $H_k(\xi)$ is an open submanifold of $H_k(E)$; in fact these $H_k(\xi)$ give a defining atlas for the differentiable structure of $H_k(E)$. When F is another smooth fiber bundle over M and $\varphi : E \to F$ is a smooth fiber bundle morphism then $\sigma \mapsto \varphi \circ \sigma$ is a smooth map $H_k(\varphi) : H_k(E) \to H_k(F)$. Thus when $k > n/2$ we can "extend" H_k to a functor from the category of smooth fiber bundles over M to the category of smooth hilbert manifolds.

In this section we will give the full details of this construction for the case $k = 1$ and $n = 1$ (so M is either the interval I or the circle S^1). A complete exposition of the general theory can be found in [Pa6].

We begin by considering the case of a trivial bundle $\xi = I \times \boldsymbol{R}^n$, so that a section of ξ is just a map of I into $\boldsymbol{R}^n$.

We will denote by $H_0(I, R^n)$ the hilbert space $L^2(I, R^n)$ of square summable maps of the unit interval I into R^n. For $\sigma, \lambda \in H_0(I, R^n)$ we denote their inner product by $\langle \sigma, \lambda \rangle_0 = \int_0^1 \langle \sigma(t), \lambda(t) \rangle \, dt$, and $\|\sigma\|_0^2 = \langle \sigma, \sigma \rangle_0$.

Recall that a continuous map $\sigma : I \to R^n$ is called *absolutely continuous* if σ' exists almost everywhere, and is in $L^1(I, R^n)$ (i.e., $\int_0^1 \|\sigma'(t)\| \, dt < \infty$). In this case $\sigma(t) = \sigma(0) + \int_0^t \sigma'(s) \, ds$—and conversely if $f \in L^1(I, R^n)$ then $t \mapsto p + \int_0^t f(s) \, ds$ is absolutely continuous and has derivative f. Since I has finite measure, by the Schwarz inequality $L^1(I, R^n) \supseteq L^2(I, R^n) = H_0(I, R^n)$. The set of absolutely continuous maps $\sigma : I \to R^n$ such that σ' is in $H_0(I, R^n)$ is called the Sobolev space $H_1(I, R^n)$.

11.1.1. Proposition. $H_1(I, R^n)$ *is a hilbert space with the inner product*

$$\langle \lambda, \sigma \rangle_1 = \langle \lambda(0), \sigma(0) \rangle + \langle \lambda, \sigma \rangle_0.$$

PROOF. This just says that the map $\lambda \mapsto (\lambda(0), \lambda')$ of $H_1(I, R^n)$ to $R^n \oplus H_0(I, R^n)$ is bijective. The inverse is $(p, \sigma) \mapsto p + \int_0^t \sigma(s) \, ds$. ∎

11.1.2. Theorem (Sobolev Inequality). *If σ is in $H_1(I, R^n)$ then*

$$\|\sigma(t) - \sigma(s)\| \leq |t - s|^{\frac{1}{2}} \|\sigma'\|_0$$
$$\leq |t - s|^{\frac{1}{2}} \|\sigma\|_1.$$

PROOF. If h is the characteristic function of the interval $[s, t]$ then $\|h\|_{L^2}^2 = \int_0^1 h^2(t) \, dt = \int_s^t 1 \, dt = |t - s|$, hence by the Schwarz inequality $\|\sigma(t) - \sigma(s)\| = \|\int_s^t \sigma'(x) \, dx\| = \|\int_0^1 h(x)\sigma'(x) \, dx\| \leq |t - s|^{\frac{1}{2}} \|\sigma'\|_0$ ∎

11.1.3. Corollary. $\|\sigma\|_\infty \leq 2\|\sigma\|_1$.

PROOF. $\|\sigma(0)\| \leq \|\sigma\|_1$, by the definition of $\|\sigma\|_1$, hence

$$\|\sigma(t)\| \leq \|\sigma(0)\| + \|\sigma(t) - \sigma(0)\|$$
$$\leq \|\sigma(0)\| + |t|^{\frac{1}{2}} \|\sigma\|_1 \leq 2\|\sigma\|_1. \quad ∎$$

11.1.4. Theorem. *The inclusion maps of $H_1(I, R^n)$ into $C^0(I, R^n)$ and into $H_0(I, R^n)$ are completely continuous.*

PROOF. Since the inclusion $C^0(I, \mathbf{R}^n) \hookrightarrow H_0(I, \mathbf{R}^n)$ is continuous, it will suffice to show that $H_1(I, \mathbf{R}^n) \hookrightarrow C^0(I, \mathbf{R}^n)$ is completely continuous. Let S be bounded in $H_1(I, \mathbf{R}^n)$. We must show that S has compact closure in $C^0(I, \mathbf{R}^n)$ or, by the Ascoli-Arzela Theorem, that S is bounded in the C^0 norm ($\| \ \|_\infty$) and is equicontinuous. Boundedness is immediate from the preceding corollary, while the Sobolev Inequality implies that S satisfies a uniform Hölder condition of order $\frac{1}{2}$ and so *a fortiori* is equicontinuous. ∎

We will denote by $S(I, \mathbf{R}^n)$ the vector space of all functions $\sigma : I \to \mathbf{R}^n$. As usual we identify $S(I, \mathbf{R}^n)$ with the vector space of all sections of the product bundle $I \times \mathbf{R}^n$. Given a smooth map $\varphi : I \times \mathbf{R}^n \to I \times \mathbf{R}^p$ of the form $(t, x) \mapsto (t, \varphi_t(x))$, with each φ_t a linear map of $\mathbf{R}^n \to \mathbf{R}^p$, we can regard φ as a smooth vector bundle morphism between the product bundle $I \times \mathbf{R}^n$ and $I \times \mathbf{R}^p$, hence it induces a linear map $\tilde{\varphi}$ of $S(I, \mathbf{R}^n)$ to $S(I, \mathbf{R}^p)$; namely $\tilde{\varphi}(\sigma)(t) = \varphi_t(\sigma(t))$. Clearly $\tilde{\varphi}$ is a continuous linear map of $C^0(I, \mathbf{R}^n)$ to $C^0(I, \mathbf{R}^p)$ and also a continuous linear map of $H_0(I, \mathbf{R}^n)$ to $H_0(I, \mathbf{R}^p)$. If $\sigma \in C^0(I, \mathbf{R}^n)$ is absolutely continuous then, if σ is differentiable at $t \in I$ so is $\tilde{\varphi}(\sigma)$, and $\tilde{\varphi}(\sigma)'(t) = \varphi_t(\sigma'(t)) + (\frac{\partial}{\partial s})_{s=t} \varphi(s, \sigma(t))$. It follows that $\tilde{\varphi}$ is also absolutely continuous and is in $H_1(\mathbf{R}^p)$ if σ is in $H_1(\mathbf{R}^n)$. Thus $\tilde{\varphi}$ is also a continuous linear map of $H_1(I, \mathbf{R}^n)$ to $H_1(I, \mathbf{R}^p)$. Of course it follows in particular that if $n = p$ and φ is a vector bundle automorphism of $I \times \mathbf{R}^n$ (i.e., each φ_t is in $GL(n, \mathbf{R})$), then $\tilde{\varphi}$ is an automorphism of $C^0(I, \mathbf{R}^n)$, of $H_0(I, \mathbf{R}^n)$, and of $H_1(I, \mathbf{R}^n)$.

Now suppose ξ is a smooth vector bundle over I. Since any bundle over I is trivial, we can find a trivialization of ξ, i.e., a vector bundle isomorphism φ of the product vector bundle $I \times \mathbf{R}^n$ with ξ. Then $\sigma \mapsto \varphi \circ \sigma$ is a bijective linear map $\tilde{\varphi}$ between the space $S(I, \mathbf{R}^n)$ of all sections of $I \times \mathbf{R}^n$ and the space $S(\xi)$ of all sections of ξ. This map $\tilde{\varphi}$ will of course map $C^0(I, \mathbf{R}^n)$ isomorphically onto $C^0(\xi)$, but moreover we can now define hilbertable spaces $H_0(\xi)$ and $H_1(\xi)$ of sections of ξ with $H_1(\xi) \subseteq C^0(\xi) \subseteq H_0(\xi)$, by specifying that $\tilde{\varphi}$ is also an isomorphism of $H_0(I, \mathbf{R}^n)$ with $H_1(\xi)$ and of $H_1(I, \mathbf{R}^n)$ with $H_1(\xi)$. By the above remarks it is clear that these definitions are independent of the choice of trivialization φ. Moreover it also follows from these remarks that H_0 and H_1 are actually functorial from the category $\mathbf{VB}(I)$ of smooth vector bundles over I and smooth vector bundle morphisms to the category $\mathbf{Hilb}$ of hilbertable Banach spaces and bounded linear maps. That is, if $\varphi : \xi \to \eta$ is a morphism of smooth vector bundles over I then $\tilde{\varphi} : S(\xi) \to S(\eta)$, $\sigma \mapsto \varphi \circ \sigma$ restricts to morphisms (i.e., a continuous linear maps) $H_0(\xi) \to H_0(\eta)$ and $H_1(\xi) \to H_1(\eta)$.

From the corollary of the Sobolev inequality we have:

11.1.5. Theorem. *For any smooth vector bundle ξ over I the inclusion maps of $H_1(\xi)$ into $C^0(\xi)$ and into $H_0(\xi)$ are completely continuous.*

Let $\mathbf{FB}(I)$ denote the category of smooth fiber bundles and smooth fiber bundle morphisms over I and let $\mathbf{Mfld}$ denote the category of smooth hilbert manifolds and smooth maps. Note that we have weakening of structure functors that "include" $\mathbf{VB}(I)$ into $\mathbf{FB}(I)$ and $\mathbf{Hilb}$ into $\mathbf{Mfld}$. Our goal is to "extend" H_1 to a functor from $\mathbf{FB}(I)$ to $\mathbf{Mfld}$. For technical reasons it is expedient to carry out this process in two steps, extending H_1 first on morphisms and only then on objects. So we introduce two "mongrel" categories, $\mathbf{FVB}(I)$ and $\mathbf{MHilb}$. The objects of $\mathbf{FVB}(I)$ are the smooth vector bundles over I, but its morphisms are fiber bundle morphisms. Similarly, the objects of $\mathbf{MHilb}$ are hilbertable Banach spaces, and its morphisms are the smooth maps between them.

11.1.6. Theorem.   *If ξ and η are smooth vector bundles over I and $\varphi : \xi \to \eta$ is a smooth fiber bundle morphism then $\sigma \mapsto \varphi \circ \sigma$ is a smooth map $C^0(\varphi) : C^0(\xi) \to C^0(\eta)$ and it restricts to a smooth map $H_1(\varphi) : H_1(\xi) \to H_1(\eta)$. Thus H_1 extends to functor from $\mathbf{FVB}(I)$ to $\mathbf{MHilb}$.*

PROOF.   As in the case of vector bundle morphisms we can assume that ξ and η are product bundles $I \times R^n$ and $I \times R^p$ respectively, so $\varphi : I \times R^n \to I \times R^p$ is a smooth map of the form $(t, x) \mapsto (t, \varphi_t(x))$. Then as above $C^0(\varphi) : C^0(I, R^n) \to C^0(I, R^p)$ is defined by $C^0(\varphi)(\sigma)(t) = \varphi_t(\sigma(t))$, and it is easy to check that $C^0(\varphi)$ is a differentiable map and that its differential is given by $DC^0(\varphi)_\sigma(\lambda)(t) = D_1\varphi_{(t,\sigma(t))}(\lambda(t))$. If σ is absolutely continuous then, for t in I such that $\sigma'(t)$ exists, $C^0(\varphi)(\sigma)'(t) = D_1\varphi_{(t,\sigma(t))}(\sigma'(t)) + (\frac{\partial}{\partial s})_{s=t}\varphi(s, \sigma(t))$. It follows that $C^0(\varphi)(\sigma)$ is also absolutely continuous and that if σ is in $H_1(R^n)$ then $C^0(\varphi)(\sigma)$ is in $H_1(R^p)$. In other words $C^0(\varphi)$ restricts to a map $H_1(\varphi) : H_1(R^n) \to H_1(R^p)$, and it is again easy to check that this map is differentiable and that its differential is given by the same formula as above. Then by an easy induction we see that $H_1(\varphi)$ is smooth and that $D^m H_1(\varphi)_\sigma(\lambda_1, \ldots, \lambda_m)(t) = D_1^m\varphi_{(t,\sigma(t))}(\lambda_1(t), \ldots, \lambda_m(t))$.   ∎

11.1.7. Remark.   Note that the same does **not** hold for the functor H_0! For example define $\varphi : I \times R \to I \times R$ by $(t, x) \mapsto (t, x^2)$. Define $\sigma : I \to R$ by $\sigma(t) = t^{-\frac{1}{4}}$, so clearly $\sigma \in H_0(I, R)$. But $\varphi_t(\sigma(t)) = t^{-\frac{1}{2}}$, which is not square summable.

Now suppose that E is a smooth fiber bundle over I. A smooth vector bundle ξ over I is called a *vector bundle neighborhood in E*, (abbreviated to VBN), if the total space of ξ is open in the total space of E and if the inclusion $\xi \hookrightarrow E$ is a fiber bundle morphism. Of course then $C^0(\xi)$ is open in $C^0(E)$, and for any σ in $C^0(\xi)$ we will say that ξ is a VBN of σ in E.

11.1.8. Proposition.   *If E is a vector bundle and ξ is a VBN in E then $C^0(\xi)$ is a smooth open submanifold of the Banach space $C^0(E)$, and* *similarly $H_1(\xi)$ is a smooth open submanifold of the Hilbert space $H_1(E)$.*

PROOF. Since the inclusion $i : \xi \hookrightarrow E$ is a fiber bundle morphism, by the preceding Theorem the inclusions $C^0(i) : C^0(\xi) \hookrightarrow C^0(E)$ and $H_1(i) :$ $H_1(\xi) \to H_1(\eta)$ are smooth, and in fact by the Inverse Function Theorem they are diffeomorphisms onto open submanifolds. ∎

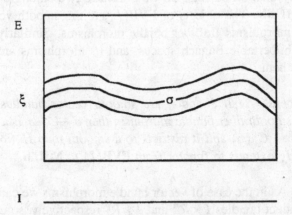

A VBN ξ of the section σ of the
fiber bundle E over the unit interval I

Given any *smooth* section σ of E we will now see how to construct a VBN in E, having σ as its zero section. We proceed as follows.

Let $TF(E)$ denote the subbundle of TE defined as the kernel of the differential of the projection of E onto I. $TF(E)$ is called the "tangent bundle along the fibers of E" since, for any fiber E_t of E, the tangent bundle, $T(E_t)$, is just the restriction of $TF(E)$ to E_t . This allows us to define a smooth "exponential" map Exp of $TF(E)$ into E such that if $e \in E_t$ then $\text{Exp}(TF(E)_e) \subseteq E_t$. Namely, choose a complete Riemannian metric for the total space of E. This induces a complete Riemannian metric on each fiber E_t, and hence an exponential map $\text{Exp}_t : TF(E)|E_t \to E_t$, and we define $\text{Exp} : TF(E) \to E$ to be equal to Exp_t on $TF(E)|E_t$. The fact that solutions of an ODE depend smoothly on parameters insures that Exp is a smooth map. (Note: Exp will *not* in general agree with the usual exponential map of E since the fibers E_t are not in general totally geodesic in E.)

Given a smooth section σ of E we define a smooth vector bundle E^σ over I by $E^\sigma = \sigma^*(TF(E))$. Note that, for $t \in I$, $E_t^\sigma = TF(E)_{\sigma(t)} = T(E_t)_{\sigma(t)}$. We define a smooth fiber bundle morphism $\mathcal{E} : E^\sigma \to E$ by $\mathcal{E}(v) = \text{Exp}(v) = \text{Exp}_t(v)$ for $v \in E_t^\sigma$. Since $\text{im}(\sigma)$ is compact we can choose an $\epsilon > 0$ less than the injectivity radius of E_t at $\sigma(t)$ for all t in I. Then $\mathcal{E}$ maps E_ϵ^σ, the open ϵ-disk bundle in E^σ, onto an open subbundle ξ of E; namely, for $t \in I$, ξ_t is the ball of radius ϵ in E_t about $\sigma(t)$. Finally let θ denote a diffeomorphism of R onto $(-\epsilon, \epsilon)$. Then we can define a fiber bundle isomorphism $\Theta : E^\sigma \approx E^\sigma_\epsilon$ by $\Theta(v) = \theta(\|v\|)(v/(1 + \|v\|))$, and composing this with $\mathcal{E}$ gives us a fiber bundle isomorphism $\mathcal{E} \circ \Theta : E^\sigma \approx \xi$. This proves:

11.1.9. VBN Existence Theorem.   *If E is any fiber bundle over I and σ is a smooth section of E then there is a vector bundle neighborhood ξ of σ in E having σ as its zero section. In more detail, given a complete Riemannian metric for E we can find such a vector bundle neighborhood structure on the open subbundle ξ of E whose fiber at t is the ball of radius ϵ about $\sigma(t)$ in E_t, provided that ϵ is chosen smaller than the injectivity radius for E_t at $\sigma(t)$ for all $t \in I$.*

11.1.10. Corollary.   *$C^0(E)$ is the union of the $C^0(\xi)$ for all VBN ξ of E. In fact, if $\sigma_0 \in C^0(E)$ and U is a neighborhood of σ_0 in E then there is a VBN ξ of σ_0 with $\xi \subseteq U$.*

PROOF.   Without loss of generality we can assume that U has compact closure. Choose $\epsilon > 0$ less than the injectivity radius of $E_{\Pi(e)}$ at e for all $e \in U$, and such that the disk of radius 2ϵ about $\sigma_0(t)$ in E_t is included in U for all $t \in I$. Choose a smooth section σ of E such that for all $t \in I$ the distance from $\sigma(t)$ to $\sigma_0(t)$ in E_t is less than ϵ. Then by the Theorem we can find a VBN ξ in E having σ as zero section and with fiber at t the ball of radius ϵ about $\sigma(t)$ in E_t. Clearly this ξ is a VBN of σ_0. ∎

11.1.11. Definition.   If E is a smooth fiber bundle over I then we define a smooth Banach manifold structure for $C^0(E)$ by requiring that, for each VBN ξ in E, the Banach space $C^0(\xi)$ is an open submanifold of $C^0(E)$. We define $H_1(E)$ to be the union of the $H_1(\xi)$ for all VBN ξ in E, and similarly we define a Hilbert manifold structure for $H_1(E)$ by requiring that, for each such ξ, the Hilbert space $H_1(\xi)$ is an open submanifold of $H_1(E)$.

11.1.12. Remark.   To see that this definition indeed makes $C^0(E)$ into a *smooth* manifold, let $\sigma \in C^0(E)$ and let ξ_1 and ξ_2 be two vector bundle neighborhoods of σ in E. It will suffice to find an open neighborhood O of σ in $C^0(E)$ that is included both in $C^0(\xi_1)$ and in $C^0(\xi_2)$, and on which both $C^0(\xi_1)$ and $C^0(\xi_2)$ induce the same differentiable structure. But by the VBN Existence Theorem there is a VBN η of σ that is included in the intersection of ξ_1 and ξ_2, and by the above Proposition the Banach space $C^0(\eta)$ is a smooth open submanifold both of $C^0(\xi_1)$ and of $C^0(\xi_2)$. A similar argument works for H_1.

11.1.13. Theorem.   *Let E and F be two smooth fiber bundles over I*

and let $\varphi : E \to F$ be a smooth vector bundle morphism. Then $\sigma \mapsto \varphi \circ \sigma$ is a smooth map $C^0(\varphi) : C^0(E) \to C^0(F)$ and restricts to a smooth map $H_1(\varphi) : H_1(E) \to H_1(F)$.

PROOF.   Given a section σ of E and a VBN η of $\varphi \circ \sigma$ in F, we can, by the VBN Existence Theorem, find a VBN ξ of ξ in E with $\varphi(\xi) \subseteq \eta$. By definition of the differentiable structures on $C^0(E)$ and $C^0(F)$ it will suffice to show that $\sigma \mapsto \varphi \circ \sigma$ maps $C^0(\xi)$ smoothly into $C^0(\eta)$. But this follows from an earlier Theorem. The same argument works for H_1. ∎

11.1.14. Remark.   We note that we have now reached our goal of extending H_1 to a functor from **FB**(I) to **Mfld**.

11.1.15. Corollary.   *If E is a smooth (closed) subbundle of F then $C^0(E)$ is a smooth (closed) submanifold of $C^0(F)$ and $H_1(E)$ is a smooth (closed) submanifold of $H_1(F)$.*

11.1.16. Remark.   It is clear from the definition of the differentiable structure on $C^0(E)$ and from the construction of VBN's above that if σ is a smooth section of a fiber bundle E, then $T(C^0(E))_\sigma$, the tangent space to $C^0(E)$ at σ, is canonically isomorphic to $C^0(E^\sigma)$, where as above E^σ denotes the vector bundle $\sigma^*(TF(E))$ over I. If $\varphi : E \to F$ is a smooth fiber bundle morphism and $C^0(\varphi)(\sigma) = \tau$ then clearly $D\varphi$ induces a vector bundle morphism $\varphi^\sigma : E^\sigma \to F^\tau$ and $C^0(\varphi^\sigma) : C^0(E^\sigma) \to C^0(F^\tau)$ is $D(C^0(\varphi))_\sigma$, the differential $C^0(\varphi)$ at σ. Similarly $T(H_1(E))_\sigma = H_1(E^\sigma)$ and $D(H_1(\varphi))_\sigma = H_1(\varphi^\sigma)$.

11.1.17. Remark.   If M is any smooth manifold then $I \times M$ is a smooth fiber bundle over I and of course we have a natural identification of $C^0(I, M)$ with $C^0(I \times M)$. Thus $C^0(I, M)$ becomes a smooth Banach manifold. Similarly $H_1(I, M)$ is well-defined and has the structure of a smooth Hilbert manifold. If M is a regularly embedded smooth (closed) submanifold of N then $I \times N$ is a smooth (closed) subbundle of $I \times N$ and hence $C^0(I, M)$ is a smooth (closed) submanifold of $C^0(I, N)$ and $H_1(I, M)$ is a smooth (closed) submanifold of $H_1(I, N)$. In particular if M is embedded as a closed submanifold of R^N then $H_1(I, M)$ is a closed submanifold of the Hilbert space $H(I, R^N)$ and so becomes a *complete* Riemannian manifold in the induced Riemannian metric. This will be important for our later applications to the calculus of variations.

11.2   Geodesics.

Let $X = I = [0,1]$, Y a complete Riemannian manifold, P and Q two points of Y, and $\mathcal{M}_0 = \mathrm{Im}(I,Y)$ the space of all immersions $\sigma : I \to Y$ such that $\sigma(0) = P$ and $\sigma(1) = Q$. We will consider two Lagrangians. The first is defined by $L(\sigma) = \|\sigma'(s)\|$, so the corresponding functional is the arc length :

$$\mathcal{L}(\sigma) = \int_0^1 \|\sigma'(s)\| \, ds,$$

and a critical point of $\mathcal{L}$ is called a geodesic of Y joining P and Q. The second Lagrangian is the energy density $E(\sigma) = \langle \sigma', \sigma' \rangle$, and $\mathcal{E}$ on $\mathcal{M}_0$ denotes the corresponding energy functional:

$$\mathcal{E}(\sigma) = \frac{1}{2} \int_0^1 \langle \sigma'(s), \sigma'(s) \rangle \, ds.$$

In what follows we will use the functional $\mathcal{E}$ as a model, illustrating the five step program of the Introduction that shows abstract Morse Theory applies to a particular Calculus of Variations problem. In the course of this we will see that $\mathcal{E}$ and $\mathcal{L}$ in a certain sense have "the same" critical points so we will rederive some standard existence theorems for geodesics. Critical points of $\mathcal{E}$ are sometimes called harmonic maps of I into Y, but as we shall see they are just geodesics parametrized proportionally to arc length.

First we compute $\nabla \mathcal{E}$ by using local coordinates. To simplify the notation a little we will adopt the so-called "Einstein summation convention". This means that **a summation is implicit over the complete range of a repeated index**. For example if T_{ij} is an $n \times n$ matrix then $\mathrm{Trace}(T) = T_{ii} = \sum_{i=1}^n T_{ii}$. Suppose $x = (x_1, \ldots, x_n)$ is a local coordinate system of M, and write $x_i(s) = x_i(\sigma(s))$, and $ds^2 = g_{ij} \, dx_i \, dx_j$. Then in this coordinate system

$$E(\sigma) = E(x, x') = g_{ij}(x) x_i' x_j'.$$

So writing $g_{ij,k} = \frac{\partial g_{ij}}{\partial x_k}$, the Euler-Lagrange equation is given by

$$g_{kl,i} x_k' x_l' = \frac{\partial E}{\partial x_i}$$
$$= \frac{d}{ds} \frac{\partial E}{\partial x_i'}$$
$$= 2(g_{ij} x_j')'$$
$$= 2g_{ij,k} x_k' x_j' + 2g_{ij} x_j''$$
$$= g_{il,k} x_k' x_l' + g_{ik,l} x_l' x_k' + 2g_{ij} x_j'',$$

i.e.,

$$g_{ij}x_j'' + \frac{1}{2}\{g_{il,k} + g_{ik,l} - g_{kl,i}\} \, x_k' x_l' = 0 \qquad (11.2.1)$$

Let (g^{ij}) denote the inverse of the matrix (g_{ij}). Multiplying both sides of (11.2.1) by g^{im} and summing over i, we get

$$x_m'' + \frac{1}{2}g^{im}\{g_{il,k} + g_{ik,l} - g_{kl,i}\} \, x_k' x_l' = 0.$$

Let Γ_{kl}^m be the Christoffel symbols associated to g, defined by:

$$\nabla_{\frac{\partial}{\partial x_k}}\left(\frac{\partial}{\partial x_l}\right) = \Gamma_{kl}^m \frac{\partial}{\partial x_m},$$

where ∇ is the Levi-Civita connection for g. Note that

$$g_{ij,k} = \frac{\partial}{\partial x_k}\langle\frac{\partial}{\partial x_i}, \frac{\partial}{\partial x_j}\rangle = \nabla_{\frac{\partial}{\partial x_k}}\langle\frac{\partial}{\partial x_i}, \frac{\partial}{\partial x_j}\rangle.$$

Then direct computation, using that that ∇ is torsion free and compatible with the metric, gives

$$\Gamma_{kl}^m = \frac{1}{2}g^{im}\{g_{il,k} + g_{ik,l} - g_{kl,i}\}.$$

So the Euler-Lagrange equation for $\mathcal{E}$ in local coordinates becomes

$$x_m'' + \Gamma_{kl}^m x_k' x_l' = 0. \qquad (11.2.2)$$

Note that if x is the geodesic coordinate centered at $\sigma(s_o)$, then (11.2.2) is the same as $\nabla_{\sigma'(s_o)}\sigma' = 0$. So the invariant formulation of (11.2.2) is $\nabla_{\sigma'}\sigma' = 0$.

The second method to compute $\nabla\mathcal{E}$ is using covariant derivatives. By the Nash isometric embedding theorem we may assume that Y is a submanifold of R^m with the induced metric. Let ∇ denote the Levi-Civita connection of Y, $u \in TY_x$, P_x the orthogonal projection of R^m onto TY_x, and ξ a tangent vector field on Y. Then $(\nabla_u\xi)(x) = P_x(d\xi(u))$. Suppose σ_t is a smooth curve in $\mathcal{M}_0$, with $\sigma_0 = \sigma$. Then

$$h = \left(\frac{d\sigma_t}{dt}\right)_{t=0} \in (T\mathcal{M}_0)_\sigma$$

is a vector field along σ and $h(0) = h(1) = 0$. We have

$$\delta\mathcal{E} = \frac{d(\mathcal{E}(\sigma_t))}{dt}\bigg|_{t=0} = \int_0^1 \langle\sigma'(s), h'(s)\rangle ds,$$

$$= -\int_0^1 \langle\sigma''(s), h(s)\rangle ds$$

$$= -\int_0^1 \langle P_\sigma(\sigma''), h(s)\rangle ds,$$

$$= -\int_0^1 \langle\nabla_{\sigma'}\sigma', h\rangle ds.$$

So the condition for σ to be a critical point of $\mathcal{E}$ is that $\nabla_{\sigma'}\sigma' = 0$, i.e., that σ' is parallel along σ. This has an elementary but important consequence;

$$(E(\sigma))' = \frac{d}{ds}\langle\sigma',\sigma'\rangle = 2\langle\nabla_{\sigma'}\,\sigma',\sigma'\rangle = 0,$$

so that *if σ is a critical point of $\mathcal{E}$, then $\|\sigma'(s)\|$ is constant.*

Next we want to discuss the relation between the two functionals $\mathcal{E}$ and $\mathcal{L}$. But first it is necessary to recall some relevant facts about reparameterizing immersions of I into Y. Let G denote the group of orientation preserving, smooth diffeomorphisms of I, i.e.,

$$G = \{u : I \to I|\ u(0) = 0, u(1) = 1, u'(t) > 0 \text{ for all } t\ \}.$$

Then G acts freely on $\mathcal{M}_0$ by $u \cdot \sigma(t) = \sigma(u(t))$, and we call elements of $\mathcal{M}_0$ belonging to the same orbit "reparameterizations" of each other. It is clear from the change of variable formula for integrals that $\mathcal{L}$ is invariant under G, i.e., $\mathcal{L}(u \cdot \sigma) = \mathcal{L}(\sigma)$. Thus $\mathcal{L}$ is constant on G orbits, or in other words reparameterizing a curve does not change its length. Let Σ denote the set of σ in $\mathcal{M}_I$ such that $\|\sigma'\|$ is some constant c, depending on σ. Recall that this is the condition that we saw above was satisfied automatically by critical points of $\mathcal{E}$. It is clearly also the condition that the length of σ between 0 and t should be a constant c times t (so in particular c is just the length of σ), and so we call the elements of Σ paths "parametrized proportionally to arc length". Recall that Σ is a cross-section for the action of G on $\mathcal{M}_0$, that is, every orbit of G meets Σ in a unique point, or equivalently, any immersion of I into Y can be uniquely reparametrized proportionally to arc length. In fact the element s of G that reparameterizes $\sigma \in \mathcal{M}_0$ proportionally to arc length is given explicitly by $s(t) = \frac{1}{\ell}\int_0^t \|\sigma'(t)\|\,dt$, where ℓ is the length of σ. This means that we can identify Σ with the orbit space $\mathcal{M}_0/G$. Now since $\mathcal{L}$ is invariant under G, it follows that if σ is a critical point of $\mathcal{L}$, then the whole G orbit of σ consists of geodesics, and in particular the point where the orbit meets Σ is a geodesic. So, in searching for geodesics we may as well restrict attention to Σ. Moreover it is clear (and we will verify this below) that to check whether $\sigma \in \Sigma$ is a critical point of $\mathcal{L}$, it suffices to check that it is a critical point of $\mathcal{L}|\Sigma$. But on Σ, $\mathcal{E} = \mathcal{L}^2/2$, so they have the same critical points. Thus:

11.2.1. Theorem.   *Let $\sigma : [0,1] \to Y$ be a smooth curve. Then the following three statements are equivalent:*

(i) σ is a critical point of the energy functional $\mathcal{E}$,

(ii) σ is parametrized proportionally to arc length and is a critical point of the arc length functional $\mathcal{L}$,

(iii) $\nabla_{\sigma'}\sigma' = 0$.

Now let's check this by direct computation. If σ is a critical point of $\mathcal{E}$ then

$$\frac{\partial E}{\partial x_i} = \left(\frac{\partial E}{\partial x_i'}\right)'.$$

But, since $L = \sqrt{E}$,

$$\frac{\partial L}{\partial x_i} = \frac{1}{2}E^{-\frac{1}{2}}\frac{\partial E}{\partial x_i},$$

$$\left(\frac{\partial L}{\partial x_i'}\right)' = \frac{1}{2}\left(E^{-\frac{1}{2}}\frac{\partial E}{\partial x_i'}\right)'$$

$$= -\frac{1}{4}E^{-3/2}(E(\sigma))'\frac{\partial E}{\partial x_i'} + \frac{1}{2}E^{-\frac{1}{2}}\left(\frac{\partial E}{\partial x_i'}\right)'$$

$$= 0 + \frac{1}{2}E^{-\frac{1}{2}}\left(\frac{\partial E}{\partial x_i'}\right)'.$$

So

$$\frac{\partial L}{\partial x_i} = \left(\frac{\partial L}{\partial x_i'}\right)',$$

i.e., σ is a critical point of $\mathcal{L}$.

To prove the converse suppose σ is a critical point of $\mathcal{L}$ parametrized proportionally to it arc length, i.e., $L(\sigma) \equiv c$, a constant, which implies that $L' = (L(\sigma))' = 0$. Since $E = L^2$, we have

$$\frac{\partial E}{\partial x_i} = 2L\frac{\partial L}{\partial x_i},$$

$$\left(\frac{\partial E}{\partial x_i'}\right)' = \left(2L\frac{\partial L}{\partial x_i'}\right)'$$

$$= 2L\left(\frac{\partial L}{\partial x_i'}\right)'.$$

Hence σ is a critical point of $\mathcal{E}$.

A very important general principle is involved here. Geometrically natural functionals J, such as length or area, tend to be invariant under "coordinate transformations". The same is true for the functionals that physicists extremalize to define basic physical laws. After all, the laws of physics should *not* depend on the size or orientation of the measuring gauges used to observe events. Following physics terminology that goes back to Hermann Weyl, a group of coordinate transformations that leave a variational problem J invariant is referred to as a "gauge group" for the problem. Now the existence of a large gauge group usually has profound and wonderful consequences. But these are not always mathematically convenient. In fact, from our point of view, there is an obvious drawback to a large gauge group G. Clearly if σ is a critical point of J then the whole gauge orbit, $G\sigma$, consists of critical points at the same level of J. But remember that, if Condition C is to be satisfied, then the set of critical points at a given level must be compact. It follows that having all the gauge group orbits not relatively compact is incompatible with Condition C. Typically, the gauge group G is a large infinite dimensional group, and the isotropy groups G_σ are finite dimensional or even compact, and this is clearly bad news for Condition C. In particular we can now see that the invariance of the length functional $\mathcal{L}$ under the group G of reparameterizations of the interval means that it cannot satisfy Condition C. The way around this problem is clear. We commonly regard immersions of the interval that differ by a reparameterization as "the same" geometric curve. Similarly, the physicist regards as "the same" two physical configurations that differ only by a gauge transformation. In general, if $f : \mathcal{M} \to \mathbf{R}$ is a variational functional invariant under a gauge group G, then we think of points of $\mathcal{M}$ belonging to the same gauge orbit as being two representations of the "the same" basic object. Thus it seems natural to carry out our analysis on the space $\mathcal{M}/G$ of gauge orbits and try to verify Condition C there. Since f is G-invariant, it gives a well defined function on $\mathcal{M}/G$. But unfortunately $\mathcal{M}/G$ is in general *not* a smooth manifold. It is possible to do a reasonable amount of analysis on the orbit space, despite its singularities; for example if M is Riemannian and G acts isometrically then ∇f is clearly a G-invariant vector field, so that the flow φ_t it generates will commute with the action of G and give a well defined flow on the orbit space. Nevertheless experience seem to show that it is usually better not to "divide out" the action of G explicitly. The following definition clearly captures the notion of f satisfying Condition C on $\mathcal{M}/G$, without actually passing to the possibly singular quotient space:

11.2.2. Definition.   Let $\mathcal{M}$ be a Riemannian manifold, G a group of isometries of $\mathcal{M}$, and $f : \mathcal{M} \to \mathbf{R}$ a smooth G-invariant function on $\mathcal{M}$. We will say that f *satisfies Condition C modulo* G if given a sequence $\{x_n\}$ in $\mathcal{M}$ such that $|f(x_n)|$ is bounded and $\|\nabla f_{x_n}\| \to 0$, there exists a sequence $\{g_n\}$ in G such that the sequence $\{g_n x_n\}$ has a convergent subsequence.

If M is complete and f is bounded below, the proof in section 9.1 that the flow generated by $-\nabla f$ is a positive semigroup generalizes easily to the case that f satisfies Condition C only modulo a gauge group G, so the First and Second Deformation Theorems are also valid in this more general context.

In actual practice, instead of showing that a functional J satisfies Condition C modulo a gauge group G directly, there are several methods for implicitly dividing out the gauge equivalence. One such method is to impose a so-called "gauge fixing condition" that defines a "cross-section" of $\mathcal{M}$, i.e., a smooth submanifold Σ of $\mathcal{M}$ that meets all the gauge orbits, and show that

J restricted to Σ satisfies Condition C. Another approach is to look for a functional $\mathcal{J}$, that "breaks the gauge symmetry" (that is, $\mathcal{J}$ is *not* G-invariant), and yet has "the same" critical points as J, in the sense that every critical point of $\mathcal{J}$ is also a critical point of J, and every G-orbit of critical points of J contains a critical point of $\mathcal{J}$. All this is elegantly illustrated by the length functional $\mathcal{L}$; the energy functional $\mathcal{E}$ breaks the reparameterization symmetry of $\mathcal{L}$, and nevertheless has the "the same" critical points as $\mathcal{L}$. The appropriate gauge fixing condition in this case is of course parameterization proportionally to arc length. But best of all, these strategies actually *succeed* in this case for, as we shall now see, the energy functional does satisfy Condition C on an appropriate Sobolev completion of $\mathcal{M}_0$.

We now begin carrying out the five steps mentioned in the Introduction. First we discuss the completion $\mathcal{M}$ of $\mathcal{M}_0$ and the extension of $\mathcal{E}$ to $\mathcal{M}$. Recall that we are assuming that Y is isometrically imbedded in $\boldsymbol{R}^m$. Since Y is complete, it is closed in $\boldsymbol{R}^m$ and it follows from Theorem 11.1.5 that $H_1(I, Y)$ is a closed submanifold of the Hilbert space $H_1(I, \boldsymbol{R}^m)$. Here the H_1-norm is defined by

$$\|\sigma\|_1^2 = \|\sigma(0)\|_0^2 + \int_0^1 \|\sigma'(t)\|^2 dt.$$

Clearly

$$\mathcal{M} = \{\sigma \in H_1(I, Y) \mid \sigma(0) = P, \ \sigma(1) = Q\}$$

is a smooth, closed codimension $2m$ submanifold of $H_1(I, Y)$ (because the map $\sigma \mapsto (\sigma(0), \sigma(1))$ of $H_1(I, Y)$ into $\boldsymbol{R}^m \times \boldsymbol{R}^m$ is a submersion), and so it is a complete Riemannian manifold. It is easily seen that $\mathcal{E}$ naturally extends to $\mathcal{M}$. For if $\sigma \in H_1(I, \boldsymbol{R}^m)$, then $\hat{\mathcal{E}}(\sigma) = \int_0^1 \langle \sigma', \sigma' \rangle ds$ is a well-defined extension of $\mathcal{E}$ to $H_1(I, \boldsymbol{R}^m)$, and $\|\sigma\|_1^2 = \|\sigma(0)\|_0^2 + \hat{\mathcal{E}}(\sigma)$. Since both $\|\sigma\|_1^2$ and $\|\sigma(0)\|_0^2$ are continuous quadratic forms on $H_1(I, \boldsymbol{R}^m)$, they are smooth, and so is $\hat{\mathcal{E}}$. Hence $\tilde{\mathcal{E}} = \hat{\mathcal{E}}|\mathcal{M}$ is smooth. $\tilde{\mathcal{E}}$ is clearly bounded from below by 0.

11.2.3. Definition. A critical point of $\tilde{\mathcal{E}}$ on $\mathcal{M}$ is called a *harmonic map* of I into Y joining P to Q.

Note that for $\sigma \in \mathcal{M}$, we have

$$T\mathcal{M}_\sigma = \{v \in H_1(I, \boldsymbol{R}^m) \mid v(t) \in TY_{\sigma(t)}, v(0) = v(1) = 0\},$$

$$d\tilde{\mathcal{E}}_\sigma(v) = \langle \sigma', v' \rangle_0 = \langle \nabla\tilde{\mathcal{E}}(\sigma), v \rangle_1,$$

where $\nabla\tilde{\mathcal{E}}(\sigma)$ is in $T\mathcal{M}_\sigma$ such that $(\nabla\tilde{\mathcal{E}}(\sigma))' = \sigma'$ as L^2 functions. If σ is smooth then by integration by parts $\nabla\tilde{\mathcal{E}}(\sigma) = -\nabla_{\sigma'}(\sigma')$.

We may assume that $0 \in Y$ and $P = 0$. Thus the elements of $\mathcal{M}$ are H_1-maps $\sigma : I \to R^n$ with $\operatorname{im}(\sigma) \subset Y$, $\sigma(0) = 0$, $\sigma(1) = Q$ and

$$\tilde{\mathcal{E}}(\sigma) = \|\sigma\|_1^2 = f_0,$$

the distance function from 0.

Next we prove that $\tilde{\mathcal{E}}$ satisfies condition C. Since smooth sections are dense in $\mathcal{M}$, it will suffice to show that given a sequence $\{\sigma_n\} \in \mathcal{M}_0$ such that

$$\mathcal{E}(\sigma_n) = \|\sigma_n\|_1^2 \le c_0 \quad \text{and} \quad \nabla\mathcal{E}(\sigma_n) \to 0 \text{ in } H_1$$

then $\{\sigma_n\}$ has a convergent subsequence in $\mathcal{M}$. Note that

$$TM_{\sigma_n} = \{h \in H_1(I, R^m) \mid h(0) = h(1) = 0, \ h(t) \in TY_{\sigma_n(t)}\},$$

$$d\mathcal{E}_{\sigma_n}(h) = -\int_0^1 \langle P_{\sigma_n}(\sigma_n''), h \rangle \, dt = \langle \nabla\mathcal{E}(\sigma_n), h \rangle_1, \quad \forall \, h \in TM_{\sigma_n}.$$

So, by the Schwarz inequality, if $h \in H_1(I, R^m)$ and $h(0) = h(1) = 0$, then we have

$$|\langle P_{\sigma_n}(\sigma_n''), h \rangle_0| \le \|\nabla\mathcal{E}(\sigma_n)\|_1 \|h\|_1. \tag{11.2.3}$$

Since $\mathcal{M}$ is closed in $H_1(I, R^m)$, it will suffice to show that some subsequence (still denoted by $\{\sigma_n\}$) satisfies

$$\|\sigma_n - \sigma_m\|_1^2 \to 0.$$

Since $\{\sigma_n\}$ is bounded in $H_1(I, R^m)$ and the inclusion of $H_1(I, R^m)$ into $C_0(I, R^m)$ is compact (Theorem 11.1.4), we can assume

$$\|\sigma_n - \sigma_m\|_\infty \to 0.$$

Note that

$$\|\sigma_n - \sigma_m\|_1^2 = \langle \sigma_n', (\sigma_n - \sigma_m)' \rangle_0 - \langle \sigma_m', (\sigma_n - \sigma_m)' \rangle_0,$$

so it will suffice to show that

$$\langle \sigma_n', (\sigma_n - \sigma_m)' \rangle_0 \to 0.$$

Since the σ_n are smooth and $\sigma_n - \sigma_m$ vanishes at 0 and 1, we can integrate by parts and the latter is equivalent to

$$\langle \sigma_n'', (\sigma_n - \sigma_m) \rangle_0 \to 0.$$

Now, since σ_n' is tangent to Y, $P_{\sigma_n}\sigma_n' = \sigma_n'$ (recall that $x \mapsto P_x$ is the Gauss map of Y, i.e., P_x is the orthogonal projection of $\boldsymbol{R}^m$ onto TY_x, and $P_\sigma\sigma'$ is the map $t \mapsto P_{\sigma(t)}\sigma'(t)$). Thus

$$\sigma_n'' = (P_{\sigma_n}\sigma_n')' = P_{\sigma_n}'\sigma_n' + P_{\sigma_n}\sigma_n''.$$

Therefore we will be finished if we can prove the following two facts:

(A) $|\langle P_{\sigma_n}'\sigma_n', (\sigma_n - \sigma_m)\rangle_0| \to 0,$

(B) $|\langle P_{\sigma_n}\sigma_n'', (\sigma_n - \sigma_m)\rangle_0| \to 0.$

As for (A), by Hölder's inequality we have

$$|\langle P_{\sigma_n}'\sigma_n', (\sigma_n - \sigma_m)\rangle_0| \le \|P_{\sigma_n}'\sigma_n'\|_{L^1}\|\sigma_n - \sigma_m\|_{L^\infty}.$$

Recalling that $\|\sigma_n - \sigma_m\|_\infty \to 0$, it will suffice to show $\|P_{\sigma_n}'\sigma_n'\|_{L^1}$ is bounded. By the Schwarz inequality, it will suffice to prove $\|P_{\sigma_n}'\|_{L^2}$ and $\|\sigma_n'\|_{L^2} = \|\sigma_n\|_1$ are bounded. The latter is true by assumption, and since $P_{\sigma_n}' = dP_{\sigma_n} \circ \sigma_n'$, and dP is bounded on a compact set, P_{σ_n}' is bounded. Since $\sigma_n - \sigma_m$ vanishes at 0 and 1 and is bounded in the H_1-norm, statement (B) follows from (11.2.3).

It remains to prove regularity. Note that equation $d\mathcal{E}_\sigma(v) = \langle \sigma', v'\rangle_0 = 0$ for $v \in T\mathcal{M}_\sigma$ is equivalent to

$$\langle \sigma', (P_\sigma v)'\rangle_0 = 0 \quad \text{for all } v \in H_1(I, \boldsymbol{R}^m). \tag{11.2.4}$$

Since $\text{Im}(P)$ is contained in the linear space of self-adjoint operators on $\boldsymbol{R}^m$, $(P_\sigma)'$ is also self-adjoint, and by chain rule, we have

$$\begin{aligned}
\langle \sigma', (P_\sigma v)'\rangle_0 &= \langle \sigma', (P_\sigma)'(v) + P_\sigma(v')\rangle_0 \\
&= \langle (P_\sigma)'(\sigma'), v\rangle_0 + \langle P_\sigma(\sigma'), v'\rangle_0 \\
&= \langle (P_\sigma)'(\sigma'), v\rangle_0 + \langle \sigma', v'\rangle_0.
\end{aligned} \tag{11.2.5}$$

Since $\sigma \in H_1(I, \boldsymbol{R}^m)$, σ is continuous and $\|\sigma\|_\infty$ is bounded.

By the chain rule $(P_\sigma)'(\sigma')$ is smooth in σ and quadratic in σ', and so it is in L^1. Then

$$\gamma(t) = \int_0^t (P_\sigma)'(\sigma')ds \tag{11.2.6}$$

is in C^0. Substituting (11.2.6) into (11.2.5) and using integration by parts, we obtain

$$\langle \gamma', \overset{\bullet}{v}\rangle_0 + \langle \sigma', v'\rangle_0 = \langle \sigma' - \gamma, v'\rangle_0 = 0.$$

It follows that σ' and γ differ by a constant. Since γ is continuous, σ is C^1. We can now "pull ourselves up by our own bootstraps". It follows from the definition of γ that if σ is C^k ($k \geq 1$), then γ' is C^{k-1}, and hence γ is C^k. Then σ' is also C^k, so σ is C^{k+1}. By induction, σ is smooth.

We can now apply our general theory of critical points to the geodesic problem.

11.2.4. Theorem. *Given any two points P and Q of a complete Riemannian manifold Y, there exists a geodesic joining P to Q whose length is the distance from P to Q. Moreover any homotopy class of paths from P to Q contains a geodesic parametrized proportionally to arc length that minimizes length and energy in that homotopy class.*

PROOF. Since it is the energy $\mathcal{E}$, rather than the length $\mathcal{L}$, that satisfies Condition C, our general theorem really only applies directly to $\mathcal{E}$. But recall that on the set Σ of paths parametrized proportionally to arc length, $\mathcal{L} = \sqrt{\mathcal{E}}$. Now since any path σ has a reparameterization $\tilde{\sigma}$ in Σ with the same length, it follows that $\inf(\mathcal{L}) = \inf(\sqrt{\mathcal{E}})$. And since we know $\mathcal{E}$ must assume its minimum at a point σ of Σ, it follows that this σ is also a minimum of $\mathcal{L}$. ∎

So far we have considered the theory of geodesics joining two fixed points. There is just as important and interesting a theory of closed geodesics. For this we take for X not the interval I, but rather the circle S^1, so our space $\mathcal{M}_0$ consists of the smooth immersions of S^1 into Y. As usual we will identify a continuous (or smooth) map of S^1 with a map of I that has a continuous (or smooth) periodic extension with period one. In this way we regard the various spaces of maps of the circle into R^m (and into Y) as subspaces of the corresponding spaces of maps of I into R^m (and into Y). This allows us to carry over all the formulas and norms defined above. In particular we have the formula:

$$\|\sigma\|_1^2 = \|\sigma(0)\|^2 + \mathcal{E}(\sigma).$$

At this point there is a small but important difference in the theory. If we consider immersions of I joining P to Q, then $\sigma(0) = P$, is constant, hence bounding the energy bounds the H_1 norm. But for the case of immersions of S^1 into Y, the point $\sigma(0)$ can be any point of Y, so if we want to insure that $\|\sigma(0)\|^2$ is bounded then **we must require that Y is bounded, and hence compact.** Once this extra requirement is made, bounding the energy again bounds the H_1 norm, and the whole development above works exactly the same for immersions of S^1 as it did for immersions of I. In particular:

11.2.5. Theorem. *If Y is a compact Riemannian manifold, then given any free homotopy class α of maps of S^1 into Y there is a representative σ of α that is a closed geodesic parametrized proportionally to arc length and that minimizes both length and energy in that homotopy class.*

The requirement that Y be compact is real, and not just an artifact of the proof. For example, consider the surface of revolution in $\mathbf{R}^3$ obtained by rotating the graph of $y = \frac{1}{x}$ about the x-axis. It is clear that the homotopy class of the circles of rotation has no representative of minimum length or energy.

11.3   Non-linear eigenvalue problem.

Let $(V, \langle \, , \, \rangle)$ be a Hilbert space. Let J, $F : V \to \mathbf{R}$ be smooth functions, 1 a regular value of F, and $\mathcal{M}$ the level hypersurface $F^{-1}(1)$ of V. Then by the Lagrange multiplier principle, $u \in V$ is a critical point of $J|\mathcal{M}$ if and only if there is a constant λ such that

$$\nabla J(u) = \lambda \nabla F(u). \tag{11.3.1}$$

If $F(u) = \langle u, u \rangle$ and J is the quadratic function defined by $J(u) = \langle P(u), u \rangle$ for some bounded self-adjoint operator P on V, then (11.3.1) becomes the eigenvalue problem for the linear operator P:

$$P(u) = \lambda u.$$

So if either F or J is quadratic, we will refer (11.3.1) as the *non-linear eigenvalue problem*.

In this section we will study a simple non-linear eigenvalue problem of this type. But first we need to review a little hard analysis.

Let X be a compact, smooth n-dimensional Riemannian manifold, ∇ the Levi-Civita connection for g, and dv the Riemannian volume element. For each p with $1 \le p < \infty$ we associate a Banach space $L^p(X)$, the space of all measureable functions $u : X \to \mathbf{R}$ such that

$$\|u\|_{L^p}^p = \int_X |u(x)|^p \, dv(x) < \infty.$$

Next we introduce the L_k^p-norm on $C^\infty(X)$ as follows:

$$\|u\|_{L_k^p}^p = \sum_{i=0}^{k} \int_X \|\nabla^i u(x)\|^p \, dv(x).$$

11.3.1. Definition.   For $1 \le p < \infty$ and each non-negative integer k, we define the Sobolev Banach space $L_k^p(X)$ to be $L^p(X)$ if $k = 0$, and to be the completion of $C^\infty(X)$ with respect to the Sobolev L_k^p-norms for positive k.

The Sobolev spaces $L^2_k(X)$ are clearly Hilbert spaces. It is not difficult to identify $L^p_k(X)$ with the space of measureable functions that have distributional derivatives of order $\leq k$ in L^p.

Another family of Banach space that will be important for us are the Hölder spaces, $C^{k,\alpha}(X)$, where k is again a non-negative integer and $0 < \alpha < 1$. It is easy to describe the space $C^{0,\alpha}(X)$; it consists of all maps $u : X \to R$ that are "Hölder continuous of order α", in the sense that

$$N_\alpha(u) = \sup_{x,y \in X} \frac{|u(x) - u(y)|}{d(x,y)^\alpha} < \infty,$$

where $d(x,y)$ is the distance of x and y in X. The norm $\| \ \|_{C^{0,\alpha}}$ for the Hölder space $C^{0,\alpha}(X)$ is defined by

$$\|u\|_{C^{0,\alpha}} = \|u\|_\infty + N_\alpha(u),$$

where as usual $\|u\|_\infty$ denotes the "sup" norm of u, $\|u\|_\infty = \max_{x \in X} |u(x)|$.

The higher order Hölder spaces can be defined in a similar manner. Let $X_1, \ldots, X_m$ be smooth vector fields on X such that $X_1(x), \ldots, X_m(x)$ spans TX_x at each point $x \in X$. We define $C^{k,\alpha}(X)$ to be the set of $u \in C^k(X)$ such that

$$N_\alpha(\nabla^k u) = \sum_{(i_1,\ldots,i_k)} N_\alpha(\nabla^k u(X_{i_1},\ldots,X_{i_k})) < \infty,$$

and we define the $C^{k,\alpha}$ norm of such a u by:

$$\|u\|_{C^{k,\alpha}} = \|u\|_{C^k} + N_\alpha(\nabla^k u),$$

where the C^k norm, $\| \ \|_{C^k}$, is as usual defined by:

$$\|u\|_{C^k} = \sum_{i=0}^k \|\nabla^i u\|_\infty.$$

Let V and W be Banach spaces with $\| \ \|_V$ and $\| \ \|_W$ respectively. If V is a linear subspace of W, then proving the inclusion $V \hookrightarrow W$ is continuous is equivalent to proving an estimate of the form $\|v\|_W \leq C\|v\|_V$ for all v in V.

There are a number of such inclusion relationships that exist between certain of the $L^p_k(X)$ and $C^{k,\alpha}(X)$. These go collectively under the name of "embedding theorems" (for proofs see [GT], [So], [Am], [Cr], [My]). They play a central role in the modern theory of PDE.

11.3.2.   Sobolev Embedding Theorems.   *Let X be a smooth, compact n-dimensional Riemannian manifold.*

(1) If $k - \frac{n}{p} \geq l - \frac{n}{q}$ and $k \geq l$, then $L_k^p(X)$ is contained in $L_l^q(X)$ and the inclusion map is continuous. If both inequalities are strict then this embedding is even compact.

(2) If $k - \frac{n}{p} \geq l + \alpha$, then $L_k^p(X)$ is contained in $C^{l,\alpha}(X)$ and the inclusion map is continuous. If the inequality is strict then this embedding is even compact.

11.3.3.   Corollary.   *If $p \leq \frac{2n}{n-2}$ then $L_1^2(X)$ is contained in $L^p(X)$. If the inequality is strict then this embedding is even compact.*

In the following we let $\|u\|_q$ denote the norm $\|u\|_{L^q}$. Since the inclusion $i : L_1^2(X) \hookrightarrow L^{\frac{2n}{n-2}}(X)$ is continuous, there is a constant C such that

$$\|u\|_{\frac{2n}{n-2}} \leq C(\|\nabla u\|_2 + \|u\|_2).$$

Let $c(X)$ denote the infimum of $A > 0$ for which there exists a $B > 0$ so that

$$\|u\|_{\frac{2n}{n-2}}^2 \leq A\|\nabla u\|_2^2 + B\|u\|_2^2$$

for all $u \in L_1^2(X)$. It turns out that $c(X)$ depends only on the dimension n of X, that is, (see [Au]):

11.3.4.   Theorem.   *There is a universal constant $c(n)$ such that for any compact Riemannian manifold X of dimension n and any $\epsilon > 0$, there is a $b(\epsilon) > 0$ for which the inequality*

$$\|u\|_{\frac{2n}{n-2}}^2 \leq (c(n) + \epsilon)\|\nabla u\|_2^2 + b(\epsilon)\|u\|_2^2$$

holds for all u in $L_1^2(X)$.

This constant $c(n)$ is referred to as the "best constant for the Sobolev Embedding Theorem".

We now state the standard *a priori* estimates for linear elliptic theory (for proofs see [Tr]):

11.3.5.   Theorem.   *Let (X, g) be a compact, Riemannian manifold, and $\Delta u = f$.*

(1) If $f \in C^{k,\alpha}(X)$ then $u \in C^{k+2,\alpha}(X)$.

(2) If $p > 1$ and $f \in L_k^p(X)$ then $u \in L_{k+2}^p(X)$.

For our discussion below, we also need the following Theorem of Brezis and Lieb [BL]:

11.3.6. Theorem.   *Suppose $0 < q < \infty$ and v_n a bounded sequence in L^q. If $v_n \to v$ pointwise almost everywhere, then $v \in L^q$ and*

$$\int_X |v_n|^q \, dv - \int_X |v_n - v|^q \, dv \to \int_X |v|^q \, dv.$$

Now suppose $2 < p \le \sigma(n) = \frac{2n}{n-2}$. Then by Corollary 11.3.3, $L^p(X)$ is continuously embedded in $L_1^2(X)$. So

$$\mathcal{M} = \{u \in L_1^2(X) \mid \int_X |u(x)|^p \, dv(x) = 1\}$$

defines a closed hypersurface of the Hilbert space $L_1^2 = L_1^2(X)$. The tangent plane of $\mathcal{M}$ at u is

$$T\mathcal{M}_u = \{\varphi \in L_1^2 \mid \langle |u|^{p-2}u, \varphi \rangle_0 = 0\},$$

where

$$\langle u, \varphi \rangle_0 = \int_X u\varphi \, dv$$

is the L^2-inner product.

Let $f : X \to R$ be a given smooth function, and define $J : \mathcal{M} \to R$ by

$$J(u) = \int_X \|\nabla u(x)\|^2 + f(x)u^2(x) \, dv. \tag{11.3.2}$$

By the Lagrange multiplier principle, the Euler-Lagrange equation of J on $\mathcal{M}$ is

$$\Delta u - fu = \lambda |u|^{p-2}u, \tag{11.3.3}$$

for some constant λ. Multiplying both sides of (11.3.3) by u and integrating over X we see that $\lambda = -J(u)$. So

$$\Delta u - fu = -J(u)|u|^{p-2}u. \tag{11.3.4}$$

The study of this equation is motivated by the following:

11.3.7. Yamabe Problem. Let (X, g) be a compact, Riemannian manifold. Is there a positive function u on X such that the scalar curvature of $\tilde{g} = u^{\frac{4}{n-2}} g$ is a constant function? Let f denote the scalar curvature function of g. Then it follows from a straight forward computation that the scalar curvature of $\tilde{g}$ is

$$\left\{ \frac{4(n-1)}{n-2} \Delta u + fu \right\} u^{-\frac{n+2}{n-2}}.$$

So the Yamabe problem is equivalent to finding a positive solution to equation (11.3.3) with $p = \sigma(n) = \frac{2n}{n-2}$ (for details see [Au],[Sc1],[Sc2]).

It is easily seen that

$$dJ_u(v) = \int_X \langle \nabla u, \nabla v \rangle + fuv \ dv \qquad \text{for } v \in T\mathcal{M}_u$$

If $v \in L_1^2$, then $v - \langle |u|^{p-2} u, v \rangle_0 u \in T\mathcal{M}_u$ and

$$dJ_u(v - \langle |u|^{p-2} u, v \rangle_0 u) = -J(u) \langle |u|^{p-2} u, v \rangle_0$$
$$+ \int_X \langle \nabla u, \nabla v \rangle + fuv \ dv \qquad (11.3.5)$$

By Hölder's inequality, for $u \in \mathcal{M}$ (i.e., $\|u\|_{L^p} = 1$), we have

$$\int_X u^2 \ dv \leq \|u^2\|_{\frac{p}{2}} \|1\|_{\frac{p}{p-2}} = (vol(X))^{\frac{p-2}{p}}. \qquad (11.3.6)$$

Let $b = \|f\|_\infty$. Then

$$J(u) \geq -b \int_M u^2 dv,$$

and J is bounded from below on $\mathcal{M}$. Note that

$$\|\nabla u\|_0^2 = J(u) - \int_X fu^2 dv \leq J(u) + \|f\|_\infty \|u\|_0^2. \qquad (11.3.7)$$

The following result and the proof are essentially in Brezis and Nirenberg [BN].

11.3.8. Theorem. Let $\sigma(n) = \frac{2n}{n-2}$.
(1) If $p < \sigma(n)$ then J satisfies condition C and critical points of J are smooth.
(2) If $p = \sigma(n)$, $c(n)$ is the best constant for the Sobolev embedding theorem, and $\alpha < 1/c(n)$, then the restriction of J to $J^{-1}((-\infty, \alpha])$ satisfies condition C and critical points of J in $J^{-1}((-\infty, \alpha])$ are smooth.

PROOF. In our discussion below $n = \dim(X)$ is fixed.
First we will prove condition C. Suppose $u_m \in L_1^2$, $J(u_m) \leq c$ and

$$\nabla J(u_m) \to 0 \quad \text{in } L_1^2. \tag{11.3.8}$$

We may assume that

$$J(u_m) \to c_0 \leq c. \tag{11.3.9}$$

Since $J(u_m)$ is bounded, it follows from (11.3.6) and (11.3.7) that $\{u_m\}$ is bounded in L_1^2. By the Sobolev embedding theorem 11.3.3, $\{u_m\}$ is bounded in $L^{\sigma(n)}$ and has a convergent subsequence in L^2. Since a bounded set in a reflexive Banach space is weakly precompact, there is a $u \in L_1^2$ and a subsequence of u_m converging weakly to u in L_1^2, so by passing to a subsequence we may assume that:

$$\|u_m - u\|_{L^2} \to 0, \tag{11.3.10}$$

$$\{u_m\} \quad \text{is bounded in } L^{\sigma(n)}, \tag{11.3.11}$$

$$u_m - u \to 0 \quad \text{weakly in } L_1^2. \tag{11.3.12}$$

$$u_m \to u \quad \text{almost everywhere.} \tag{11.3.13}$$

It follows that

$$\langle u_m - u, u \rangle_{L_1^2} = \int_X \langle \nabla(u_m - u), \nabla u \rangle + (u_m - u)u \, dv \to 0,$$

$$\int_X (u_m - u)u \, dv \to 0.$$

So we have $\langle \nabla(u_m - u), \nabla u \rangle_0 \to 0$, which implies that

$$\langle \nabla u_m, \nabla u \rangle_0 \to \|\nabla u\|_{L^2}^2,$$

$$\|\nabla(u_m - u)\|_{L^2}^2 - \left\{ \|\nabla u_m\|_{L^2}^2 - \|\nabla u\|_{L^2}^2 \right\} \to 0.$$

In particular, we have

$$\|\nabla u_m\|_2^2 - \langle \nabla u_m, \nabla u \rangle_0 - \|\nabla(u_m - u)\|_2^2 \to 0. \tag{11.3.14}$$

Since $C^\infty(X)$ is dense in L_1^2, we may further assume the u_m are smooth. Using (11.3.8), we have

$$dJ_{u_m}(v_m) \to 0, \quad \text{if } \{v_m\} \text{ is bounded in } L_1^2.$$

Since $u_m - u$ is bounded in L_1^2, by (11.3.5) and the above condition we have

$$-J(u_m)\langle|u_m|^{p-2}u_m, (u_m - u)\rangle_0 + \langle\nabla u_m, \nabla(u_m - u)\rangle_0$$
$$+ \langle fu_m, (u_m - u)\rangle_0 \to 0 \qquad (11.3.15)$$

It follows from (11.3.9) (11.3.10) and (11.3.14) that

$$-c_0\langle|u_m|^{p-2}u_m, (u_m - u)\rangle_0 + \|\nabla(u_m - u)\|_{L^2}^2 \to 0. \qquad (11.3.16)$$

If $p < \sigma(n)$, then by Sobolev embedding theorem we may assume that $u_m \to u$ in L^p. By Hölder's inequality,

$$|\langle|u_m|^{p-2}u_m, u_m - u\rangle_0| \leq \|u_m^{p-1}\|_{L^{\frac{p}{p-1}}} \|u_m - u\|_{L^p} = \|u_m - u\|_{L^p} \to 0.$$

So $\|\nabla(u_m - u)\|_{L^2} \to 0$, which implies that $u_m \to u$ in L_1^2 and $u \in \mathcal{M}$. This proves condition C for case (1).

If $p = \sigma(n)$, then we want to prove that for $\alpha < \frac{1}{c(n)}$, J restricts to $J^{-1}((-\infty, \alpha])$ satisfies condition C. So we may assume $c_0 \leq \alpha$. Since $\{|u_m|^{p-2}u_m\}$ is a bounded sequence in $L^{\frac{p}{p-1}}$, by passing to a subsequence we may assume that $\{|u_m|^{p-2}u_m\}$ converges weakly to $|u|^{p-2}u$ in $L^{\frac{p}{p-1}}$. So for $u \in L^p = (L^{\frac{p}{p-1}})^*$ we have

$$\int_X |u_m|^{p-2}u_m u \, dv \to \int_X |u|^{p-2}uu \, dv = \int_X |u|^p \, dv,$$

$$\langle|u_m|^{p-2}u_m, (u_m - u)\rangle_0 = \int_X |u_m|^p - u_m^{p-1}u \, dv \to 1 - \int_X u^p dv,$$

By Theorem 11.3.6, we have

$$\|u_m - u\|_{L^p}^p \to 1 - \|u\|_{L^p}^p \leq 1. \qquad (11.3.17)$$

So (11.3.16) gives

$$\epsilon_m = -c_0\|u_m - u\|_{L^\sigma}^p + \|\nabla(u_m - u)\|_{L^2}^2 \to 0. \qquad (11.3.18)$$

Choose α_0 such that $\alpha < \alpha_0 < \frac{1}{c(n)}$. Set $x_m = \|u_m - u\|_{L^p}$. Using Theorem 11.3.4 and (11.3.10), we obtain

$$-c_0 x_m^p + \alpha_0 x_m^2 \leq \epsilon_m \to 0.$$

Since $c_0 \leq \alpha < \alpha_0$, there is a constant $\delta > 1$ (only depends on $c(n)$ and α) such that

(i) $\varphi(x) = -c_0 x^p + \alpha_0 x^2 > 0$ on $[0, \delta]$,

(ii) if $t_n \in [0, \delta]$ and $\varphi(t_m) \leq b_m \to 0$, then $t_m \to 0$.

Because of (11.3.17), we may assume that $x_m \in [0, \delta]$. So

$$\|u_m - u\|_{L^p} \to 0.$$

It then follows from (11.3.18) that $u_m - u \to 0$ in L_1^2, which proves Condition C for case (2).

Next we prove regularity for case (1). For simplicity we will discuss the special case $n = 3$ and $p = 4 < \sigma(3) = 6$. Suppose $u \in L_1^2$ is a critical point of J then $\triangle u = fu + \lambda u^3 \in L^2$, where $\lambda = -J(u)$. So by Theorem 11.3.5 (2), $u \in L_2^2$. Since $2 - 3/2 = 1/2$, $u \in C^\alpha$ if $\alpha \leq \frac{1}{2}$. Applying Theorem 11.3.5 (2), $u \in C^{2+\alpha}$. Applying the same estimate repeatedly implies that u is smooth. The general case is similar (for example see [Au]).

Regularity for case (2) follows from Trudinger's Theorem ([Tr]). ∎

Now $J(u) = J(-u)$, and $\mathcal{M}/Z_2$ is diffeomorphic to the infinite dimensional real projective space RP^∞. As a consequence of the above theroem and Lusternik-Schnirelman theory (Corollary 9.2.11) that we have:

11.3.9. Theorem.   If $p < \frac{2n}{n-2}$, then there are infinitely many pairs of smooth functions u on (X^n, g) such that

$$\triangle u = fu + \lambda |u|^{p-2} u,$$

where $\lambda = -J(u)$.

Appendix

We review some basic facts and standard definitions and notations from the theory of differentiable manifolds and differential topology. Proofs will be omitted and can be found in [La] and [Hi].

Manifold will always mean a paracompact, smooth (meaning C^∞) manifold satisfying the second axiom of countability, and modeled on a hilbert space of finite or infinite dimension. Only in the final chapters do we deal explicitly with the infinite dimensional case, and before that the reader who feels more comfortable in the finite dimensional context can simply think of all the manifolds that arise as being finite dimensional. In particular when we assume that the model hilbert space is V, with inner product $\langle\ ,\ \rangle$ then the reader can assume $V = R^n$ and $\langle x, y \rangle = x \cdot y = \sum_{i=1}^n x_i y_i$.

The tangent space to a smooth manifold X at x is denoted by TX_x, and if $F : X \to Y$ is a smooth map and $y = F(x)$ then $DF_x : TX_x \to TY_y$ denotes the differential of F at x. If Y is a hilbert space then as usual we canonically identify TY_y with Y itself. With this identification we denote the differential of F at x by $dF : TF_x \to Y$. In particular if $f : X \to R$ is a smooth real valued function on X then, for each x in X its differential $df_x : TX_x \to R$ is an element of T^*X_x, the cotangent space to X at x. Also if X is modelled on V and $\Phi : O \to V$ is a chart for X at p, we have an isomorphism $d\Phi_p : TX_p \to V$. A Riemannian structure for X is an assignment to each x in X of a continuous, positive definite inner product $\langle\ ,\ \rangle_x$ on TX_x, such that the associated norm is complete. If $\Phi : O \to V$ is a chart as above then for each x in O there is a uniquely determined bounded, positive, self-adjoint operator $g(x)$ on V such that for $u, v \in TX_x$,

$$\langle u, v, \rangle_x = \langle g(x)d\Phi(u), d\Phi(v) \rangle,$$

where $\langle\ ,\ \rangle$ is the inner product in V. The Riemannian structure is smooth if for each chart Φ the map $x \mapsto g(x)$ from O into the Banach space of self-adjoint operators on V is smooth. (When $V = R^n$ this just means that the matrix elements $g_{ij}(x)$ are smooth functions of x.)

For a Riemannian manifold X there is a norm preserving duality isomorphism $\ell \mapsto \hat{\ell}$ of T^*X_x with TX_x, characterized by $\ell(u) = \langle u, \hat{\ell} \rangle_x$. In particular if $f : X \to R$ is a smooth function, then the dual $(df_x)\hat{}$ of df_x is called the gradient of f at x and is denoted by ∇f. The vector field ∇f plays a central rôle in Morse theory, and we note that its characteristic property is that for each Y in TX_x, $Yf \overset{\text{def}}{=} df(Y)$, the directional derivative of f at x in the direction Y, is given by $\langle Y, \nabla f_x \rangle$. It follows from the Schwarz inequality that if $df_x \neq 0$ then, among all the unit vectors Y at x, the directional derivative of f in the direction Y assumes its maximum, $\|\nabla f\|$, uniquely for $Y = \frac{1}{\|\nabla f\|}\nabla f$.

We recall that given a smooth map $F : X \to Y$ a point x of X is called a *regular point* of F if $DF_x : TX_x \to TY_{F(x)}$ is surjective. Other points of X are called *critical points* of F. A point y of Y is called a *critical value* of F if $F^{-1}(y)$ contains at least one critical point of F. Other points of Y are called *regular values*. (Note that if y is a non-value of F, i.e., if $F^{-1}(y)$ is empty, then y is nevertheless considered to be a "regular value" of F.) By the Implicit Function Theorem if x is a regular point of F and $y = F(x)$, then there is a neighborhood O of x in X such that $O \cap F^{-1}(y)$ is a smooth submanifold of X (of dimension $\dim(X) - \dim(Y)$ when $\dim(X) < \infty$). Thus if y is a regular value of F then $F^{-1}(y)$ is a (possibly empty) closed, smooth submanifold of X.

If X is an n–dimensional smooth manifold, then a subset S of X is said to have measure zero in X if for each chart $\Phi : O \to R^n$ for X, $\Phi(S \cap O)$ has Lebesque measure zero in R^n. Note that it follows that S has no interior.

Morse-Sard Theorem. *[DR, p.10] If X and Y are finite dimensional smooth manifolds and $F : X \to Y$ is a smooth map, then the set of critical values of F has measure zero in Y and in particular it has no interior.*

Corollary. *If X is compact then the set of regular values of F is open and dense in Y.*

If $f : X \to R$ is a smooth function and $df_x \neq 0$, then since R is one-dimensional, $df_x : TX_x \to R$ must be surjective, i.e., x is a regular value of f. Thus for a real valued smooth function the critical points are exactly the points where df_x is zero. Of course when X is Riemannian we can equally well characterize the critical points of f as the zeros of the vector field ∇f.

Let X be a smooth Riemannian manifold, and M a smooth submanifold of X with the induced Riemannian structure. If $F : X \to R$ is a smooth function on X and $f = F|M$ is its restriction to M then, at a point x of M, df_x is the restriction to TM_x of dF_x, and it follows from this and the characterization of the gradient above that ∇f_x is the orthogonal projection onto TM_x of ∇F_x. Thus x is a critical point of f if and only if ∇f_x is orthogonal to TM_x. Now suppose c is a regular value of some other smooth, real valued function $G : X \to R$ and $M = G^{-1}(c)$. Then $TM_x = \ker(dG_x) = \nabla G_x^\perp$, hence in this case $TM_x^\perp$ is spanned by ∇G_x. This proves:

Lagrange Multiplier Theorem. *Let F and G be two smooth real valued functions on a Riemannian manifold X, c a regular value of G, and $M = G^{-1}(c)$. Then x in M is a critical point of $f = F|M$ if and only if $\nabla F_x = \lambda \nabla G_x$ for some real λ.*

Let Y be a smooth vector field on a manifold X. A *solution curve* for Y is a smooth map σ of an open interval (a, b) into X such that $\sigma'(t) = Y_{\sigma(t)}$ for all $t \in (a, b)$. It is said to have *initial condition* x if $a < 0 < b$ and $\sigma(0) = x$, and it is called *maximal* if it is not the restriction of a solution curve with properly larger domain. An equivalent condition for maximality is the following: either $b = \infty$ or else $\sigma(t)$ has no limit points as $t \to \infty$, and similarly either $a = -\infty$ or else $\sigma(t)$ has no limit points as $t \to -\infty$

Global Existence and Uniqueness Theorem for ODE. *If Y is a smooth vector field on a smooth manifold X, then for each x in X there is a unique maximal solution curve of Y, $\sigma_x : (\alpha(x), \beta(x)) \to X$, having x as initial condition.*

For $t \in R$ we define $D(\varphi_t) = \{x \in X \mid \alpha(x) < t < \beta(x)\}$ and $\varphi_t : D(\varphi_t) \to X$ by $\varphi_t(x) = \sigma_x(t)$. Then $D(\varphi_t)$ is open in X and φ_t is a difeomorphism of $D(\varphi_t)$ onto its image. The collection $\{\varphi_t\}$ is called the flow generated by Y, and we call the vector field Y *complete* if $\alpha \equiv -\infty$ and $\beta \equiv \infty$. In this case $t \mapsto \varphi_t$ is a one parameter group of diffeomorphisms of X (i.e., a homomorphism of R into the group of diffeomorphisms of X).

References

[Ab] Abresh, U., Isoparametric hypersurfaces with four or six distinct principal curvatures, *Math. Ann.* **264** (1983), 283–302.

[Ad] Adams, J.F., *Lectures on Lie groups*. Benjamin Inc., New York, 1969.

[Am] Adams, R.A., *Sobolev Spaces*. Academic Press, New York, 1972.

[Al] Almgren, F.J. Jr., Some interior regularity theorems for minimal surfaces and an extension of Bernstein's theorem, *Ann. of Math.* **85** (1966), 277–292.

[At] Atiyah, M.F., Convexity and commuting Hamiltonians, *Bull. London Math. Soc.* **14** (1982), 1–15.

[Au] Aubin, T., *Non-linear analysis on manifolds, Monge-Ampère Equations*. Springer-Verlag, New York, 1982. vol. 252 Grundlehren Series

[Bä] Bäcklund, A.V., *Concerning surfaces with constant negative curvature*. New Era Printing Co. Lancaster Pa., 1905.

[Ba] Banchoff, T.F., The spherical two-piece property and tight surfaces in spheres, *J. Differential Geometry* **4** (1970), 193–205.

[BG] Benson, C.T. and Grove, L.C., *Finite Reflection Groups*. Bogden S. Quigley Inc., 1971.

[Bé] Bérard, P.H., From vanishing theorems to estimating theorems, the Bochner technique revisted, preprint

[BG] Bernstein, I., and Ganea, T., Homotopical nilpotency, *Illinois J. Math.* **5** (1961), 99–130.

[Be] Besse, A., *Einstein manifolds*. Springer-Verlag, New York, 1987. Ergeb. Math. Ihrer Grenz., 3 Folge Band 10

[B] Bochner, S., Vector fields and Ricci curvature, *Bull. Amer. Math. Soc.* **52** (1946), 776–797.

[BY] Bochner,S. and Yano,K., *Curvature and Betti numbers*. Annals Math. Studies 32, Princeton Univ. press, 1953.

[Bb] Bombieri, E., *Seminar on Minimal Submanifolds*. Annals Math. Studies 103, Princeton Univ. press, 1983.

[Bo] Borel, A., *Seminar on Transformation Groups*. Annals Math. Studies 6, Princeton Univ. press, 1960.

[Bt] Bott, R., Non-degenerate critical manifolds, *Ann. of Math.* **60** (1954), 248–261.

[BS] Bott, R and Samelson, H., Applications of the theory of Morse to symmetric spaces, *Amer. J. Math.* **80** (1958), 964–1029.

[Bu] Bourbaki, N., *Groupes et Algebres de Lie*. Hermann, Paris, 1968.

[Br] Bredon, G.E., *Introduction to Compact Transformation Groups*. Academic Press, New York, 1972.

[BL] Brezis, H. and Lieb, E., A relation between pointwise convergence of functions and convergence of integrals, *Proc. Amer. Math. Soc.* **88** (1983), 486–490.

[BN] Brezis, H. and Nirenberg, L., Positive solutions of non-linear elliptic equations involving critical Sobolev exponents, *Comm. Pure Appl. Math.* **36** (1983), 437–477.

[CM] Cairns S., and Morse, M., *Critical Point Theory in Global Analysis and Differential Topology*. Academic Press, New York, 1969.

[Cb] Calabi, E., Minimal immersions of surfaces in euclidean spheres, *J. Differential Geometry* **1** (1967), 111-125.

[Cr] Calderón, A.P., Lebesgue spaces of differentiable functions and distributions, *Proc. Symp. Pure Math., Amer. Math. Soc.* **4** (1961), 33–49.

[Ca1]   Cartan, É., Sur les variétés de courbure constante dans l'éspace euclidien ou non euclidien I, *Bull. Soc. Math. France* **47** (1919), 125–160.

[Ca2]   Cartan, É., Sur les variétés de courbure constante dans l'éspace euclidien ou non euclidien II, *Bull. Soc. Math. France* **48** (1920), 132–208.

[Ca3]   Cartan, É., Familles des surfaces isoparamétriques dans les éspace à courbure constante, *Annali di Mat.* **17** (1938), 177–191.

[Ca4]   Cartan, É., Sur des familles remarquables d'hypersurfaces isoparametriques dans les éspaces spheriques, *Math.Z.* **45** (1939), 335–367.

[Ca5]   Cartan, É., Sur quelques familles remarquables d'hypersurfaces des éspaces spheriques à 5 et à 9 dimensions, *Univ. Nac. Tucuman Revista A* **1** (1940), 5–42.

[Ca6]   Cartan, É., *Leçons sur la Géométrie des Éspaces de Riemann.* Gauthier Villars, Paris, 1946.

[CW1]   Carter,S., and West, A., Tight and taut immersions, *Proc. London Math. Soc.* **25** (1972), 701-720.

[CW2]   Carter,S., and West, A., Isoparametric systems and transnormality, *Proc. London Math. Soc.* **51** (1985), 521-542.

[CW3]   Carter, S., and West, A., Generalized Cartan polynomials, *J. London Math. Soc.* **32** (1985), 305-316.

[CR1]   Cecil, T.E., and Ryan, P.J., Tight spherical embeddings, *Springer Verlag Lecture Notes in Math. no. 838* (1981), 94–104..

[CR2]   Cecil, T.E., and Ryan, P.J., *Tight and Taut Immersions of Manifolds.* Research notes in Math. vol. 107, Pitman Publ. Inc. Boston, 1985.

[Ch1]   Chern, S.S., A simple intrinsic proof of the Gauss-Bonnet formula for the closed Riemannian manifolds, *Ann. of Math.* **45** (1944), 747–752.

[Ch2]   Chern, S.S., On isothermic coordinates, *Comment. Math. Helv.* **28** (1954), 301–309.

[Ch3]   Chern, S.S., Simple proofs of two theorems on minimal surfaces, *Extrait de L'Enseignement mathématique* **15** (1969), 53–61.

[Ch4]   Chern, S.S., *Curves and Surfaces in Euclidean Spaces*, Univ. of California, Berkeley Lecture Notes

[Ch5]   Chern, S.S., *Minimal submanifolds in a Riemannian manifold.* Kansas University, 1968. Lecture notes

[CC]   Chern, S.S. and Chevalley, C., Élie Cartan and his mathematical work, *Bull. Amer. Math. Soc.* **58** (1952), 217–250.

[CdK]   Chern, S.S., do Carmo, M., and Kobayashi, S., Minimal submanifolds of a sphere with second fundamental form of constant length, *Functional analysis and related fields* (1970), 59–75. Springer-Verlag

[CL1]   Chern, S.S. and Lashof, R.K., On the total curvature of immersed manifolds 1, *Amer. Jour. Math.* **79** (1957), 306–318.

[CL2]   Chern, S.S. and Lashof, R.K., On the total curvature of immersed manifolds 2, *Mich. Math. Jour.* **5** (1958), 5–12.

[Cv]   Chevalley, C., Invariants of finite groups generated by reflections, *Amer. J. Math.* **77** (1955), 778–782.

[Co1]   Conlon L., Variational completeness and K-transversal domains, *J. Differential Geometry* **5** (1971), 135–147.

[Co2]   Conlon L., A class of variationally complete representations, *J. Differential Geometry* **7** (1972), 149–160.

[De]   De Almeida, S.C., Minimal hypersurfaces of S^4 with non-zero Gauss-Kronecker curvature, *Bol. Soc. Bras. Mat.* **14** (1983), 137–146.

[Da]   Dadok J., Polar coordinates induced by actions of compact Lie groups, *Transactions Amer. Math. Soc.* **288** (1985), 125–137.

[Dv]   Davis, M., Smooth G-manifolds as collections of fiber bundles, *Pacific. Jour. Math.* **77** (1978), 315–363.

[DR]   de Rham, G., *Variétés Différentiables.* Hermann & Cie., Paris, 1955.

[Do]   Do Carmo, M.P., *Differential geometry of curves and surfaces.* Prentice Hall Inc. NJ, 1976.

[Dy]   Dynkin, E. B., The structure of semisimple Lie algebras, *Uspeki Mat. Nauk. (N.S.)* **2** (1981), 59–127. AMS Translations, No. 17 (1950).

[Ee]   Eells, J., On equivariant harmonic maps, *Proc. of 1981 Symp. Diff. Geom. and Diff. Equations* (1985), 55–74.

[EL]   Eells, J., and Lemaire, A report on harmonic maps, *Bull. London. Math. Soc.* **10** (1978), 1–68.

[Ei]   Eisenhart, L.P., *A Treatise on the Differential Geometry of Curves and Surfaces.* Boston, Ginn., 1909.

[Fe]   Ferus, D., Symmetric submanifolds of Euclidean space, *Math. Ann.* **247** (1980), 81–93.

[FKM]   Ferus, D., Karcher, H. and Münzner, H. F., Cliffordalgebren und neue isoparametrische hyperflächen, *Math. Z.* **177** (1981), 479–502.

[FK]   Ferus, D., and Karcher, H., Non-rotational minimal spheres and minimizing cones, *Comment. Math. Helv.* **60** (1985), 247–269.

[GT]   Gilbarg, D., and Trudinger, N.S., *Elliptic partial differential equations of second order.* Springer-Verlag, New York, 1977. Grundlehren der mathematischen Wissenschaften no. 224

[GS]   Guillemin, V., and Sternberg, S., Convexity properties of the moment mapping, *Invent. Math.* **67** (1982), 491–513.

[He]   Helgason, S., *Differential Geometry and Symmetric Spaces.* Academic Press, 1978.

[Hk]   Hicks, N.J., *Notes on Differential Geometry.* Van Nostrand Co. NJ, 1965. Math. Studies Series, no. 3

[Hi]   Hirsch, M., *Differential Topology.* Springer-Verlag, New York, 1982. Graduate Texts in Math. Series, no. 33

[Ho]   Hopf, H., *Differential Geometry in the Large.* Springer-Verlag, 1983. Lecture notes in Mathematics

[Hr]   Horn, A., Doubly stochastic matrices and the diagonal of a rotation matrix, *Amer. J. Math.* **76** (1954), 620–630.

[Hs1]   Hsiang, W.Y., On the compact homogeneous minimal submanifolds, *Proc. Nat. Acad. Sci. USA* **56** (1966), 5–6.

[Hs2]   Hsiang, W.Y., Minimal cones and the spherical Berstein problem I, *Ann. of Math.* **118** (1983), 61–73.

[Hs3]   Hsiang, W.Y., Minimal cones and the spherical Berstein problem II, *Invent. Math.* **74** (1983), 351–369.

[HL]   Hsiang, W.Y., and Lawson B.H.Jr., Minimal submanifolds of low cohomogeneity, *J. Differential Geometry* **5** (1971), 1–38.

[HPT1]   Hsiang, W.Y., Palais R.S., and Terng C.L., The topology and geometry of isoparametric submanifolds in Euclidean spaces, *Proc. Nat. Acad. Sci. USA* **82** (1985), 4863–4865.

[HPT2]   Hsiang, W.Y., Palais R.S., and Terng C.L., The topology of isoparametric submanifolds, *J. Differential Geometry* **27** (1988), 423–460.

[KW]   Kazdan, J.L., and Warner, F. W., Existence and conformal deformation of metrics with prescribed Gaussian and scalar curvatures, *Annals of Math.* **101** (1975), 317–331.

[Ko]   Kobayashi, S., Imbeddings of homogeneous spaces with minium total curvature, *Tohoku Math. J.* **19** (1967), 63–70.

[KN]   Kobayashi, S., and Nomizu, K., *Foundations of Differential Geometry I, II.* Interscience New York, 1963.

[KT]   Kobayashi, S. and Takeuchi,M., Minimal imbeddings of R-spaces, *J. Differential Geometry* **2** (1968), 203–215.

[Ks]   Kostant, B., On convexity, the Weyl group and the Iwasawa decomposition, *Ann. Sci. Ec. Norm. Sup.* **6** (1973), 413–455.

[Ku1]   Kuiper, N.H., The homotopy type of the unitary group of Hilbert space, *Topology* **3** (1965), 19–30.

[Ku2]   Kuiper, N.H., Minimum total absolute curvature for immersions, *Invent. Math.* **10** (1970), 209–238.

[Lb]   Lamb, G.L., *Elements of Soliton Theory.* John Wiley, New York, 1980.

[La]   Lang, S., *Introduction to Differentiable Manifolds.* Interscience, New York, 1966.

[Lw1]   Lawson, H.B. Jr., Complete minimal surfaces in S^3, *Ann. of Math.* **90** (1970), 335–374.

[Lw2]   Lawson, H.B. Jr., *Lectures on Minimal Submanifolds, I.* Publish or Perish, Inc., 1980. Math. Lecture series 9

[Lc]   Levi-Civita, T., Famiglie di superfici isoparametriche nell' ordinario spazio euclideo, *Atti Accad. naz. Lincei. Rend. U. Sci. Fis. Mat. Natur.* **26** (1937), 355–362.

[Li]   Lions, P.L., On the existence of positive solutions of semilinear elliptic equations, *SIAM Rev.* **24** (1982), 441–467.

[Mi1]   Milnor, J.W., *Morse Theory.* Annals Math. Study 51, Princeton Univ. Press, 1963.

[Mi2]   Milnor, J.W., and Stasheff, J.D., *Characteristic Classes.* Annals Math. Study 76, Princeton Univ. Press, 1974.

[Mo]   Moore, J.D., Isometric immersions of space forms in space forms, *Pacific J. Math.* **40** (1972), 157–166.

[My]   Morrey, C.B. Jr., *Multiple Integrals in the Calculus of Variations.* Springer-Verlag, New York, 1966. vol. 130 Grundlehren Series.

[Mü1]   Münzner, H.F., Isoparametrische Hyperflächen in Sphären, I, *Math. Ann.* **251** (1980), 57–71.

[Mü2]   Münzner, H.F., Isoparametrische Hyperflächen in Sphären, II, *Math. Ann.* **256** (1981), 215–232.

[NY]   Nagano, T., and Yano, K., Einstein spaces admitting a one-parameter group of conformal transformations, *Ann. of Math.* **69** (1959), 451-461.

[Ne]   Newell, A.C., *Solitons in Mathematics and Physics.* SIAM, 1985. 48 CBMS

[No]   Nomizu, K., Élie Cartan's work on isoparametric families of hypersurfaces, *Proc. Symp. Pure Math., Amer. Math. Soc.* **27** (1975), 191–200.

[NR]   Nomizu, K., and Rodriguez, L.L., Umbilical submanifolds and Morse functions, *Nagoya Math. J.* **48** (1972), 197–201.

[Ob]   Obata, M., Conformal transformations of compact Riemannian manifolds, *Illinois J. Math.* **6** (1962), 292-295.

[On]   O'Neill, B., The fundamental equations of a submersions, *Michigan Math. J.* **13** (1966), 459–469.

[Os]   Osserman, R., *A Survey of Minimal Surfaces.* Van Nostrand, New York, 1969.

[OT1]   Ozeki, H. and Takeuchi, M., On some types of isoparametric hypersurfaces in spheres,1, *Tohoku Math. J.* **127** (1975), 515–559.

[OT2]   Ozeki, H. and Takeuchi, M., On some types of isoparametric hypersurfaces in spheres,2, *Tohoku Math. J.* **28** (1976), 7–55.

[Pa1] Palais, R.S., On the existence of slices for actions of non-compact groups, *Ann. of Math.* **73** (1961), 295–323.

[Pa2] Palais, R.S., Morse theory on Hilbert manifolds, *Topology* **2** (1963), 299–340.

[Pa3] Palais, R.S., *Seminar on the Atiyah-Singer Index Theorem*. Annals of Math. Study 4, Princeton University press, 1964.

[Pa4] Palais, R.S., Homotopy theory of infinite dimensional manifolds, *Topology* **5** (1966), 1–16.

[Pa5] Palais, R.S., Lusternik-Schnirelman theory on Banach manifolds, *Topology* **5** (1966), 115–132.

[Pa6] Palais, R.S., *Foundations of Global Non-linear Analysis*. Benjamin Co, New York, 1968.

[Pa7] Palais, R.S., *Critical point theory and the minimax principle*. Proc. Symp. on Global Analysis, Amer. Math. Soc., 1968.

[Pa8] Palais, R.S., A topological Gauss-Bonnet theorem, *J. Differential Geom.* **13** (1978), 385-398.

[Pa9] Palais, R.S., The principle of symmetric criticality, *Comm. Math. Phys.* **69** (1979), 19–30.

[Pa10] Palais, R.S., Applications of the symmetric critical principle in mathematical physis and differential geometry, *Proc. of 1981 Symp. Diff. Geom. and Diff. Equations* (1985), 247–302.

[PS]   Palais, R.S., and Smale, S., A generalized Morse theory, *Bull. Amer. Math. Soc.* **70** (1964), 165–172.

[PT1] Palais, R.S. and Terng, C.L., Reduction of variables for minimal submanifolds, *Proceedings, Amer. Math. Soc.* **98** (1986), 480-484.

[PT2] Palais, R.S. and Terng, C.L., A general theory of canonical forms , *Transaction Amer. Math. Soc.* **300** (1987), 771–789.

[Ra]   Rabinowitz, P., *Minimax Methods in Critical Point Theory with Applications to Differential Equations*. Amer. Math. Soc., 1986. CBMS Regional Conference No. 65

[SU]   Sacks, J. and Uhlenbeck, K., Minimal immersions of two spheres, *Ann. of Math.* **113** (1981), 1–24.

[Sa]   Samelson, H., *Notes on Lie algebra*. Van Nostrand Co. NJ, 1969. Math. Studies series no. 23

[Sc1] Schoen, Richard, Conformal deformation of a Riemannian metric to a constant scalar curvature, *J. Differential Geom.* **20** (1984), 479-495.

[Sc2] Schoen, Richard, Recent progress in geometric partial differential equations, *Proc. International Cong. Math. Berkeley* (1985), 121-130.

[Sh]   Schwarz, G., Smooth functions invariant under the action of a compact Lie group, *Topology* **14** (1975), 63-68.

[Se]   Segré, B., Famiglie di ipersuperfici isoparametriche negli spazi euclidei ad un qualunque numero di dimensioni, *Atti Accad. naz Lincie Rend. Cl. Sci. Fis. Mat. Natur.* **27** (1938), 203-207.

[Su]   Schur, I., Über eine Klasse von Mittelbildungen mit Anwendungen auf der Determinanten theorie, *Sitzungsberichte der Berliner Mathematischen Gesellschaft* **22** (1923), 9-20.

[Si]   Simons, J., Minimal varieties in Riemannian manifolds , *Ann. of Math.* **88** (1968), 62-105.

[Sm1] Smale, S., Morse theory and a non-linear generalization of the Dirichlet problem, *Ann. of Math.* **80** (1964), 382-396.

[Sm2] Smale, S., An infinite dimensional version of Sard's theorem, *Amer. J. Math.* **87** (1965), 861-866.

[So]   Sobolev, S.L., *Applications of Functional Analysis in Mathematical Physics*. Amer. Math. Soc., 1963. Translations of Math. Monographs No. 7

[Sp]   Spivak, M., *A Comprehensive Introduction to Differential Geometry, vol. I.* Publish or Perish Inc., 1970.

[TT]   Tenenblat, K., and Terng, C.L., Bäcklund's theorem for n-dimensional submanifolds of R^{2n-1}, *Ann. of Math.* **111** (1980), 477–490.

[Te1]   Terng, C.L., A higher dimensional generalization of the Sine-Gordon equation and its soliton theory, *Ann. of Math.* **111** (1980), 491–510.

[Te2]   Terng, C.L., Isoparametric submanifolds and their Coxeter groups, *J. Differential Geometry* **21** (1985), 79–107.

[Te3]   Terng, C.L., A convexity theorem for isoparametric submanifolds, *Invent. Math.* **85** (1986), 487–492.

[Te4]   Terng, C.L., Some geometric development related to Kovalewski's work, *The Legacy of Sonya Kovalevskaya, Contemporary Math., Amer. Math. Soc.* **64** (1986), 195–205.

[Te5]   Terng, C.L., Proper Fredholm submanifolds of Hilbert spaces, to appear in J. Diff. Geometry

[Tr]   Trudinger, N., Remarks concerning the conformal deformation of Riemannian structures on compact manifolds, *Ann. Sc. Norm. Sup. Pisa* **22** (1968), 265–274.

[U]   Uhlenbeck, K., Harmonic maps: A direct method in calculus of variations, *Bull. Amer. Math. Soc.* **76** (1970), 1082–1087.

[Wa]   Warner, G., *Harmonic analysis on semi-simple Lie groups, I.* Springer-Verlag, Berlin, 1972.

[We]   Wente, H.C., Counterexample to a conjecture of H. Hopf, *Pacific J. Math.* **121** (1986), 193-243.

[Wu1]   Wu, H., A remark on the Bochner technique in differential geometry , *Proc. Amer. Math. Soc.* **78** (1980), 403–408.

[Wu2]   Wu, H., The Bochner technique, *Proc. 1980 Symp. on Differential Geom.and Diff. Equations* (1982), –. Science press and Gordon Breach

[Ya]   Yamabe, H., On a deformation of Riemannian structures on compact manifolds, *Osaka J. Math.* **12** (1960), 21-27.

[Yn]   Yano, K., *Integral Formulas in Riemannian Geometry.* Marcel Dekker, Inc., New York, 1970.

[Yau]   Yau, S.T., *Seminar on Differential Geometry.* Annals Math. Studies 102, Princeton University press, 1982.

Index

This book was typeset by the authors with Professor Donald E. Knuth's TEX typesetting system, using the TEXtures software for the Apple Macintosh family of computers, developed by Kellerman & Smith. The camera copy was prepared on an Apple LaserWriter using its native Times-Roman family of typefaces for the main text fonts, Knuth's Computer Modern Symbol font for the mathematical symbols, and a slanted version of Times-Roman, created with Kellerman & Smith's EdMetrics program, for theorem statements.

In what follows all references to monographs, are applicable also to multiauthorship volumes such as seminar notes.

§1. Lecture Notes aim to report new developments - quickly, informally, and at a high level. Monograph manuscripts should be reasonably self-contained and rounded off. Thus they may, and often will, present not only results of the author but also related work by other people. Furthermore, the manuscripts should provide sufficient motivation, examples and applications. This clearly distinguishes Lecture Notes manuscripts from journal articles which normally are very concise. Articles intended for a journal but too long to be accepted by most journals, usually do not have this "lecture notes" character. For similar reasons it is unusual for Ph.D. theses to be accepted for the Lecture Notes series.

Experience has shown that English language manuscripts achieve a much wider distribution.

§2. Manuscripts or plans for Lecture Notes volumes should be submitted either to one of the series editors or to Springer-Verlag, Heidelberg. These proposals are then refereed. A final decision concerning publication can only be made on the basis of the complete manuscripts, but a preliminary decision can usually be based on partial information: a fairly detailed outline describing the planned contents of each chapter, and an indication of the estimated length, a bibliography, and one or two sample chapters - or a first draft of the manuscript. The editors will try to make the preliminary decision as definite as they can on the basis of the available information.

§3. Lecture Notes are printed by photo-offset from typed copy delivered in camera-ready form by the authors. Springer-Verlag provides technical instructions for the preparation of manuscripts, and will also, on request, supply special staionery on which the prescribed typing area is outlined. Careful preparation of the manuscripts will help keep production time short and ensure satisfactory appearance of the finished book. Running titles are not required; if however they are considered necessary, they should be uniform in appearance. We generally advise authors not to start having their final manuscripts specially tpyed beforehand. For professionally typed manuscripts, prepared on the special stationery according to our instructions, Springer-Verlag will, if necessary, contribute towards the typing costs at a fixed rate.

The actual production of a Lecture Notes volume takes 6-8 weeks.

. . ./. . .

§4. Final manuscripts should contain at least 100 pages of mathematical text and should include
- a table of contents
- an informative introduction, perhaps with some historical remarks. It should be accessible to a reader not particularly familiar with the topic treated.
- a subject index; this is almost always genuinely helpful for the reader.

§5. Authors receive a total of 50 free copies of their volume, but no royalties. They are entitled to purchase further copies of their book for their personal use at a discount of 33.3 %, other Springer mathematics books at a discount of 20 % directly from Springer-Verlag.

Commitment to publish is made by letter of intent rather than by signing a formal contract. Springer-Verlag secures the copyright for each volume.

LECTURE NOTES

ESSENTIALS FOR THE PREPARATION
OF CAMERA-READY MANUSCRIPTS

Springer

Springer-Verlag
Berlin Heidelberg New York
London Paris Tokyo Hong Kong

The preparation of manuscripts which are to be reproduced by photo-offset require special care. Manuscripts which are submitted in technically unsuitable form will be returned to the author for retyping. There is normally no possibility of carrying out further corrections after a manuscript is given to production. Hence it is crucial that the following instructions be adhered to closely. If in doubt, please send us 1 - 2 sample pages for examination.

General. The characters must be uniformly black both within a single character and down the page. Original manuscripts are required: photocopies are acceptable only if they are sharp and without smudges.

On request, Springer-Verlag will supply special paper with the text area outlined. The standard TEXT AREA (OUTPUT SIZE if you are using a 14 point font) is 18 x 26.5 cm (7.5 x 11 inches). This will be scale-reduced to 75% in the printing process. If you are using computer typesetting, please see also the following page.

Make sure the TEXT AREA IS COMPLETELY FILLED. Set the margins so that they precisely match the outline and type right from the top to the bottom line. (Note that the page number will lie outside this area). Lines of text should not end more than three spaces inside or outside the right margin (see example on page 4).

Type on one side of the paper only.

Spacing and Headings (Monographs). Use ONE-AND-A-HALF line spacing in the text. Please leave sufficient space for the title to stand out clearly and do NOT use a new page for the beginning of subdivisons of chapters. Leave THREE LINES blank above and TWO below headings of such subdivisions.

Spacing and Headings (Proceedings). Use ONE-AND-A-HALF line spacing in the text. Do not use a new page for the beginning of subdivisons of a single paper. Leave THREE LINES blank above and TWO below headings of such subdivisions. Make sure headings of equal importance are in the same form.

The first page of each contribution should be prepared in the same way. The title should stand out clearly. We therefore recommend that the editor prepare a sample page and pass it on to the authors together with these instructions. Please take the following as an example. Begin heading 2 cm below upper edge of text area.

MATHEMATICAL STRUCTURE IN QUANTUM FIELD THEORY

John E. Robert
Mathematisches Institut, Universität Heidelberg
Im Neuenheimer Feld 288, D-6900 Heidelberg

Please leave THREE LINES blank below heading and address of the author, then continue with the actual text on the same page.

Footnotes. These should preferable be avoided. If necessary, type them in SINGLE LINE SPACING to finish exactly on the outline, and separate them from the preceding main text by a line.

Symbols. Anything which cannot be typed may be entered by hand in BLACK AND ONLY BLACK ink. (A fine-tipped rapidograph is suitable for this purpose; a good black ball-point will do, but a pencil will not). Do not draw straight lines by hand without a ruler (not even in fractions).

Literature References. These should be placed at the end of each paper or chapter, or at the end of the work, as desired. Type them with single line spacing and start each reference on a new line. Follow "Zentralblatt für Mathematik"/"Mathematical Reviews" for abbreviated titles of mathematical journals and "Bibliographic Guide for Editors and Authors (BGEA)" for chemical, biological, and physics journals. Please ensure that all references are COMPLETE and ACCURATE.

IMPORTANT

Pagination. For typescript, number pages in the upper right-hand corner in LIGHT BLUE OR GREEN PENCIL ONLY. The printers will insert the final page numbers. For computer type, you may insert page numbers (1 cm above outer edge of text area).

It is safer to number pages AFTER the text has been typed and corrected. Page 1 (Arabic) should be THE FIRST PAGE OF THE ACTUAL TEXT. The Roman pagination (table of contents, preface, abstract, acknowledgements, brief introductions, etc.) will be done by Springer-Verlag.

If including running heads, these should be aligned with the inside edge of the text area while the page number is aligned with the outside edge noting that right-hand pages are odd-numbered. Running heads and page numbers appear on the same line. Normally, the running head on the left-hand page is the chapter heading and that on the right-hand page is the section heading. Running heads should not be included in proceedings contributions unless this is being done consistently by all authors.

Corrections. When corrections have to be made, cut the new text to fit and paste it over the old. White correction fluid may also be used.

Never make corrections or insertions in the text by hand.

If the typescript has to be marked for any reason, e.g. for provisional page numbers or to mark corrections for the typist, this can be done VERY FAINTLY with BLUE or GREEN PENCIL but NO OTHER COLOR: these colors do not appear after reproduction.

COMPUTER-TYPESETTING. Further, to the above instructions, please note with respect to your printout that
- the characters should be sharp and sufficiently black;
- it is not strictly necessary to use Springer's special typing paper. Any white paper of reasonable quality is acceptable.

If you are using a significantly different font size, you should modify the output size correspondingly, keeping length to breadth ratio 1 : 0.68, so that scaling down to 10 point font size, yields a text area of 13.5 x 20 cm (5 3/8 x 8 in), e.g.

Differential equations.: use output size 13.5 x 20 cm.

Differential equations.: use output size 16 x 23.5 cm.

Differential equations.: use output size 18 x 26.5 cm.

Interline spacing: 5.5 mm base-to-base for 14 point characters (standard format of 18 x 26.5 cm).
If in any doubt, please send us 1 - 2 sample pages for examination. We will be glad to give advice.

Vol. 1259: F. Cano Torres, Desingularization Strategies for Three-Dimensional Vector Fields. IX, 189 pages. 1987.

Vol. 1260: N.H. Pavel, Nonlinear Evolution Operators and Semigroups. VI, 285 pages. 1987.

Vol. 1261: H. Abels, Finite Presentability of S-Arithmetic Groups. Compact Presentability of Solvable Groups. VI, 178 pages. 1987.

Vol. 1262: E. Hlawka (Hrsg.), Zahlentheoretische Analysis II. Seminar, 1984–86. V, 158 Seiten. 1987.

Vol. 1263: V.L. Hansen (Ed.), Differential Geometry. Proceedings, 1985. XI, 288 pages. 1987.

Vol. 1264: Wu Wen-tsün, Rational Homotopy Type. VIII, 219 pages. 1987.

Vol. 1265: W. Van Assche, Asymptotics for Orthogonal Polynomials. VI, 201 pages. 1987.

Vol. 1266: F. Ghione, C. Peskine, E. Sernesi (Eds.), Space Curves. Proceedings, 1985. VI, 272 pages. 1987.

Vol. 1267: J. Lindenstrauss, V.D. Milman (Eds.), Geometrical Aspects of Functional Analysis. Seminar. VII, 212 pages. 1987.

Vol. 1268: S.G. Krantz (Ed.), Complex Analysis. Seminar, 1986. VII, 195 pages. 1987.

Vol. 1269: M. Shiota, Nash Manifolds. VI, 223 pages. 1987.

Vol. 1270: C. Carasso, P.-A. Raviart, D. Serre (Eds.), Nonlinear Hyperbolic Problems. Proceedings, 1986. XV, 341 pages. 1987.

Vol. 1271: A.M. Cohen, W.H. Hesselink, W.L.J. van der Kallen, J.R. Strooker (Eds.), Algebraic Groups Utrecht 1986. Proceedings. XII, 284 pages. 1987.

Vol. 1272: M.S. Livšic, L.L. Waksman, Commuting Nonselfadjoint Operators in Hilbert Space. III, 115 pages. 1987.

Vol. 1273: G.-M. Greuel, G. Trautmann (Eds.), Singularities, Representation of Algebras, and Vector Bundles. Proceedings, 1985. XIV, 383 pages. 1987.

Vol. 1274: N. C. Phillips, Equivariant K-Theory and Freeness of Group Actions on C*-Algebras. VIII, 371 pages. 1987.

Vol. 1275: C.A. Berenstein (Ed.), Complex Analysis I. Proceedings, 1985–86. XV, 331 pages. 1987.

Vol. 1276: C.A. Berenstein (Ed.), Complex Analysis II. Proceedings, 1985–86. IX, 320 pages. 1987.

Vol. 1277: C.A. Berenstein (Ed.), Complex Analysis III. Proceedings, 1985–86. X, 350 pages. 1987.

Vol. 1278: S.S. Koh (Ed.), Invariant Theory. Proceedings, 1985. V, 102 pages. 1987.

Vol. 1279: D. Ieşan, Saint-Venant's Problem. VIII, 162 Seiten. 1987.

Vol. 1280: E. Neher, Jordan Triple Systems by the Grid Approach. XII, 193 pages. 1987.

Vol. 1281: O.H. Kegel, F. Menegazzo, G. Zacher (Eds.), Group Theory. Proceedings, 1986. VII, 179 pages. 1987.

Vol. 1282: D.E. Handelman, Positive Polynomials, Convex Integral Polytopes, and a Random Walk Problem. XI, 136 pages. 1987.

Vol. 1283: S. Mardešić, J. Segal (Eds.), Geometric Topology and Shape Theory. Proceedings, 1986. V, 261 pages. 1987.

Vol. 1284: B.H. Matzat, Konstruktive Galoistheorie. X, 286 pages. 1987.

Vol. 1285: I.W. Knowles, Y. Saitō (Eds.), Differential Equations and Mathematical Physics. Proceedings, 1986. XVI, 499 pages. 1987.

Vol. 1286: H.R. Miller, D.C. Ravenel (Eds.), Algebraic Topology. Proceedings, 1986. VII, 341 pages. 1987.

Vol. 1287: E.B. Saff (Ed.), Approximation Theory, Tampa. Proceedings, 1985–1986. V, 228 pages. 1987.

Vol. 1288: Yu. L. Rodin, Generalized Analytic Functions on Riemann Surfaces. V, 128 pages, 1987.

Vol. 1289: Yu. I. Manin (Ed.), K-Theory, Arithmetic and Geometry. Seminar, 1984–1986. V, 399 pages. 1987.

Vol. 1290: G. Wüstholz (Ed.), Diophantine Approximation and Transcendence Theory. Seminar, 1985. V, 243 pages. 1987.

Vol. 1291: C. Mœglin, M.-F. Vignéras, J.-L. Waldspurger, Correspondances de Howe sur un Corps p-adique. VII, 163 pages. 1987

Vol. 1292: J.T. Baldwin (Ed.), Classification Theory. Proceedings, 1985. VI, 500 pages. 1987.

Vol. 1293: W. Ebeling, The Monodromy Groups of Isolated Singularities of Complete Intersections. XIV, 153 pages. 1987.

Vol. 1294: M. Queffélec, Substitution Dynamical Systems – Spectral Analysis. XIII, 240 pages. 1987.

Vol. 1295: P. Lelong, P. Dolbeault, H. Skoda (Réd.), Séminaire d'Analyse P. Lelong – P. Dolbeault – H. Skoda. Seminar, 1985/1986. VII, 283 pages. 1987.

Vol. 1296: M.-P. Malliavin (Ed.), Séminaire d'Algèbre Paul Dubreil et Marie-Paule Malliavin. Proceedings, 1986. IV, 324 pages. 1987.

Vol. 1297: Zhu Y.-l., Guo B.-y. (Eds.), Numerical Methods for Partial Differential Equations. Proceedings. XI, 244 pages. 1987.

Vol. 1298: J. Aguadé, R. Kane (Eds.), Algebraic Topology, Barcelona 1986. Proceedings. X, 255 pages. 1987.

Vol. 1299: S. Watanabe, Yu.V. Prokhorov (Eds.), Probability Theory and Mathematical Statistics. Proceedings, 1986. VIII, 589 pages. 1988.

Vol. 1300: G.B. Seligman, Constructions of Lie Algebras and their Modules. VI, 190 pages. 1988.

Vol. 1301: N. Schappacher, Periods of Hecke Characters. XV, 160 pages. 1988.

Vol. 1302: M. Cwikel, J. Peetre, Y. Sagher, H. Wallin (Eds.), Function Spaces and Applications. Proceedings, 1986. VI, 445 pages. 1988.

Vol. 1303: L. Accardi, W. von Waldenfels (Eds.), Quantum Probability and Applications III. Proceedings, 1987. VI, 373 pages. 1988.

Vol. 1304: F.Q. Gouvêa, Arithmetic of p-adic Modular Forms. VIII, 121 pages. 1988.

Vol. 1305: D.S. Lubinsky, E.B. Saff, Strong Asymptotics for Extremal Polynomials Associated with Weights on ℝ. VII, 153 pages. 1988.

Vol. 1306: S.S. Chern (Ed.), Partial Differential Equations. Proceedings, 1986. VI, 294 pages. 1988.

Vol. 1307: T. Murai, A Real Variable Method for the Cauchy Transform, and Analytic Capacity. VIII, 133 pages. 1988.

Vol. 1308: P. Imkeller, Two-Parameter Martingales and Their Quadratic Variation. IV, 177 pages. 1988.

Vol. 1309: B. Fiedler, Global Bifurcation of Periodic Solutions with Symmetry. VIII, 144 pages. 1988.

Vol. 1310: O.A. Laudal, G. Pfister, Local Moduli and Singularities. V, 117 pages. 1988.

Vol. 1311: A. Holme, R. Speiser (Eds.), Algebraic Geometry, Sundance 1986. Proceedings. VI, 320 pages. 1988.

Vol. 1312: N.A. Shirokov, Analytic Functions Smooth up to the Boundary. III, 213 pages. 1988.

Vol. 1313: F. Colonius, Optimal Periodic Control. VI, 177 pages. 1988.

Vol. 1314: A. Futaki, Kähler-Einstein Metrics and Integral Invariants. IV, 140 pages. 1988.

Vol. 1315: R.A. McCoy, I. Ntantu, Topological Properties of Spaces of Continuous Functions. IV, 124 pages. 1988.

Vol. 1316: H. Korezlioglu, A.S. Ustunel (Eds.), Stochastic Analysis and Related Topics. Proceedings, 1986. V, 371 pages. 1988.

Vol. 1317: J. Lindenstrauss, V.D. Milman (Eds.), Geometric Aspects of Functional Analysis. Seminar, 1986–87. VII, 289 pages. 1988.

Vol. 1318: Y. Felix (Ed.), Algebraic Topology – Rational Homotopy. Proceedings, 1986. VIII, 245 pages. 1988

Vol. 1319: M. Vuorinen, Conformal Geometry and Quasiregular Mappings. XIX, 209 pages. 1988.